ENVIRONMENTAL AND CLIMATIC IMPACT
OF COAL UTILIZATION

ACADEMIC PRESS RAPID MANUSCRIPT REPRODUCTION

PROCEEDINGS OF THE SYMPOSIUM ON ENVIRONMENTAL AND CLIMATIC
IMPACT OF COAL UTILIZATION HELD IN WILLIAMSBURG, VIRGINIA,
APRIL 17–19, 1979

ENVIRONMENTAL AND CLIMATIC IMPACT OF COAL UTILIZATION

edited by

JAG J. SINGH

NASA Langley Research Center
Hampton, Virginia

ADARSH DEEPAK

Institute for Atmospheric Optics and
Remote Sensing
Hampton, Virginia

1980
ACADEMIC PRESS
A Subsidiary of Harcourt Brace Jovanovich, Publishers
New York   London   Toronto   Sydney   San Francisco

ACADEMIC PRESS, INC.
111 Fifth Avenue, New York, New York 10003

United Kingdom Edition published by
ACADEMIC PRESS, INC. (LONDON) LTD.
24/28 Oval Road, London NW1 7DX

Library of Congress Cataloging in Publication Data

Symposium on Environmental and Climatic Impact of
Coal Utilization, Williamsburg, Virginia, 1979.
Environmental and climatic impact of coal utilization.

1. Coal—Environmental aspects—Congresses.
I. Singh, Jag J. II. Deepak, Adarsh. III. Title.
TD195.C58S97 1979   628.5'32   79-28681
ISBN 0-12-646360-3

PRINTED IN THE UNITED STATES OF AMERICA

80 81 82 83   9 8 7 6 5 4 3 2 1

CONTENTS

v

Pollutant Dispersal Modeling—K. Demerjian, Chairman

CLIMATIC IMPACT OF COAL PLANT EMISSIONS
R. Reck and M. Mitchell, Chairmen

ENVIRONMENTAL IMPACT OF COAL PLANT EMISSIONS
A. Green and A. Deepak, Chairmen

PARTICIPANTS

MARY ANNE ALVIN, Westinghouse Research and Development Center, 1310 Beulah Road, Pittsburgh, Pennsylvania 15235

JOHN A. AVALOS, Los Angeles Department of Water and Power, P. O. Box 111, Los Angeles, California 90051

CARMEN E. BATTEN, NASA–Langley Research Center, MS 404B, Hampton, Virginia 23665

ROBERT N. BEADLE, Department of Energy/Office of Environment, Washington, D. C. 20545

MICHAEL A. BOX, Institute for Atmospheric Optics and Remote Sensing, P. O. Box P, Hampton, Virginia 23666

FRANK BRESCIA, 637 Hillsdale Avenue, Hillsdale, New Jersey 07642

THOMAS A. CAHILL, Department of Physics, University of California, Davis, California 95616

ROGER W. CARLSON, University of Illinois, 289 Morrill Hall, Urbana, Illinois 61801

NORBERT C. CHEN, Oak Ridge National Laboratory, P. O. Box Y, Bldg 9204-1, MS-3, Oak Ridge, Tennessee 37830

WAI K. CHENG, Aerodyne Research Inc., Crosby Drive, Bedford, Massachusetts 01730

KWO-SUN CHU, Department of Physics, Talladega College, Talladega, Alabama 35160

PETR CHYLECK, Center for Earth and Planetary Physics, Harvard University, Cambridge, Massachusetts 02138

WESLEY R. COFER III, NASA–Langley Research Center, MS 404B, Hampton, Virginia 23665

ADARSH DEEPAK, Institute for Atmospheric Optics and Remote Sensing, P. O. Box P, Hampton, Virginia 23666

KENNETH DEMERJIAN, Environmental Sciences Research Laboratory, Meteorology and Assessment Division, MD 80, Research Triangle Park, North Carolina 27711

EUGENE R. DUFRESNE, Jet Propulsion Laboratory, 4800 Oak Grove Drive, Pasadena, California 91103

WILLIAM J. EADIE, Battelle–Pacific NW Lab, Richland, Washington 99352

THOMAS L. EDDY, Mechanical Engineering, West Virginia University, Morgantown, West Virginia 26506

HOWARD B. EDWARDS, NASA–Langley Research Center, MS 235, Hampton, Virginia 23665

ROBERT N. ELI, West Virginia University, Morgantown, West Virginia 26506

JAMES P. FRIEND, Drexel University, Department of Chemistry, Philadelphia, Pennsylvania 19104

GERALD L. GREGORY, NASA–Langley Research Center, MS 325, Hampton, Virginia 23665

ALEX E. S. GREEN, University of Florida, 221 SSRB, Gainesville, Florida 32611

PATRICK J. HAMILL, Systems and Applied Sciences, 17 Research Drive, Hampton, Virginia 23666

KWANG S. HAN, Department of Chemistry and Physics, Hampton Institute, Hampton, Virginia 23668

ROBERT C. HARRISS, NASA–Langley Research Center, MS 270, Hampton, Virginia 23665

ROBERT V. HESS, NASA–Langley Research Center, MS 401A, Hampton, Virginia 23665

HSIAO-MING HSU, University of Michigan, Ann Arbor, Michigan 48109

JAMES G. HUDSON, Atmospheric Sciences Center, Desert Research Institute, Box 60770, Reno, Nevada 89503

RUDOLF B. HUSAR, Department of Mechanical Engineering, Washington University, Box 1185, St. Louis, Missouri 63130

ALI IMAM, Geoscience Department, Elizabeth City State University, Elizabeth City, North Carolina 27909

RONALD J. ISAACS, AER, Inc., 872 Massachusetts Avenue, Cambridge, Massachusetts 02139

CASIMIR JACHIMOWSKI, NASA–Langley Research Center, MS 404B, Hampton, Virginia 23665

MALCOLM K. W. KO, AER, Inc., 872 Massachusetts Avenue, Cambridge, Massachusetts 02139

FRANK C. KORNEGAY, Oak Ridge National Laboratory, P. O. Box X, Oak Ridge, Tennessee 37830

MARK L. KRAMER, Meteorological Evaluation Service, 134 Broadway, Amityville, New York 11701

THOMAS F. LAVERY, Environmental Analysis Division, ERT, 2625 Townsgate Road, Westlake, California 91361

GORDON LERFALD, Wave Propagation Laboratory, NOAA/Environmental Research Laboratory, 325 Broadway, Boulder, Colorado 80302

JOEL S. LEVINE, NASA–Langley Research Center, MS401B, Hampton, Virginia 23665

STEVEN E. LINDBERG, Oak Ridge National Laboratory, P. O. Box X, Oak Ridge, Tennessee 37830

THOMAS E. LUKOW, Bureau of Land Management, 2515 Warren Avenue, P. O. Box 1828, Cheyenne, Wyoming 82001

LUCIEN B. MCDONALD, State Air Pollution Control BD, 448 Goodspeed Road, Virginia Beach, Virginia 23451

MICHAEL MCELROY, Pierce Hall, Harvard University, Cambridge, Massachusetts 02138

AXEL MATTSON, NASA–Langley Research Center, MS 110, Hampton, Virginia 23665

CHARLES T. MILLER, DOE/Fossil Fuel Processing, MS E 333/P. O. Box 369, Olney, Maryland 20545

J. MURRAY MITCHELL, NOAA–EDIS/Code Dx6, 8060 13th Street, Silver Spring, Maryland 20910

RICHARD A. MORRISON, Talladega College, Talladega, Alabama 35160

GERALD R. NORTH, NASA–Goddard Space Flight Center, Code 915, Greenbelt, Maryland 20771

T. NOVAKOV, Lawrence Berkeley Laboratory, University of California, Berkeley, California 94720

JOHN M. ONDOV, Lawrence Livermore Laboratory, P. O. Box 5508, Livermore, California 94550

DONALD H. PACK, EPA/Scientific Advisory Board, 1826 Opalocka Drive, McLean, Virginia 22101

DAVID E. PATTERSON, Washington University, Box 1185, St. Louis, Missouri 63130

RALPH PERHAC, Environmental Assessment Department, Electric Power Research Institute, P. O. Box 10412, Palo Alto, California 94303

DONALD H. PHILLIPS, NASA–Langley Research Center, MS 283, Hampton, Virginia 23665

GERALD L. POTTER, Lawrence Livermore Laboratory, P. O. Box 808, Livermore, California 94546

ED PRIOR, NASA–Langley Research Center, MS 401B, Hampton, Virginia 23665

RUTH A. RECK, General Motors Research Laboratory, Warren, Michigan 48090

ROBERT S. ROGOWSKI, NASA–Langley Research Center, MS 283, Hampton, Virginia 23665

LES ROSE, NASA–Langley Research Center, MS 139A, Hampton, Virginia, 23665

GLEN SACHSE, NASA–Langley Research Center, MS 235A, Hampton, Virginia 23665

JOHN SAMOS, NASA–Langley Research Center, MS 139A, Hampton, Virginia 23665

DAVID R. SCHRYER, NASA–Langley Research Center, MS 283, Hampton, Virginia 23665

DANIEL I. SEBACHER, NASA–Langley Research Center, MS 404B, Hampton, Virginia 23665

HENRY SHAW, Exxon Research and Engineering Company, P. O. Box 8, Linden, New Jersey 07036

JAG J. SINGH, NASA–Langley Research Center, MS 235, Hampton, Virginia 23665

THOMAS STEMPORA, Commonwealth Edison Co., P. O. Box 767, Room 1700, Chicago, Illinois 60690

JAMES E. STITT, NASA–Langley Research Center, MS 117, Hampton, Virginia 23665

GERALD M. STOKES, Battelle–Pacific NW Laboratory, Richland, Washington 99352

NIEN DAK SZE, AER, Inc., 872 Massachusetts Avenue, Cambridge, Massachusetts 02139

HARRY J. SVEC, Iowa State University, DOE/Ames Laboratory, Ames, Iowa 50011

D. R. TAYLOR, Department of Chemistry, Ft. Collins, Colorado 80523

RICHARD P. TURCO, R & D Associates, 720 El Medio Avenue, Pacific Palisades, California 90272

CHARLES VAN VALIN, NOAA/US Department of Commerce, Boulder, Colorado 80303

N. T. WAKELYN, NASA–Langley Research Center, MS 404B, Hampton, Virginia 23665

KENNETH T. WHITBY, University of Minnesota, Department of Mechanical Engineering, Minneapolis, Minnesota 55455

CHARLES H. WILSON, NASA–Langley Research Center, MS 404B, Hampton, Virginia 23665

GEORGE M. WOOD, NASA–Langley Research Center, MS 234, Hampton, Virginia 23665

GLENN K. YUE, Institute for Atmospheric Optics and Remote Sensing, P. O. Box P, Hampton, Virginia 23666

PREFACE

This volume contains the technical proceedings of the Symposium on Environmental and Climatic Impact of Coal Utilization held in Williamsburg, Virginia, April 17–19, 1979.

It is imperative to undertake a continual review and discussion of the impact of increasing coal utilization on terrestrial environment and climate because of dwindling world supplies of liquid and gaseous fossil fuels. This symposium was organized to provide an interdisciplinary forum for such a review. Fifty-five scientists from the industrial community, universities, government agencies, and research laboratories attended the symposium in which thirty seven papers were presented. Complete texts of thirty-two of these papers, and their ensuing discussions, are included in this volume.

The symposium was divided into twelve sessions, each devoted to a specific aspect of the impact problem, ably chaired by the following scientists: J. J. Singh, Aerosol Characteristics; K. T. Whitby, Aerosol Formation and Dynamics; T. A. Cahill, Pollutant Measurement Techniques; R. M. Perhac, Aerosol Optical Effects; T. Novakov, M. McElroy, Emission of Sulfur and Carbon Compounds; J. Friend, Atmospheric Sulfur Budget; K. Demerjian, Pollutant Dispersal Modeling; R. Reck and M. Mitchell, and A. E. S. Green and A. Deepak, Environmental Impact of Coal Plant Emissions. Each session was organized to include one or two reviews and a similar number of contributed papers to present and discuss the current state of knowledge and the results of the latest investigations in selected areas of research. Ample time was allowed for discussions following each paper. Discussions presented were recorded and the transcripts postedited.

Mr. James E. Stitt, Director for Electronics, NASA–LaRC, welcomed the participants of the symposium and spoke of NASA's role in the development of remote sensing instrumentation for the measurement of air and water pollutants and their impact on climate. He also explained that NASA had the chartered role of assisting other governmental agencies in the application of remote measurement techniques in solving their problems related to environmental laws and regulations. Dr. Ralph M. Perhac, EPRI, in his introductory remarks, drew attention to the seriousness of the energy problem in the United States. To ensure proper representation of major disciplines involved, a symposium program committee composed of the following scientists was set up: Drs. J. J. Singh (co-Chairman), NASA-LaRC; A. Deepak (co-Chairman) and Glenn K. Yue, IFAORS: Thomas A. Cahill, University of California; William P. Elliot, NOAA/Air Resources Laboratory; James P. Friend, Drexel University; Ralph M. Perhac, EPRI; and Ruth A. Reck, General Motors Research Laboratory.

ACKNOWLEDGMENTS

The editors wish to acknowledge the cooperation of the participants, members of the Technical Program Committee, session chairmen, and speakers for making this a successful and valuable symposium for everyone. Special thanks are due the authors for their cooperation in enabling a prompt publication of the symposium proceedings. It is a pleasure to acknowledge the valuable assistance of Sue Crotts and Sherry Allen, IFAORS, in organizing the symposium and in preparing and typing the final manuscripts. Special thanks are due Dr. Glenn K. Yue, IFAORS, for assisting with all aspects of the symposium.

The Symposium was cosponsored by the Electric Power Research Institute (EPRI), NASA–Langley Research Center (NASA–LaRC), National Oceanic and Atmospheric Administration (NOAA), Virginia State Air Pollution Control Board (VSAPCB), and the Institute for Atmospheric Optics adn Remote Sensing (IFAORS).

Symposium Speakers and Chairmen (Left to Right): J. P. Friend, Drexel U.; K. Demerjian, Environ. Sci. Res. Lab.; M. McElroy, Harvard U.; N. D. Sze, Atmos. & Environ. Res. (AER); R. P. Turco, R & D Assoc.; K. T. Whitby, U. of Minnesota; D. E. Patterson, Washington U.; C. VanValin, NOAA/Environ. Res. Lab.; R. A. Reck, General Motors Res. Lab.; J. J. Singh, NASA–LaRC; T. F. Lavery, Environ. Res. & Tech.; T. Novakov, U. of California/Berkeley; P. J. Hamill, Systems & Applied Sci.; D. R. Schryer, NASA–LaRC; A. Deepak, IFAORS: H-M Hsu, U. of Michigan; D. H. Pack, EPA/Sci. Adv. Board; G. M. Wood, NASA–LaRC; T. E. Lukow, Bureau of Land Managmt; M. K. W. Ko, AER. Raised Row (Left to Right): G. R. North, NASA-Goddard; A. E. S. Green, U. of Florida; G. L. Potter, Lawrence Livermore Lab.; F. C. Kornegay, Oak Ridge Nat'l Lab.; R. A. Morrison, Talladega College; G. K. Yue, IFAORS; T. A. Cahill, U. of California.

PHYSICAL AND CHEMICAL CHARACTERIZATION
OF AEROSOL EMISSIONS FROM
COAL-FIRED POWER PLANTS[1]

John M. Ondov and Arthur H. Biermann

Lawrence Livermore Laboratory
University of California
Livermore, California

Based on the 1977 EPA National Emissions Report, conventional
coal-fired power plants are the largest single anthropogenic
source of atmospheric fine particles and sulfur oxides, and the
second largest source of nitrogen oxides. In addition to these
criteria pollutants, potentially toxic trace elements and heavy
metals, naturally occurring radionuclides and potentially carcinogenic organic and inorganic compounds have been identified
in atmospheric discharges from coal-fired power plants. The
atmospheric emission of fine particles is of special importance
because they contain toxic substances, have long atmospheric
residence times, and are efficiently deposited in the lung.
Results of the detailed chemical and physical characterization
of aerosol particles from coal combustion in several utility scale
coal-fired power plants are reported. The work includes classification of morphology and measurement of distribution parameters
of aerosol particles discharged into the atmosphere as well as
comprehensive trace element heavy metal and radionuclide analyses
of size fractionated aerosol particles. Results of these studies
indicate that potentially toxic trace elements and radionuclides
are concentrated on fine particles by vapor deposition and
mechanical enrichment mechanisms. The origin of particles and
the relationship between concentration and particle size are
discussed in view of current theoretical models. Finally, the
effects of various emission control strategies are discussed.

[1]Work performed under the auspices of the U.S. Department of
Energy by the Lawrence Livermore Laboratory under contract
number W-7405-ENG-48.

I. INTRODUCTION

The National Coal Association forecasts an increase in the
use of coal by utilities from 444 million tons in 1976 to approximately 850 million tons in 1985. Most of the coal is burned in
large, centrally located, conventional power plants, which,
according to the 1977 Environmental Protection Agency National
Emissions Report, discharged into the atmosphere some 2×10^9 kg
of fly ash, 1.5×10^{10} kg of SO_2, and 4.2×10^9 kg of NO_x. This
corresponds to 17 percent of the total anthropogenic, atmospheric,
particulate-emission inventory. Thus, conventional power plants,
fired by pulverized coal, are the largest single anthropogenic
source of atmospheric fine particles and sulfur oxides and the
second largest source of nitrogen oxides. Associated with the
atmospheric emission of fine particulate matter and oxides of
sulfur and nitrogen is the concurrent release of potentially toxic
trace elements and heavy metals, naturally occurring radionuclides,
and potentially carcinogenic organic and inorganic compounds.

Consideration of studies of trace element emissions from coal-
fired power plants equipped with cold-side electrostatic precipitators (ESPs) by this laboratory (1-3), as well as those of
references 4 to 7 and others indicate that emissions of Se, Hg,
As, W, and U may be large compared with other sources in typical
urban areas and with their natural fluxes into the environment.

Physical and chemical properties of particulate emissions
have been examined from eight different boilers, all burning pulverized coal and equipped with either hot- or cold-side electrostatic precipitators or venturi wet scrubbers. The results of
these studies, including classification of morphology, measurement
of distribution parameters of aerosol particles, and comprenehsive
analyses of fly-ash particles for trace elements, heavy metals,
and radionuclides are discussed. Reference to a company or product name does not imply approval of recommendation of the product

by the University of California or the U.S. Department of Energy
to the exclusion of others that may be suitable.

II. EXPERIMENTAL METHODS

Total aerosol and size-segregated fly-ash samples were collected in-stack at each of the plants using a modified EPA-Method-
5-type sampling system. With this system, particle collectors
are mounted at the in-stack end of the sampling probe to eliminate
wall losses incurred with the Method-5 system. Filter samples
were obtained with 1-μm-pore fluoropore or 0.4 μm Nucleopore filters. Size segregated fly-ash samples were obtained with 8- and
12-stage University of Washington Mark III and Mark V inertial
cascade impactors. Polycarbonate or polyimide impaction sub-
strates were coated with grease to improve collection efficiency.

At some of the plants, aerosol-distribution parameters were
monitored in near-real time using a Thermo System's electric
aerosol analyzer and a Royco optical, particle-size analyzer,
which were operated with a dilution system connected to a second
isokinetic sampling probe. Generating load, coal feed rate, and
boiler and control-device parameters were monitored during stack-
fly-ash collection periods. Pulverized coal, bottom-ash, and
precipitator fly-ash streams were also sampled during stack
sampling.

Coal, stack fly ash, bottom ash, and fly ash collected by the
particulate control devices were analyzed for up to 43 elements
by instrumental neutron activation analysis. Measurements of
Ni and Pb in bulk coal and fly-ash samples were made by energy
dispersive x-ray fluorescence analysis and of Cd and Be by atomic
absorption spectroscopy, using a heated graphite analyzer.
Details of these analyses were previously described (8,9).

III. DISCUSSION

A. *Distribution Modes of Fly-Ash Aerosols*

The bulk of the by-products fly ash from coal combustion represents the residual mineral matter in coal. During combustion,
the carbon matrix is burned away, and the minerals, such as clays,
shales, sulfides, carbonates, chlorides, and various assessory
minerals are exposed to temperatures of 1500° to 1600°C. The
resultant mineral phases are typically unreacted parent minerals
and their thermally transformed daughter phases, including
unburned or partially burned coal. At combusion temperatures,
most of the nonvolatile mineral matter fuses, and results in
rather distinct, highly spherical particles having the composition
of amorphous mullite. The mass median aerodynamic diameter of
aerosol particles emitted from furnaces at utilities burning
pulverized coal is typically around 150 μm.

Fisher *et al.* carefully characterized the morphology of stack-
emitted aerosol particles in four fly-ash fractions with volume
median diameters ranging from 2.2 to about 20 μm (10). These
studies indicate that most of the particles that escape collection
by emission-control devices are solid glass spheres. A greater
abundance of hollow spheres and spheres filled with particles,
termed pleurospheres, were found in the larger size fractions.
These, however, were rare in isokinetically collected aerosol-
particle samples.

In addition to these larger particles, a distinct peak in the
distribution of particles occurs at submicron particle sizes, as
illustrated by the volume compared with size distribution in
Fig. 1. These smaller particles result in part from the condensation of volatile metallic and nonmetallic oxides. Padia *et al.*
suggested that as much as 1 to 4 percent of the silica in fly ash
may be volatilized by the formation of volatile silicon monoxide
from reactions of SiO$_2$ and carbon (11). The subsequent cooling

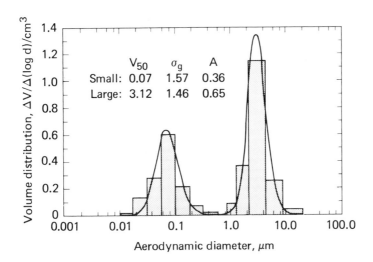

*FIGURE 1. Particle distributions of aerosols from plants
burning pulverized coal are bimodal. This distribution was
obtained instrumentally with electric aerosol and optical aerosol
analyzers connected to an isokinetic sampling system.*

of silicon and other oxide vapors results in supersaturation and
the formation of condensation nuclei, which rapidly coagulate to
form particles of diameters on the order of 0.1 μm.

Another mechanism that may be responsible for producing fine
particles is bubble bursting. At high combusion temperatures, the
suspended mineral grains begin to melt at the surface.
Interstitial water and CO_2 from decomposition outgas from the
unmelted core and form bubbles. If the bubbles break, smaller
particles may form in a fashion analogous to that proposed by
Junge in the formation of sea-salt aerosols (12).

Much of the mineral matter present in United States coals is
typically distributed in the carbon matrix as fine mineral
particles with characteristic dimensions of 1 to 15 μm (13). The
much larger particle sizes that are generally observed are probably formed by coagulation of molten mineral grains as the surfaces of burning coal particles recede.

B. *Chemical Enrichment of Aerosol Particles*

    1. Models of Vapor Deposition on Particle Surfaces. However
they are formed, these particles serve as surfaces for the condensation or adsorption of substances volatilized during combustion. Davison *et al.* demonstrated an inverse size dependence
of the concentration of many elements in coal fly ash (14). This
is indeed the behavior one would expect if some portion of the
element resided in a relatively thin surface layer that was
deposited from the gas phase. From purely geometric considerations, they derived the expression for the relationship
between the concentration of an element (C_e) and the particle
radius (r) as follows:

$$C_e = C_v + \frac{3\,C_s{}'}{r} \tag{1}$$

where the first term C_v is the concentration of the element in
the volume component of the particle and the second term is the
concentration of the element on the surface. Because the mass of
the particle is proportional to r^3, it is clear that the 1/r
dependence of concentration predicted by the model assumes,
therefore, that the mass transfer to the surface of the particle
is proportional to r^2, i.e., dependent on the surface area.

    This expression, however, is only valid for small surface
thicknesses, and for particles in the free molecular region of
aerosol physics, i.e., particles of diameters very much less than
the mean free path of gas molecules. Data were recently obtained
in 12 size ranges with a University of Washington MKV cascade
impactor (15).

    In Fig. 2, concentrations of three trace elements in μg/g
thought to be volatilized by combustion are plotted compared agains
particle size. These data fit quite well with the $1/r^2$ curve
rather than the 1/r curve. Therefore, particles measured in
this work follow aerosol mechanics for the slip-flow region of
aerosol physics. In this region, the mass flux to the surface of t

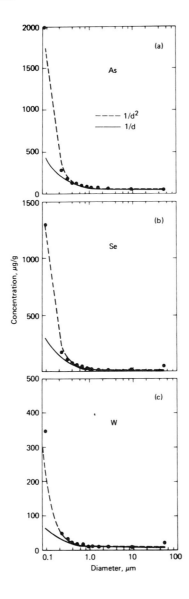

FIGURE 2. The data on concentration compared with particle size of three trace elements fit more closely a $1/r^2$ function. Therefore, mass transfer to the particle surface in the size region seems to be a function of the particle radius rather than of the radius squared. This behavior is consistent with slip-flow aerosol mechanics.

particle suspended in a stagnant gas is proportional to the
radius of the particle as indicated in

$$\frac{\Delta M}{\Delta T} = 4\pi D \ \frac{(n_i^{\ s} - n_i)}{(1 + Kn)} \tag{2}$$

This then leads to an inverse square relationship between concentration and particle size given in

$$C_e = C_v + C_s \ \frac{K}{r^2} \tag{3}$$

    2. Concentration Profiles. Given the dependency on inverse
radius expressed in Eq. (3), log plots of elemental concentrations
compared with size should yield a straight line with a negative
slope, the magnitude of which is proportional to the surface concentration. In Fig. 3, concentration profiles of several elements in aerosol particles collected downstream from a cold-side
electrostatic precipitator were plotted. Several elements,
including Se, As, Sb, W, U, V, and Ba, presumably have fairly
large surface components and were probably volatilized in the
furnace. Concentrations of Fe, Al, Na, and Sc show no size
dependence. These elements are mainly associated with the volume
component of the fly ash. Considerable deviation from linearity
is observed in the 2- to 9-μm region of the curves, especially
by W. Preliminary calculations indicate that the increase in
concentration in this region is the result of coagulation
between particles in the two predominant aerosol-particle-
distribution modes shown in Fig. 1.

    It is clear from Fig. 3 that fine aerosol particles may be
enormously enriched in concentrations of many potentially toxic
elements. Because of their small size, these particles tend to
escape collection by control devices more readily, have long
atmospheric residence times, and deposit efficiently in the human
lung. Furthermore, because large fractions of these elements

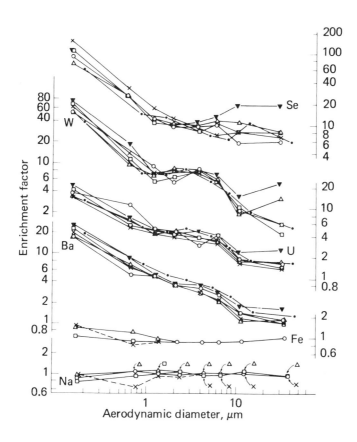

FIGURE 3. Aerosol particles from coal combustion are highly enriched in concentrations of many potentially toxic trace elements. Enrichments are greatest on smaller particles.

are concentrated on the surface of particles, their biologic
availability is likely to be greater than if distributed in the
silicate matrix.

C. *Effects of Control Devices*

   1. Hot-Side Compared with Cold-Side ESPs. Depending on the
type of control device used, both the quantity and the composition
of particulate emissions may be drastically altered. Hot-side
electrostatic precipitaters, for example, are often used at
western plants that burn low-sulfur coal. Hot-side ESPs are
located on the high temperature side of the intake air heater.
Consequently, the temperature of the gas passing through a hot-
side ESP is generally about 600° or $700^{\circ}F$, as compared with $250^{\circ}F$
in ESPs located on the low temperature side. At the higher
temperatures at which the hot-side precipitators are operated,
somewhat larger quantities of some inorganic species volatilized
in the combustion process will remain in the vapor phase, and,
therefore, escape precipitation.

   In Fig. 4 are plotted concentrations compared with size
curves of several elements in particles emitted from a plant with
a hot-side ESP. The profiles of As, as well as Ba, Ga, U, V, and
In were rather flat compared with their profiles from the cold-
side ESP in Fig. 3. This uniformity with respect to size indicates that the surface component is small relative to the volume
component. This might occur if the vapor-phase component had not
yet been deposited.

   2. Venturi Scrubbers Compared with Cold-Side ESPs.
Electrostatic precipitation and wet scrubbing are two major processes employed to control particulate emissions at coal-fired
electrical utility stations. Commercial wet scrubbers are highly
efficient in removing supermicron particles and reduce plume

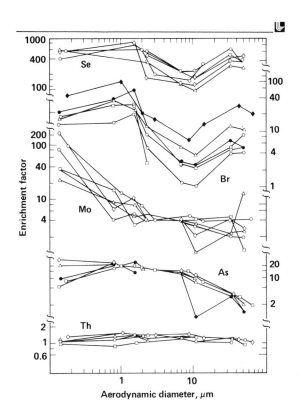

FIGURE 4. *The concentrations of several elements associated with particles collected downstream from a plant burning pulverized coal with a hot-side ESP are less highly enriched than in particles from a cold-side ESP. This may indicate that gas phase components had not yet deposited on the particles.*

visibility greatly. In addition to the control of particulate
emissions, the units simultaneously reduce SO_2 emissions (16,17).

The venturi wet scrubber, however, collects particles primarily by impaction and interception mechanisms. As shown in
Fig. 5, the collection efficiency of the unit for supermicron
particles is quite high, i.e., > 99 percent, but below 1 μm drops
rapidly with decreasing particle size. The 50 percent cut-off for
the unit was about 0.75 μm (aerodynamic diameter) (16-18).

Electrostatic precipitators collect particles by both diffusion and impaction/interception mechanisms; hence, these curves
are characterized by high collection efficiencies for both supermicron and submicron particles, with a minimum for particles of
diameters in the range of 0.1 to 1.0 μm.

Measurements were made of elemental emissions and particle-
size distributions from two power units equipped with high-energy
variable-throat, venturi wet scrubbers and a unit equipped with a
cold-side ESP. Because each of the units was in use at a single,
western, coal-fired power plant at which the same low-sulfur
high-ash coal was burned in all units and fly-ash elutriation and
particle-size characteristics of the boilers were similar, the
relative effectiveness of the two types of control devices can be
evaluated for trace element removal by comparing their emissions.

The more highly enriched, fine aerosol particles are more difficult to collect than larger, less enriched particles in both
kinds of control devices. Thus, the aerosol penetrating each of
the control devices seems further enriched. Particle-size distributions of several elements from two venturi, scrubber-
equipped units and from one ESP-equipped unit are shown in Fig. 6.
In this figure, emission factors, i.e., ng of element emitted per
joule of heat input to the boiler are plotted to account for differences in the size of the boilers. The emission factors are
plotted against the aerodynamic diameters of particles on
individual impactor stages and back-up filters determined from

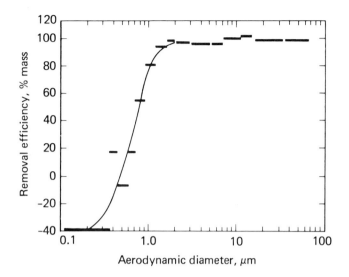

FIGURE 5. The particle removal efficiency of a venturi wet scrubber system was negative in the submicron region. Flash volatilization and mist-entrainment mechanisms are thought to be the cause of the formation of these small particles.

scanning electron microscopy (SEM) measurements (19). Because of the very low efficiency of the venturi scrubber for collecting submicron particles, the emission of many trace elements were actually greater than from the ESP-equipped unit despite the fact that the overall removal efficiency of particulate mass of the scrubber (99.8%) was about 11 times greater than that of the ESP ($\approx$ 97%). Furthermore, a significant negative efficiency was observed and suggests that the scrubber actually generated fine particles (see Fig. 5). This is probably the result of flash-volatilization of water droplets containing dissolved and sus-pended solids by the hot flue gas and in part mist entrainment. In addition, the concentrations of many metals, such as Mn, Zn, Co, Cr, and Cu, were further enriched, probably resulting from corrosion of metal surfaces inside the scrubber. Finally, par-ticulate emissions from the scrubber units occurred in particle

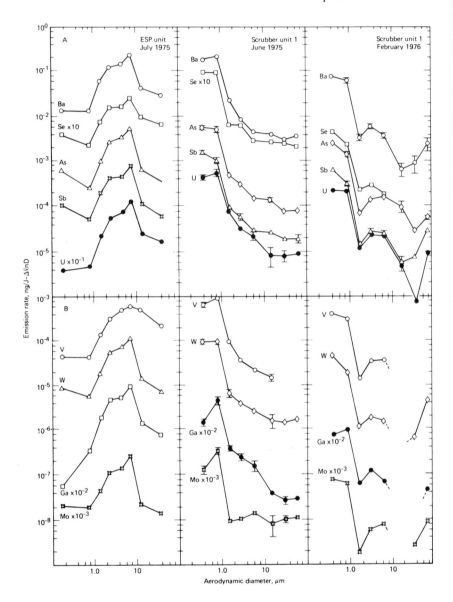

FIGURE 6. Emissions of several trace elements from a coal utility unit equipped with a venturi wet scrubber system were often larger than those from a similar unit equipped with an ESP. In addition, aerosols emitted from the scrubber were comprised of finer particles.

sizes that are more efficiently deposited in the pulmonary region
of the human respiratory system. It is concluded that the wet
scrubber system tested would be less effective in reducing the
potential inhalation hazard of those particulate emissions discussed than an ESP of comparable overall efficiency.

LIST OF SYMBOLS

C_e relative concentration of element e ($\mu g/g$)

C_v, C_s relative concentration ($\mu g/g$) distributed in the volume
             and surface of the particle, respectively

D   diffusion coefficient

K_n Knudsen number

m   mass of the species transferred

n_i, n_i^s concentration and saturation concentration (number/unit
             volume)

r   particle radius

α' discontinuity ("jump") coefficient

ρ density of the particle

ACKNOWLEDGMENT

     The authors gratefully acknowledge the assistance of
R. E. Heft, R. W. Wikkerink, D. G. Garvis, and K. O. Hamby in the
collection and analysis of samples; W. H. Martin and R. L. Tandy
for assistance in data reduction; and B. K. Ishida for assistance
in preparing the manuscript.

REFERENCES

1. Ondov, J. M., Ragaini, R. C., and Biermann, A. H., *Environ.
    Sci. Technol. 13,* 598 (1979).

2. Ondov, J. M., Ragaini, R. C., and Biermann, A. H., *Environ.
    Sci. Technol.* (Lawrence Livermore Laboratory, Livermore,
    California (UCRL-80110).

3. Coles, D. G., Ragaini, R. C., and Ondov, J. M., *Environ. Sci. Technol. 12,* 442 (1978).

4. Klein, D. H., Andren, A. W., Carter, J. A., Emery, J. F., Feldman, C., Fulkerson, W., Lyon, W. S., Ogle, J. C., Talm, Y., Van Hook, R. I., and Bolton, N., *Environ. Sci. Technol. 9,* 973-939 (1975).

5. Gladney, S., Small, J. A., Gordon, G. E., and Zoller, W. H., *Atmos. Environ. 10,* 1071-1077 (1976).

6. Andre, A. W., and Klein, D. H., *Environ. Sci. Technol. 9,* 856-858 (1975).

7. Kaakinen, J. W., Jorden, R. M., Lawasani, M. H., and West, R. E., *Environ. Sci. Technol. 9,* 862-869 (1975).

8. Heft, R. C., Absolute Instrumental Neutron Activation Analysis at Lawrence Livermore Laboratory, presented at the Third International Conference on Nuclear Methods in Environmental and Energy Research, Columbia, Missouri, October 10-13, 1977 (James R. Vogt, ed.), NTIS, Springfield, Virginia, p. 170.

9. Bonner, N. A., Bazan, F., and Camp, D. C., *Chem. Instr. 6,* 1-36 (1975).

10. Fisher, G. L., Chang, P. V., and Brummer, M., *Science, 192,* 555-556 (1976).

11. Padia, A. S., Sarafim, A. O., and Howard, J. B., The Behavior of Ash in Pulverized Coal under Simulated Combustion Conditions, presented at the Combustion Institute's Central States Section, April 1976.

12. Junge, C. E., "Air Chemistry and Radioactivity," p. 158, Academic Press, New York (1970).

13. Ulrich, G. D., An Investigation of the Mechanism of Fly-Ash Formation in Coal-Fired Utility Boilers, Interim Report USERDA FE-2205-1, May 28, 1976.

14. Davison, R. L. Natusch, D. F. S., Wallace, J. R., and Evans, E. A., Jr., *Environ. Sci. Technol. 8,* 1107-1113 (1974).

15. Biermann, A. H., and Ondov, J. M., *Atmos. Environ.*
     (1980).

16. "McIlvaine Scrubber Manual," Vol. II, pp. 5-11, McIlvaine
     Co., Northbrook, Illinois (1974).

17. Stern, A. C., "Air Pollution, Vol. III, pp. 437-495,
     Academic Press, New York (1968).

18. Calvert, S., Goldshmide, J., Leith, D., and Mehta, D., *in*
     "Wet Scrubber System Study," Scrubber Handbook, Vol. I,
     pp. 5-81, Environmental Protection Agneyc, Report R2-118a,
     (1975).

19. Ondov, J. M., Ragaini, R. C., and Biermann, A. H., *Atmos.*
     Environ. 12, 1175 (1978).

DISCUSSION

Miller: Did any of the coal plants studied do a coal washing to remove ash?

Ondov: No, these were predominantly western coal plants, strip mining coal, located in states such as New Mexico and Utah. They actually burned coal containing about 24 percent ash in most of these cases.

Harriss: I would like to make one comment on your enrichments of some of the trace metals in fine particles. In looking across several matrix materials that undergo high temperature combustion, I found that in addition to the vapor pressures in the elements, one must go back and look at the geochemistry of the raw material. Even though some trace metals have relatively low vapor pressures as oxides, they are associated with an organic binding material in the original matrix which results in high volatility.

Ondov: I am sure that is true. In addition to these--many of these elements were termed calcofilic geochemically and form sulfides that are quite volatile at combustion temperatures. There may be some organic association here. I didn't mean to imply that the elements occurred as free metals or metal oxides necessarily. A lot of the metal oxides are quite volatile too. We think uranium, for instance, might be volatilized as a hydroxide.

Brescia: What is the level of the radioactivity?

Ondov: Radioactivity ranges from 2 to 17 picocuries per gram, per nuclide for ^{40}K, ^{228}Th, ^{228}Ra, ^{210}Pb, ^{226}Ra, ^{238}U. It turns out that lead and uranium, are elements that we found enriched in aerosol particles from combustion. Depending on the control device, it could be enriched by a factor of 15 or more, in fly ash over its concentration in the coal. There will be different levels, depending on the size fraction and of course, depending on the kind of control device.

Unidentified Speaker: Do you have any feeling about the efficiency and the enrichment capability of various bag-house types of abatement devices. It seems they may really come into their own in the next few years over western coals.

Ondov: Yes. Well they ostensibly are quite good, that is, they have efficiencies as high as 99.99 percent. My understanding is that they have not really caught on so well, possibly becuase of engineering problems. But I am sure that as time goes on, they probably will be used more extensively. I see no reason, theoretically, why they wouldn't be the best thing yet.

Singh: Did you find any evidence that some of the fine aerosols
were all essentially solid particles of one element, such as
sulphur?

Ondov: No, we have no evidence of that. It is very hard to do
single particle work on individual submicron particles. We have
analyzed particles on backup filters of cascade impactors by
activation analysis and by x-ray fluorescence and x-ray microprobe. But, to my knowledge,there are no, what you might call,
primary condensation nuclei--that is to say, a single particle
composed of a single element. Larger particles, however, often
contain very high concentrations of iron or carbon.

Singh: I noted that you didn't have anything about elements
lighter than calcium. Sulfur is one of the culprits that most of
the people worry about and you did not have any data about
sulfur. Would you care to comment?

Ondov: We have found that sulfur is enriched in aerosol particles
emitted from the unit equipped with the wet venturi scrubber.
Electron spectroscopy of coal fly ash indicates that greater than
95 percent of the sulfur is present as sulfate.

Kramer: Have you done any work on eastern coal, or high sulfur
coal?

Ondov: No.

Kramer: Would you expect consideraly different results?

Ondov: Well, it is certainly possible that you can get some
different results for plants burning high sulfur coal.
Especially if you look at a situation where you have some kind
of scrubber. If you have more sulfur, you're going to have higher
acid levels which may exacerbate problems related to leaching.
We'll be looking at the fluidized bed combuster at Rivesville,
West Virginia in the near future. The sulfur level in that coal
will be quite a bit higher. But it's a completely different
kind of system. We would like to sample aerosols from eastern
plants, but have made no arrangements to do so at this time.

Brescia: You mentioned a nucleation mechanism for the formation
of aerosols. As I recall, this type of mechanism generally leads
to a highly homogeneous aerosol. Do you have evidence on the
homogeniety of the aerosol?

Ondov: Well, no. The problem here is that condensation would
lead to a homogeneous aerosol if there were nothing else that
could coprecipitate. Something like silicon monoxide might
nucleate and actually form some particles just after the gas

gets out of the boiler. As combustion gases pass through the
primary heat exchangers to make steam for the turbine, the
temperature drops. At some point, just after the combustion
chamber, you might get nucleation of say, silicon monoxide
particles. But then as the gas cools further, any primary
particles will be covered with condensable species that deposit
on particles at lower temperatures. So, even though you might
have some particles that are formed by condensation, you might
see them covered with surface deposits because of material condensing or absorbing later on, so I don't expect to see primary
particles of the kind you described. Further, it would be
difficult to differentiate particles formed by nucleation from
those formed by other mechanisms, such as by bubble-bursting, and
fragmentation. Finally, the fine particles coagulate rapidly,
so that the particles we collect are likely to be aggregates.
Therefore the particles we observe are really groups of particles
which now contain a mixture of different particles and elements.

SOME PROPERTIES OF PARTICULATE PLUMES FROM
COAL-FIRED POWER PLANTS[1]

Thomas E. Lukow[2]
William A. Cooper

University of Wyoming
Cheyenne, Wyoming[a]

An instrumented aircraft was used to study properties of the particulate plumes downwind of two modern coal-fired power plants, the Jim Bridger plant in Wyoming (7 flights) and the Colstrip plant in Montana (6 flights). Measurements included total (Aitken) particle concentrations, particle size distributions, concentrations of cloud condensation nuclei (CCN) and of ice nuclei, and plume locations and dimensions. These measurements were made at distances of from 4 to 50 km downwind of the plants. CCN concentrations of about three times the background concentration were observed in the plumes, during both summertime and wintertime studies. Aitken particle concentrations in these plumes were often measured to be in excess of $20,000$ cm^{-3}, and total particle fluxes were $>10^{16}$ s^{-1}. Particle size spectra showed that 10^3 to 10^4 cm^{-3} particles with sizes 0.01-0.1 μm diameter were produced by the plants, and increases at sizes larger than 1.0 μm were also measured. Total CCN production rates were found to be on the order of 10^{15} CCN s^{-1}, active at 1% supersaturation. In some cases, evidence was obtained that the CCN concentration increased downwind of the plants while the total particle concentration decreased; thus, CCN were being produced in the plumes downwind of the power plants.

[1]Research sponsored by the Old West Regional Commission.

[2]Present address: Bureau of Land Management, Cheyenne, WY 82001

I. SAMPLING PROCEDURES

The particulate emissions from coal-fired electric power
generating plants were studied in a series of 13 aircraft flights
during both the summer and winter of 1976. The power plants
studied were the Jim Bridger plant in southwestern Wyoming and
the Colstrip plant in southern Montana. At the time of these
studies the Jim Bridger plant had a peak operating capacity of
1000 MW, but was operating at about 650 MW. The Colstrip plant,
then with a peak capacity of 360 MW, was operating at about
250 to 300 MW. Electrostatic precipitators were operating at
both plants and, in addition, scrubbers were in operation at the
Colstrip plant.

An instrumented aircraft was used to map the plumes from the
power plants and to collect aerosol samples for these studies.
Measurements recorded by equipment on the aircraft included
total Aitken particle concentration (measured by an Environment
One condensation nucleus monitor), particle size distributions
(measured by an active scattering aerosol spectrometer and an
axially scattering spectrometer probe, both manufactured by
Particle Measuring Systems), temperature, pressure, dewpoint,
wind direction and speed, and intensity of turbulence. The
aircraft also was equipped to collect filter samples of particles
and to bag samples of the aerosol so that other instruments not
onboard the aircraft could be used to study the particles in the
plume.

The normal flight procedure began with detailed plume mapping
using Aitken nucleus concentrations, so that subsequent studies
and aerosol collections could be made near the center of the
plumes. In most cases, visual observations by the aircraft crew
would have been sufficient to determine the proper position for
the aerosol samples, but this positioning was always confirmed
by measured increases in Aitken particle concentrations. The

real-time data display capability of the aircraft data system
facilitated accurate positioning within the plume and permitted
collection of upwind (background) aerosol samples at a potential
temperature corresponding to the plume altitude.

Aerosol samples were collected in aluminized Mylar bags for
subsequent processing by instruments for the measurement of
cloud condensation nuclei (CCN), contact and deposition ice
nuclei, Aitken nuclei, and particle size spectra. The use of
storage bags ensured that the spectrum of CCN as a function of
supersaturation could be measured, a test that otherwise would
have been difficult because of the large variations in particle
concentration in different parts of the plume and the short time
that the aircraft remained within the plume. All bag samples
were processed within about 1 hour after collection, and CCN
decay losses in the bags were measured to be only about 15 percent per hour. Additional details regarding the equipment, the
measurement techniques, and the results are available in papers
by Cooper and Lukow and by Lukow (1, 2).

II. CLOUD CONDENSATION NUCLEI

The CCN detector was a conventional thermal diffusion chamber.
The droplets that formed in the supersaturated region of this
chamber were illuminated by a small laser beam and were photo-
graphed for subsequent counting and analysis. This instrument
has been compared with several other similar CCN counters, and it
produces comparable results.

Table 1 summarizes the CCN measurements. In each case, the
spectrum was fitted by the equation $N = CS^K$ where N is the
cumulative number of CCN active at supersaturation S (expressed
in percent) and C and K were adjusted to obtain the best fit to
the measurements. Figure 1 shows some representative spectra,
measured at about 0800 MDT at an altitude of 300 m AGL. The

TABLE 1.   CCN Measurement Summary

Date Code	Plant		Location	C	K
760211	B	1	7 km downwind in plume	350+100	1.19+0.22
"	"	3	5 km upwind	260+60	0.98+0.22
760212	"	4	5 km downwind	500+90	0.90+0.17
"	"	5	4 km downwind in plume	440+90	0.79+0.18
"	"	6	5 km upwind	270+70	1.27+0.25
760213A	"	8	8 km downwind in plume	330+140	0.70+0.38
- "	"	9	6 km downwind in plume	320+270	0.77+0.69
"	"	10	6 km upwind	480+160	0.67+0.31
760213B	"	7	4 km downwind in plume	510+120	0.98+0.18
"	"	11	20 km downwind in plume	460+100	0.57+0.17
"	"	12	11 km upwind	340+150	0.90+0.41
760213C	"	15	14 km downwind in plume	230+60	1.35+0.15
"	"	16	9 km downwind in plume	260+60	0.66+0.21
"	"	17	6 km upwind	120+60	0.87+0.40
760223	C	19	11 km downwind in plume	480+80	0.44+0.14
"	"	21	6 km upwind	120+40	0.52+0.28
"	"	22	18 km upwind	120+40	0.70+0.29
760224	"	23	23 km downwind	270±60	0.83±0.22
"	"	26	18 km downwind	270±70	1.01±0.23
"	"	27	4 km upwind	400±80	0.90+0.18
760225	"	28	24 km downwind in plume	770±110	0.97+0.13
"	"	29	19 km downwind in plume	470+100	1.19+0.17
"	"	30	14 km downwind in plume	380+70	0.93+0.18
"	"	31	9 downwind, edge of plume	220+40	0.76±0.22
"	"	33	8 upwind	210+50	1.00+0.25
76092	B	1	25 km downwind in plume	930+130	1.39+0.27
"	"	3	upwind	860+140	1.33+0.28
76093	"	1	16 km downwind in plume	2200+200	0.69+0.11
"	"	2	7 km downwind in plume	1770+170	0.94+0.14
"	"	3	upwind	590±110	1.82±0.37
760831A	C	1	10 km downwind	3350±270	0.97±0.12
"	"	2	upwind	1160±150	1.03+0.20
760831B	C	3	12000' MSL, above plant	890±160	1.19±0.25
"	"	4	12000' MSL, upwind	860±130	1.03±0.25
760901	"	1	4 km downwind, in plume	1200±160	0.75±0.18
"	"	2	in dust near surface	1900±200	0.72±0.14
"	"	3	upwind	1260±166	0.76±0.17

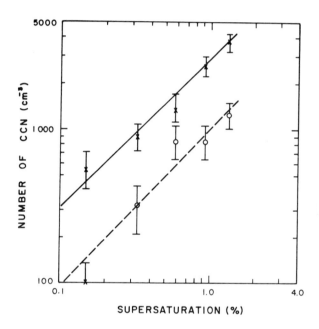

FIGURE 1. CCN concentration as a function of supersaturation, for the in-plume (X) and background (O) samples of
31 August 1976.

upwind and the downwind samples were both collected within about
10 km of the plant, from air having the same potential temperature. An inversion was still present within about 150 m of the
surface, and the atmosphere was stably stratified for at least
100 m above that inversion. The wind speed was only about
3 m s^{-1}.

Table I shows that significant CCN production was detected
at both plants on some occasions, but that on other occasions

little difference between in-plume and background samples was found. (Most samples were collected during periods of stable stratification, when higher concentrations would be expected.)

Total CCN fluxes could be estimated by assuming that the size of the Aitken nucleus plume was the same size as the CCN plume. An equivalent production rate for the power plants was determined by subtracting the corresponding background flux from the measured in-plume CCN flux. (This rate is termed an "equivalent" production rate, because substantial CCN production apparently occurred in the plumes downwind of the plants, as discussed in the next section.) Table II lists the highest equivalent production rates determined during each of the four field trips. Typical rates, for CCN active at 1% supersaturation, were about 10^{15} CCN s^{-1}. These CCN production rates are similar to those reported by Pueschel and Van Valin, who measured rates of 10^{14} s^{-1} at 10 km downwind and 10^{16} s^{-1} at 100 km downwind of the Four Corners power plant (3). The CCN production rates are somewhat lower than those measured by Hobbs *et al.* who measured production rates of 8 x 10^{15} s^{-1} at the Colstrip power plant (4).

The studies at Colstrip on 25 February 1976 and at Jim Bridger on 3 September 1976 provided opportunities to follow the plume for long distances downwind, because of the high stability of the atmosphere at plume altitude. The plumes were trapped

TABLE II. *CCN Fluxes or Equivalent Production Rates*

Date Code	Plant	Time after emission	CCN flux at 1%
760212	Bridger	10 min	9×10^{14} sec^{-1}
760225	Colstrip	40 min	3×10^{15} sec^{-1}
760831	Colstrip	80 min	2×10^{15} sec^{-1}
760903	Bridger	40 min	3×10^{14} sec^{-1}

within vertical regions only about 30 m deep. Figure 2 shows the
CCN concentration active at 1% supersaturation as a function of
time for the two days. In each case, the sample collected
farthest downwind from the plant contained the highest concentration of CCN.

These results suggest that CCN were being produced in the
plumes downwind of the power plants. The different parts of
the plume were emitted from the power plants over periods of
about 1 hour, during which time the operating power of the plant
varied less than 10%, and the wind speed was almost constant
during this period; thus, different emitted concentrations or
dilution factors probably cannot explain the observations.
Another possibility is that the downwind samples were obtained
from a portion of the plume with higher CCN concentrations, and
that the samples closer to the plant were obtained at the edge
of the plume and missed the highest CCN concentrations. A

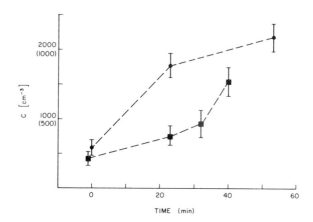

*FIGURE 2. CCN concentration, active at 1% supersaturation,
as a function of transport time in the plume from the power
plant. The dots refer to the 3 September 1976 case, and the
squares to the 25 February 1976 case; the ordinate labels in
parentheses refer to the 25 February case.*

comparison of the total particle concentrations in the bag
samples used for the CCN determination eliminates this possibility. These particle size spectra were measured by using an
electrical mobility analyzer, which covered the size range from
0.01-0.3 µm diameter. The particle size spectra for the samples
of 25 February are shown in Fig. 3. The highest CCN concentration was measured in the sample taken farthest downwind, but
that sample did not have the highest total particle concentration. The relative particle concentrations in the samples of
3 September were similar. In both cases, the CCN concentration
increased downwind as the total particle concentration decreased.
These two studies suggest that substantial CCN production can
occur within the plumes downwind of the power plants. This
phenomenon was only observed when the plume could be traced for
about 1 hour in a relatively stable layer. CCN concentrations
were observed to increase as much as 50% while the total
particle concentrations were decreasing.

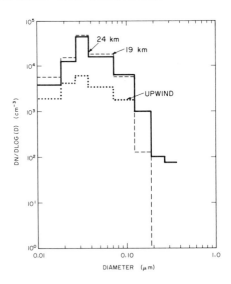

*FIGURE 3. Particle size spectra measured for the aerosol
samples upwind, 19 km downwind, and 24 km downwind of the
Colstrip plant on 25 February 1976.*

III.   AITKEN PARTICLE MEASUREMENTS

Aitken particle measurements were used both to map the
dimensions of the plume and to determine the total particle concentrations emitted from the power plants.   Figure 4 shows the
Aitken particle cross section measured at the Jim Bridger plant
on 3 September 1976, at a time when a strong inversion was
present.   The cross section is perpendicular to the wind
direction at a distance of 5 km downwind from the plant.   The

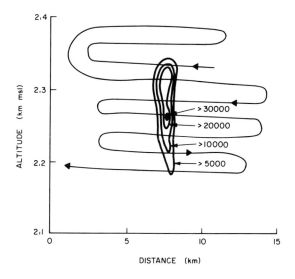

FIGURE 4.   A cross section of the Aitken particle plume of
3 September 1976.   The plume is viewed along the wind, and the
concentrations are shown in units of cm^{-3}, and the flight track
is also shown (arrows).

stack elevation was about 2.16 km MSL. Figure 5 shows a cross
section of the plume as seen in a plane parallel to the wind
direction. These plume dimensions and profiles were representative of stable conditions during both summer and winter studies.
(The measured value of the eddy dissipation rate, deduced from
cross-wind horizontal fluctuations, was 1-10 $cm^2 s^{-3}$ and the
time available for plume spreading was about 1500 s for the
example of Fig. 5.)

Increases in total particle concentration of as much as
$3 \times 10^4 cm^{-3}$, or about ten times background values, were detected
in many instances. The total particle flux in these plumes could
be determined from measurements such as those shown in Fig. 4
and from the measured wind speed; these fluxes were about 10^{16}
particles s^{-1}. No evidence of increasing particle flux with
distance downwind was found for the total particle population.

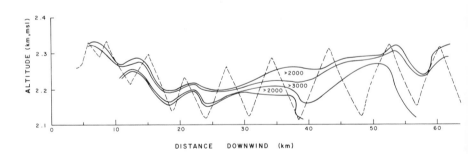

FIGURE 5. The cross section of the Aitken particle plume of
3 September 1976, as viewed across the wind. Concentrations are
shown in units of cm^{-3}, and the flight track is also shown
(dashed line).

IV. ICE NUCLEUS MEASUREMENTS

Filter samples were collected during the aircraft flights, in
some cases by flying for long periods in the plumes and in other
cases by exhausting bag samples through filters after the flights.
These filter samples were processed in the Wyoming filter pro-
cessor, described by Gordon and Vali (6). All measured ice
nucleus concentrations were less than 0.3/liter, effective at
-16° C and at 99% relative humidity. Although there was a
tendency for the in-plume samples to yield concentrations greater
than the background samples, the difference did not exceed a
factor of two and, in all but two cases, appeared to be statistically insignificant. The interpretation of these results
presents difficulty, since the addition of CCN to filters has
been shown to increase the apparent ice nucleus concentrations
measured by the filter technique; it is therefore not clear
whether even the minor detected increases are due to real ice
nucleus differences or are a result of the CCN increases (7).

In addition, 10 aerosol bag samples were processed through
an instrument for the detection of contact nuclei, described by
Vali (8). The ice nucleus concentrations from these 10 bags
varied from 0.2/liter to 1.0/liter effective at -17° C, a range
similar to that found elsewhere in Wyoming (8). Five of the
samples were in-plume samples and five were corresponding background samples collected upwind of the plants. There was no
detectable difference between the in-plume and background ice
nucleus concentrations.

A recent study of the Four Corners Plant found no difference
between in-plume and background filter samples of ice nuclei
(9). These results and the present results suggest that current
power plants are not significant sources of ice nuclei.

REFERENCES

1. Cooper, W. A., and Lukow, T. E., "Weather Modification Potential of Coal-Fired Power Plants." Final Report to the Old West Regional Commission from the University of Wyoming, Laramie, Wyoming (1977).

2. Lukow, T. E., M.S. Thesis, University of Wyoming, Laramie Wyoming (April 1977).

3. Pueschel, R. F., and Van Valin, C. C., *Trans. Am. Geophys. Union, 57,* 925 (1976).

4. Hobbs, P. V., Radke, L. F., and Stith, J., *in* "Sixth Conference on Planned and Inadvertent Weather Modification," pp. 73-74. American Meteorological Society, Boston, Massachusetts (1977).

5. Knudsen, E. O., and Whitby, K. T., *J. Aerosol Sci. 6,* 6 (1975).

6. Gordon, G., and Vali, G., *in* "The Third International Workshop on Ice Nucleus Measurements," pp. 91-122. University of Wyoming, Laramie, Wyoming (1976).

7. Vali, G., *in* "International Conference on Cloud Physics," p. 43. American Meteorological Society, Boston, Massachusetts (1976).

8. Vali, G., *in* "Conference on Cloud Physics," p. 34. American Meteorological Society, Boston, Massachusetts (1974).

9. Schnell, R. G., Van Valin, C. C., and Pueschel, R. F., *Geophys. Res. Lett. 3,* 657 (1976).

DISCUSSION

Ondov: Did you actually try to calculate the dilution factor or measure some stable gas or something to get an idea what that was? You must know the dilution factor of the plume to actually see generation of condensation nuclei.

Lukow: Well, that is correct. But, we did measure, or map the plume at various distances downwind and determined the spread of the plume. However, under these stable conditions, the plume pretty much stayed the 30 meters thick, but spread in the horizontal direction. There was very, very little vertical dispersion going on.

Ondov: But it did spread horizontally.

Lukow: Oh yes.

Ondov: In plume sampling we found dilution factors of a thousand or more, within the first few miles of that same plant. I was curious about another point. What about aerosol coagulation? In your computation of the nuclei formation rate, did you account for the removal rate of fine particles by aerosol coagulation? It is possible that the true rate of nuclei number formation would be greater if coagulation were neglected.

Lukow: That is possible.

Whitby: What time of day were these measurements made? You showed two sets: one for September and one for February.

Lukow: Right. Most of the numbers shown today were early morning measurements, the stable conditions. We did take some afternoon measurements and it was much more difficult--the plume was breaking up more -- it was a more unstable environment.

Whitby: One of the things we have found looking at a lot of plume data is that if one is going to correlate anything that is growing with the time of day, the solar insolation must be considered. Our ignoring this has been kind of foolish, I guess, up to this point, when looking at plume data because, if the sun is driving the chemistry, then obviously one has to take account the time of day. I also agree that you really cannot make much sense out of growth data at all without bringing the dilution and the aerosol dynamics into it. Another question, do you know what the effective cutoff of the CCN counter is that you were using? It is an integral measurement and growth could result either from mass deposition or from coagulation? Do you know what the effective cutoff was?

Lukow: We did some early measurements with the instrument in an effort to determine some of that. It becomes a very difficult thing to work with. I would guess that we are looking at the range of 0.02 µm radius, something like that, but I cannot give you that with any certainty, no.

Brescia: What is the nature of the material that is condensing?

Lukow: That I cannot tell you. We did not look at the material itself. We did not do any electron microscopy on it or anything like that. It was strictly an instrumental measurement looking at particle size, etc. We did not look at what the constituents were.

SULFATE AND NITRATE IN PLUME AEROSOLS
FROM A POWER PLANT NEAR COLSTRIP, MT

C. C. Van Valin
R. F. Pueschel
F. P. Parungo

Atmospheric Physics and Chemistry Laboratory
Environmental Research Laboratories
National Oceanic and Atmospheric Administration
Boulder, Colorado

Aerosol size distributions are typically bimodal with modes at less than 0.1 µm and larger than 1.0 µm radius. Elemental particle analysis by scanning electron microscope with energy dispersive X-ray analysis shows that the small particle mode in the plume contains more sulfur than the background aerosol. This is postulated because of a secondary sulfate aerosol from a gas-to-particle conversion mechanism and subsequent coagulation with the plume aerosol. The incidence of sulfur in particles of the larger size mode in the plume is relatively unchanged with increasing time from the stacks. The larger size mode represents sulfur-containing primary (fly ash) aerosols and is affected only to a small extent by absorption of SO_2 and/or coagulation of secondary sulfate particles. Particles collected outside of the power plant plume, and therefore presumed to be representative of the unpolluted atmosphere, were found to contain about 43% of particles that indicated no elements of atomic number greater than 10, i.e., that were probably composed of carbonates, nitrates, soot, or organic matter. Some of the aerosol samples were found to contain a small amount of uranium, as well as the more commonly encountered metals such as iron, chromium, and nickel. Lanthanum and cerium were also encountered occasionally.

35

I. INTRODUCTION

Power production from coal-fired electric generating plants
in the Rocky Mountain states, with the attendant emissions of
gases and particles to the atmosphere, carries the implication
for long-term impact upon the environment and agriculture of this
region. In addition, long range transport, especially of sulfur
dioxide and sulfates, may contribute to the degradation of air
quality and the occurrence of acid precipitation east of the
Mississippi River.

The research group has been engaged in the study of the
effects of power plant operation in the west, and especially
in terms of cloud physics (1-7). As part of this continuing
program, field programs have been conducted in the vicinity
of the Montana Power Co. generating plant at Colstrip, MT,
using both our research-instrumented aircraft and the mobile
laboratory.

The mobile laboratory observations have been reported
previously (8, 9); this paper will deal with airborne aerosol
analyses and with electron microscope analyses of aircraft-
collected samples.

The Colstrip power plant achieved full operational status in
1976; it has two pulverized-coal burning units, each with a gross
design output capability of 358 megawatts. Coal consumption at
maximum load is rated at 1.76×10^5 kg/hr/unit, the use of coal
with a heating value of 4.86×10^6 gram calories/kg being
assumed. Combustion effluents are directed through alkaline
fly ash-lime scrubbers that operate at typical efficiencies of
ca. 80% for SO_2 removal; the maximum particulate emission rate
is ca. 83 kg/hr. The two stacks are 500 ft (152.4 m) tall,
inside diameters are 16.5 ft (5 m), tops are 3745 ft (1141 m)
MSL. Exit velocity is 100 ft/sec (30 m/sec); emission temperature is 80° C.

II. POWER-PLANT-PLUME MEASUREMENTS

Plume cross-sectional dimensions were measured with the instrumented research aircraft during the three field projects of 1977. The plume cross-sectional measurements were done by flying across the plume at a right angle to the wind direction, starting at an altitude above the plume, and descending 100 or 200 ft for each successive plume crossing. Parameters measured included light scattering coefficient, Aitken nuclei, SO_2, NO_x, and O_3 concentrations, temperature and relative humidity. Nuclepore filter samples were collected in and near the plume and were returned to our laboratory for ice nuclei measurements and for scanning electron microscope/X-ray spectrometer elemental analysis of individual particles. Table I lists the plume dimensions and Aitken nuclei (AN) flux at two downwind distances, and the apparent net particle production rate, after the losses due to settling out, coagulation, or impingement on stationary surfaces.

The large differences between the February and August/ September measurements, both in terms of plume dimension and of particle flux, are probably a consequence of atmospheric stability and solar energy input, respectively, during a period of time when plant operating conditions remained stable. February 16 was an overcast winter day, with significant tem- perature stratification at the time of the plume crossings. August 31 and September 1 were warm summer days with moderate instability and ample sunshine. The 16 Feburary 1977 measured and calculated SO_2 and aerosol fluxes and formation rates are listed in Table II.

Aitken nuclei measurements include particles of radius less than 10^{-6} cm, while particles larger than 1 μm radius are subject to settling during the measurement; the smallest particles are

*TABLE I. Plume Parameters Near the Coal-fired Power Plant
At Colstrip, MT*

	16 Feb 77	31 Aug 77	1 Sep 77
Downwind distance (km)			
first	0.5	12.0	12.0
second	25.0	43.0	45.0
Plume width (km)			
first	56m (est)	6.0	5.1
second	2.8	12.0	7.4
Wind velocity (m sec^{-1})	6.0	8.0	6.0
Cross-sectional area (cm^2)			
first	1.06×10^7	2.8×10^{10}	1.38×10^{10}
second	5.30×10^9	5.6×10^{10}	2.01×10^{10}
AN flux (sec^{-1})			
first	1.02×10^{14}	4.0×10^{16}	2.31×10^{16}
second	4.10×10^{15}	5.1×10^{16}	3.37×10^{16}
Net production (sec^{-1})			
(disregarding losses)	4.00×10^{15}	1.1×10^{16}	1.06×10^{16}
Maximum AN			
concentrations (cm^{-3})			
first	16.0×10^3	5.0×10^3	7.7×10^3
second	3.8×10^3	3.0×10^3	7.7×10^3

usually predominant in terms of numbers, but of less importance
in terms of mass than particles of r $\geq 0.1 \times 10^{-5}$ cm. Thus, the
computation of mean particle radius is inexact and the aerosol
formation rate, as measured, could be satisfied by considerably
less net sulfur dioxide consumption than that postulated here.

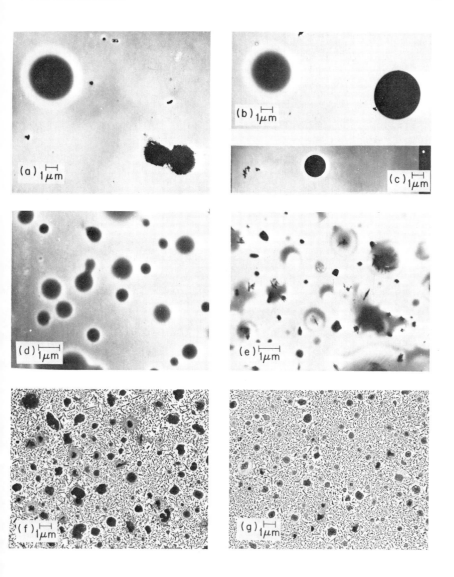

FIGURE 1. Electron photomicrographs of grids from a cascade impactor sample collected 27 February 1979, in the power plume 5 mi (8 km) from the source. (a) through (e): nitron post-coated. (f) and (g): BaCl$_2$ coated. Nitron = 1,4-diphenyl-3,5-endo-anilino-4,5-dihydro-1,2,4-triazole.

TABLE II.   *Sulfur Dioxide Depletion and Sulfate Aerosol*
Formation Rates on 16 February 1977

SO_2 stack emission, estimated	$351 \ g \ sec^{-1}$
SO_2 flux at 25 km, measured	228
SO_2 settling flux, $V_G = 1 \ cm \ sec^{-1}$	64
Net SO_2 consumption	59
H_2SO_4 formation rate	89
Aerosol formation rate	$4 \times 10^{15} \ sec^{-1}$
Mean particle radius	$1.4 \times 10^{-5} \ cm$

III.   ELECTRON MICROSCOPE ANALYSIS

Samples consisting of plume aerosol particles of diameter 0.2 to 2.0 μm were collected on electron microscope screens mounted in front of the impactor jet nozzles of a four-stage Casella cascade impactor carried on the research aircraft. The screens were vacuum coated with $BaCl_2$ or with nitron for analysis of the collected sample for the presence of sulfate (10, 11) or of nitrate (12, 13), respectively, with the transmission electron microscope.

Parts (f) and (g) of Fig. 1 are photomicrographs of the particles collected on a $BaCl_2$ coated grid that was mounted in the fourth (final) stage of the cascade impactor. The network of small, elongated particles of uniform size and distribution is the $BaCl_2$ substrate. In part (f), the black spots represent mineral or fly ash particles; the grayish areas surrounding the spots are the areas in which a sulfate coating reacted with the $BaCl_2$. A few spots here, and many in part (g), lack the central aerosol particle, as indicated by the lack of a totally black area; these spots represent sulfate particles. Where the center of the spot is darker than the surroundings, but not completely opaque, the reaction of sulfate with the substrate has not gone to completion. The mineral spots that lack the grayish halo are those that had not acquired a sulfate coating. The photomicrograph of part (f) was taken near the center of the impaction line, and that of part (g) was closer to the edge, thus accounting for the observed difference of particle size. It also shows that the small end of the particle size spectrum is dominated by sulfate particles formed from gas phase oxidation of SO_2 followed by homogeneous nucleation and condensation. The diameters of the spots of $BaSO_4$ following reaction have been found to be roughly double those of the original particles; thus, the particle size indicated here is in satisfactory agreement with that derived in Table II. Parts (a) through (e) are photomicrographs of impactors that were vacuum coated with nitron following collection. Parts (a), (b) and (c) are from the third stage of the cascade impactor; parts (d) and (e) are from the fourth stage. The nitron coating produces a specific indication of nitrates by the appearance of fibrous masses where nitrate-containing drops or particles had impinged (12); this method also indicates (but is not specific for) sulfate by a clear area around a particle, or, in the absence of an unreactive particle, of a spot that is clear at the edge and shading into greater density at the center. With nitron the diameter of the resulting spot is roughly three times the diameter of the original sulfate particle. Part (a) shows a sulfate particle (left) and two nitrate particles (right). Part (b) shows a sulfate particle (left) and an uncoated fly ash particle (right). Part (c) shows a fly ash particle coated with sulfate. Part (d) includes many sulfate particles but none containing detectable amounts of nitrate; part (e) is a mixed aerosol collection, including many fly ash and mineral particles with sulfate coating, a number of nitrate particles, and some particles that could be mixtures of sulfate and nitrate.

In most, if not all, of the samples collected during the field operations, the nitrate reaction products were found

in spots representing droplets of 1 µm or larger in diameter,
while the sulfate particles or drops are much smaller, most
often in the range 0.1 to 0.5 µm diameter. This indicates that
nitrogen dioxide, being much more water soluble and hygroscopic
than sulfur dioxide, dissolves in water droplets or absorbs
water, and then is oxidized in solution to nitrate, which is
also quite hygroscopic. Sulfur dioxide is thought to undergo
oxidation in the vapor phase followed by either homogeneous or
heterogeneous nucleation to form sulfuric acid droplets or
sulfuric acid coatings on the surfaces of inert particles.

On occasions when samples were collected at two or more
distances downwind in the plume, the fraction of particles containing sulfur was usually found to increase with increasing
distance (and time) downwind. Table III provides a summary of
those examples.

From this consistent increase with increasing time and
distance downwind, it is inferred that either coagulation between
sulfuric acid droplets and plume particles or condensation of
sulfuric acid vapor on the particles is occurring. This postula-
tion is certainly consistent with the appearance of the sulfate-
coated particles found in the electron photomicrographs of
Fig. 1. In these instances, comparison of numbers is valid
only between samples of the same day, and not between days, by
reason of changing meteorological and solar energy parameters.

IV. SUMMARY AND CONCLUSIONS

Measurements of sulfur dioxide and aerosol flux through the
plume of the coal-fired power plant near Colstrip, MT, permit
the calculated estimation of the mean sulfate particle diameter
of ca. 0.14 µm and a particle generation rate of 10^{15} to 10^{16}
sec^{-1}. This particle size estimation is consistent with the
diameters of sulfate or sulfuric acid particles collected in a
multistage cascade impactor and measured by means of transmission

TABLE III. Percent of Particles, as Determined by SEM-EDX, That Were Found to Contain Sulfur

Date	Particle Diameter, μm	Percent Containing Sulfur	Distance, km	Percent Containing Sulfur	Distance, km
2/15/77	0.1 to 0.4	28	8	40	16
	0.6 to 1.0	44	8	52	16
5/29/77	0.1 to 0.4	68	1.6	100	12
	0.6 to 1.0	34	1.6	74	12
	1.5 to 10.0	54	1.6	54	12
8/29/77	0.1 to 0.4	11	4	24	16
	0.6 to 1.0	39	4	27	16
9/1/77	0.1 to 0.4	14	25	17	50
	0.6 to 1.0	25	25	43	50

electron microscope imaging. As the distance (and time) downwind increased the proportion of particles containing sulfur appears to have increased, thus pointing to coagulation of a sulfate aerosol and plume particles or heterogeneous nucleation of sulfuric acid droplets following vapor phase oxidation of SO_2. Nitrate particles collected with the cascade impactor are predominantly of size greater than 1 μm diameter. The small size of the sulfate particles as compared with the nitrate aerosol indicates disparate mechanisms for oxidation of SO_2 and of NO_2, with the former undergoing vapor phase reactions, while the latter is oxidized in the liquid (water droplet) phase.

REFERENCES

1. Parungo, F. P., Allee, P. A., and Weickmann, H. K.,
    Geophys. Res. Lett. 5, 515-517 (1978).

2. Parungo, F. P., Ackerman, E., Proulx, H., and Pueschel, R.,
    Atmos. Environ. 12, 929-935 (1978).

3. Pueschel, R. F., and Van Valin, C. C., *Atmos. Environ. 12,*
    307-312 (1978).

4. Pueschel, R. F., *Geophys. Res. Lett. 3,* 651-653 (1976).

5. Schnell, R. C., Van Valin, C. C., and Pueschel, R. F.,
    Geophys. Res. Lett. 3, 657-660 (1976).

6. Pueschel, R. F., Van Valin, C. C., and Parungo, F. P.,
    Geophys. Res. Lett. 1, 51-54 (1974).

7. Van Valin, C. C., Pueschel, R. F., Parungo, F. P., and
    Proulx, R. A., *Atmos. Environ. 10,* 27-31 (1976).

8. Abshire, N. L., Derr, V. E., Lerfald, G. M., McNice, G. T.,
    Pueschel, R. F., and Van Valin, C. C., "Aerosol
    Characterization at Colstrip, Montana, Spring and Fall,
    1975." NOAA Technical Memorandum ERL WPL-33 (1978).

9. Van Valin, C. C., Pueschel, R. F., Wellman, D. L., Abshire,
    N. L., Lerfald, G. M., and McNice, G. T., *in* "The
    Bioenvironmental Impact of a Coal-fired Power Plant,
    Fourth Interim Report, Colstrip, Montana" (E. M. Preston
    and T. L. Gullett, eds.), pp 2-52. EPA Technical Report
    600/3-79-044 (1979).

10. Mamane, Y., and de Pena, R. G., *Atmos. Environ. 12,*
    69-82 (1978).

11. Mamane, Y., and Pueschel, R. F., *Geophys. Res. Lett. 6,*
    109-112 (1979).

12. Bigg, E. K., Ono, A., and Williams, J. A., *Atmos. Environ.
    8,* 1-13 (1974).

13. Mamane, Y., and Pueschel, R. F., "A quantitative Method
     for the Detection of Individual Nitrate Particles,"
     Atmos. Environ. (1980).

DISCUSSION

Perhac: I have two questions. First, what compounds do you
think comprise the nitrate particles; and second--since you refer
to sulfuric acid--how do you know that the sulfate particles are
sulfuric acid as opposed to other sulfate compounds?

Van Valin: My response to the first question concerning the
nitrates is that we have not done a characterization of the
elemental composition of those fibrous particles, so we do not
know the answer. The anser to the second question is that by
showing sulfuric acid in the table, and by mentioning sulfate
and sulfuric acid, it was not my intention to claim the existence
of sulfuric acid, but only to include this as an intermediate in
the formation of sulfate salts. If sulfuric acid droplets had
been collected, tiny spots would be expected to be present in
the halos surrounding the main particles, as seen in the photo-
micgrographs of the particles collected on a nitron-coated
impactor; these spots are absent. In examination of collected
particles with the scanning electron microscope with energy
dispersive X-ray analysis, we often find Ca associated with S
in the close-in samples ($\lesssim$ 5 μm), but with S appearing alone in
samples from greater downwind distances. The energy dispersive
X-ray analysis does not include emissions from elements lighter
than Na, and the fairly low excitation energy as customarily
employed in our analyses leaves some uncertainty regarding identification of trace amounts of some of the heavier elements. Thus,
although the evidence is not conclusive, there is the strong
indication that ammonium sulfate or ammonium bisulfate particles
predominate at the greater downwind distances.

Perhac: That's the big difference from the health standpoint--
whether it is ammonium sulfate or sulfuric acid is a serious
consideration. Which one is it?

Van Valin: Yes, I recognize that.

Ondov: What kind of control device was on this power plant?

Van Valin: The Colstrip plant has a wet scrubber that uses
alkaline fly ash with lime added to control pH. It is not an
electrostatic precipitator. I am not familiar with the unit
beyond that.

Ondov: Two possibilities occur to me as to what these sulfur
particles might be. Because you do have entrainment of mist
from scrubbers and the liquid contains quite a bit of sulfur in
the form of soluable sulfate, its possible that the water will

evaporate leaving fine sulfate contianing aerosols. Do you think there is a possibility that you are actually making particles by evaporating water droplets in the atmosphere? The other thing that occurs to me is that it may be condensation of sulfuric acid.

Van Valin: I would be very surprised if it were a case of simply reaching a dewpoint for sulfuric acid. The flux of particles at 25 or 30 or more miles downwind is equal to or greater than that at the shorter distances. Since the temperature of the plume quickly reaches ambient, and since the plume continues to expand, condensation would have to occur at very short distances if it were to take place at all. By way of additional response to Dr. Perhac's question, as well as replying to the part about evaporation of water from sulfuric acid droplets, I repeat that at close-in distances we find particles giving response for S and Ca. At greater distances the Ca is not found. The behavior of S-containing particles in the scanning electron microscope indicates they are not sulfuric acid, but are probably ammonium sulfate or bisulfate. Sulfuric acid droplets will boil away when hit with the electron beam. Of course, one cannot say unequivocally that neutralization did not take place between the time of sample collection and analysis; this is a strong argument for analyzing the samples as soon as possible.

Hamill: I just wanted to ask you a question about one of your tables. You had SO_2 consumption at 59 grams per second and H_2SO_4 formation at 89 grams per second. I was wondering if that was just assuming all the SO_2 wnet into H_2SO_4. How did you do that?

Van Valin: We assumed a loss for SO_2 due to settling out from the plume with a settling velocity of 1 cm sec^{-1}; the remaining SO_2 lost went to H_2SO_4. We recognize that this is an incomplete accounting for losses, but in this context there is not much point in detailed material balance since the variation of the rate of emission of SO_2 is large, and there is also uncertainty in the SO_2 flux at 25 km downwind, as measured. The emission rate estimate of 351 sec^{-1} could be on the high side; the SO_2 flux measurements were done in as methodical a way as was possible, considering problems connected with flying an airplane back and forth through a plume that is invisible and is detectable only by instrument response. The point is that the particle generation rate could be accomplished with less SO_2 consumption-H_2SO_4 generation simply by assuming a smaller mean particle radius.

Singh: What were the meteorological differences between the
August and September data?

Van Valin: Glad you mentioned that. I had intended to, but I
forgot about it. August 31 and September 1 were warm summer days
but with some stratification at the time of the measurements, and
plenty of sunlight. The February day was cold and overcast and
with a strong stratification at the time of measurement. If it
could be said that the oxidation of SO_2 was dependent upon the
input of solar energy, this example would support it.

Singh: That's probably what Dr. Whitby had in mind about the
role played by the input solar radiation.

THE FORMATION OF SULFATE AEROSOLS THROUGH
HETEROMOLECULAR NUCLEATION PROCESS

G. K. Yue

Institute for Atmospheric Optics and Remote Sensing
Hampton, Virginia

P. Hamill

Systems and Applied Sciences Corporation
Hampton, Virginia

*The formation of sulfate aerosols through homogeneous
nucleation, ion-induced nucleation, and heterogeneous nucleation
in a binary system of H_2SO_4- H_2O mixture is studied. Contours
of critical composition, critical radius and the nucleation rate
of sulfate aerosols formed through the homogeneous nucleation
process are plotted as a function of H_2SO_4 concentration and H_2O
concentration. The effect of the uncertainty of the saturated
vapor pressure of H_2SO_4 on the nucleation rate is discussed.
It is shown that the sulfate aerosols formed through the ion-
induced nucleation process can be classified as stable, transitory, and unstable according to the environmental conditions in
which they are formed. Finally, the effect of size and other
properties of the condensation nuclei on the heterogeneous
heteromolecular nucleation of sulfate particles is considered.*

I.   INTRODUCTION

Aerosols in the atmosphere are due to either primary or
secondary sources. Primary sources are those which emit
particulates directly into the atmosphere, whereas secondary

sources are those which lead to the formation of atmospheric aerosol particles through nucleation processes. A large-scale increase in coal combustion will certainly increase both the primary and secondary aerosols. In the absence of interaction with water clouds, the secondary aerosols formed through gas-to-particle conversion mechanisms are usually very small in size, but the total number concentration of secondary aerosols is several orders of magnitude higher than that of the primary aerosols. Under certain circumstances, the growth of these small secondary particles to larger ones through coagulation and condensation processes may cause severe environmental and climatic problems. An understanding of the formation of the secondary aerosols is therefore essential in assessing the impact of atmospheric particulates on our environment and in devising improved antipollution strategies for utilizing coal to meet our energy needs.

Among the toxic substances released into the atmosphere by burning coal, sulfur dioxide is probably the dominant gaseous precursor for the formation of secondary airborne particulates. Elemental analysis of differentiated aerosols collected downstream from a coal-fired plant indicates that sulfur is very strongly concentrated in the smaller size aerosols (1). Despite the increasing emphasis on the study of aerosols, the origin and behavior of these secondary sulfate particulates suspended in the air are still poorly understood.

In this paper, the classical liquid-drop model is applied to investigate the formation of sulfate aerosols through homogeneous nucleation, ion-induced nucleation, and heteromolecular nucleation in a binary system of H_2SO_4-H_2O gases under different environmental conditions. Conclusions from this study can be applied to kinetic aerosol models simulating the formation and growth of sulfate aerosols in air.

In Section II, homogeneous heteromolecular nucleation theory
is reviewed. Contours of critical composition, critical radius,
and the nucleation rate of sulfate aerosols at 25°C and 0°C are
plotted as a function of H_2SO_4 concentration and H_2O concentration. The effect of the uncertainty of the saturated vapor
pressure of H_2SO_4 on the nucleation rate is discussed. In
Section III a recently developed ion-induced nucleation theory
applicable to the H_2SO_4-H_2O binary system is presented. It is
shown that the sulfate aerosols formed in the presence of an
ionization source can be classified as stable, transitory, or
unstable according to the environmental conditions in which they
are formed. Nucleation rates for sulfate aerosols formed
through the ion-induced nucleation process under different
ambient conditions are calculated. In Section IV, the classical
heterogenous-homomolecular nucleation theory is extended to
heteromolecular systems to study the formation of H_2SO_4-H_2O
droplets on the surface of pre-existing solid particles. The
effect of size and other properties of the condensation nuclei
(or substrate) on the heterogeneous heteromolecular nucleation of
sulfate particles is elucidated. Some conclusions are given in
the last section. A list of symbols follows Section V.

II. HOMOGENEOUS NUCLEATION THEORY

The Gibbs free energy of formation of a liquid embryo in a
binary mixture of vapors is given by the expression:

$$\Delta G = n_A (\mu_A^\ell - \mu_B^g) + n_B (\mu_B^\ell - \mu_B^g) + 4\pi r^2 \sigma \qquad (1)$$

where n_i = number of moles of species i, μ_i^ℓ and μ_i^g = chemical
potential of species i in the liquid and gas phase, respectively,

r = radius of the embryo and σ = microscopic surface tension. For the special system H_2SO_4- H_2O under consideration in this paper, A refers to water and B refers to sulfuric acid.

The change of chemical potential is the transformation from gas phase to liquid phase for each component is given by the following relationship:

$$\mu_i^{\ell} - \mu_i^{g} = RT \ln \frac{P_i}{P_i^{sol}} \tag{2}$$

where R = universal gas constant, T = absolute temperature, P_i = ambient vapor pressure of species i and P_i^{sol} = partial vapor pressure of i over a flat surface of the solution.

As suggested by Yue the following quantities are defined:

$$S_A = \frac{P_A}{P_A^o} = \text{relative humidity}$$

$$S_B = \frac{P_B}{P_B^o} = \text{relative acidity}$$

$$a_A = \frac{P_A^{sol}}{P_B^o} = \text{water activity}$$

$$a_B = \frac{P_B^{sol}}{P_B^o} = \text{acid activity}$$

where P_A^o and P_B^o are equilibrium vapor pressures of A, B over a flat surface of pure substance A and B, respectively (2). Equation (1) can then be rewritten as:

$$\Delta G \, (n_A, n_B) = -n_A RT \ln \frac{S_A}{a_A} - n_B RT \ln \frac{S_B}{a_B} + 4\pi r^2 \sigma \tag{3}$$

and r is given by

$$\frac{4}{3} \pi r^3 \rho = n_A M_A + n_B M_B$$

where ρ is the density of the solution and M_A and M_B are molecular weights of A and B, respectively.

The critical composition n_A^* and n_B^* at the saddle point (where an embryo becomes stable) can be calculated by solving the following simultaneous equations:

$$\left(\frac{\partial \Delta G}{\partial n_A} \right) n_B = 0$$

$$\left(\frac{\partial \Delta G}{\partial n_B} \right) n_A = 0 \qquad (4)$$

Nair and Vohra (3) substituted Eq. (3) into Eq. (4) and obtained the following generalized Kelvin equation:

$$\ln \frac{S_A}{a_A} = \frac{2}{RT} \frac{M_A}{\rho} \frac{\sigma}{r} \left(1 + \frac{x'}{\rho} \frac{d\rho}{dx'} - \frac{3}{2} \frac{x'}{\sigma} \frac{d\sigma}{dx'} \right) \qquad (5)$$

$$\ln \frac{S_B}{a_B} = \frac{2}{RT} \frac{M_B}{\rho} \frac{\sigma}{r} \left(1 - \frac{(1 - x')}{\rho} \frac{d\rho}{dx'} + \frac{3}{2} \frac{(1 - x')}{\sigma} \frac{d\sigma}{dx'} \right) \qquad (6)$$

where x' is the fraction of the mass of acid B in the solution

$$x' = \frac{n_B M_B}{n_A M_A + n_B M_B} \qquad (7)$$

When the environmental conditions S_A, S_B, and T are given, the critical composition x'^* and the critical radius r^* at the saddle point can be calculated by solving the simultaneous Eqs. (5) and (6).

The values of x'^* and r^* at 25° C are plotted as a function of H_2SO_4 concentrations in Figs. 1 and 2. These figures are very useful in estimating the critical composition and radius under a variety of ambient conditions.

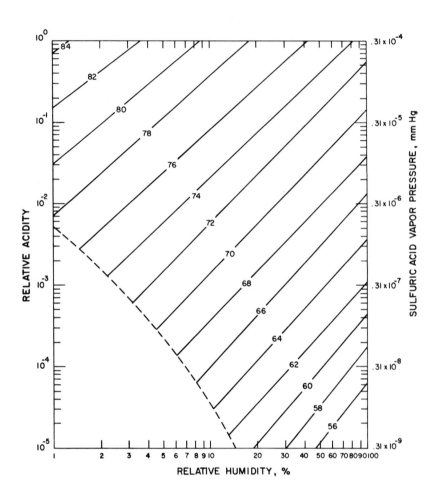

FIGURE 1. Critical composition of sulfate aerosols at 25° C
for different relative humidities and acidities. Number
represents weight percentage of H_2SO_4 in the droplet.

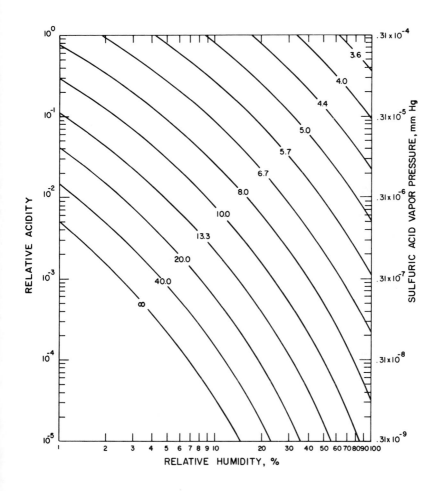

FIGURE 2. Critical size of sulfate aerosols at 25° C for different relative humidities and acidities. Number represents the critical radius in Å.

Substituting Eqs. (5) and (6) into Eq. (3) yields a simple expression of the Gibbs free energy at the saddle point

$$\Delta G^* = \frac{4}{3} \pi \sigma r^{*2} \tag{8}$$

The homogeneous nucleation rate (embryos formed per cm^3 per sec) in a binary mixture of vapors with $N_A \gg N_B$ can be written as

$$J = 4\pi r^{*2} \beta_B N_A \exp\left(-\frac{\Delta G^*}{KT}\right) \tag{9}$$

where N_A is the concentration of water vapor, K is the Boltzman constant, and β_B is the impinging rate of gaseous sulfuric acid molecules on a surface of unit area

$$\beta_B = N_B \left(\frac{KT}{2\pi M_B}\right)^{\frac{1}{2}} \tag{10}$$

where N_B is the concentration of H_2SO_4 and M_B is the mass of one molecule of H_2SO_4 (4,5).

The homogeneous nucleation rates of H_2SO_4-H_2O solution droplets at 25° C are plotted as a function of H_2SO_4 concentration in Fig. 3.

It should be noted that the value of P_B^o, the sulfuric acid vapor pressure, is still uncertain. The value of P_B^o at 23° C was evaluated by Roedel as 2.5×10^5 torr. By extrapolating, it is estimated that P_B^o at 25° C should be about 3.1×10^5 torr (6). Any future changes in the H_2SO_4 vapor pressure estimates will affect the numbers on the sulfuric acid vapor pressure scale of the plots; however, since the left-hand side is scaled in terms of relative acidity, there are no changes in the values of x'^* and r^* in Figs. 1 and 2 and the only changes in nucleation rate is to multiply by a factor equal to the change in the vapor pressure of pure H_2SO_4 in Fig. 3.

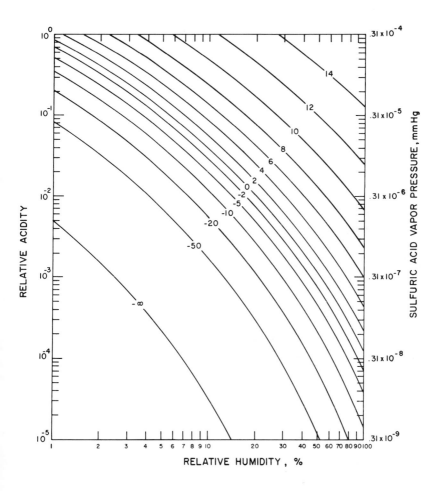

FIGURE 3.   Nucleation rate of sulfate aerosols at 25° C for
different relative humidities and acidities.   Integer on each
curve is the power to the base 10 of the nucleation rate
(no. cm^{-3} sec^{-1}).

In order to study the effect on temperature on the critical
composition, radius and nucleation rates of sulfate aerosols,
calculations are repeated with temperature at 25° C. Results
are presented in Figs. 4 to 6.

III. ION-INDUCED NUCLEATION THEORY

The Gibbs free energy of formation of a liquid embryo in a
binary mixture of vapor in the presence of an ion source is
given by the modified Thomson's theory as

$$\Delta G_I = n_A (\mu_A^\ell - \mu_A^g) + n_B (\mu_B^\ell - \mu_B^g) + 4\pi r^2 \sigma$$

$$+ \frac{Q^2}{2} (1 - \frac{1}{\varepsilon}) (\frac{1}{r} - \frac{1}{r_o}) \tag{11}$$

where the additional term accounts for the change in field
energy due to the condensation of liquid molecules of dielectric
constant ε about an ion core with charge Q and radius r_o. The
pressure of an ion will cause a decrease in the Gibbs free
energy and provide an attractive force between the foreign
center and the bulk-nucleating phase, and thus lowers the
supersaturation level necessary for a transition to a condensed
phase.

Substituting Eq. (2) into Eq. (11) yields

$$\Delta G_I = -n_A RT \ln \frac{S_A}{a_A} - n_B RT \ln \frac{S_B}{a_B} + 4\pi r^2 \sigma$$

$$+ \frac{Q^2}{2} (1 - \frac{1}{\varepsilon}) (\frac{1}{r} - \frac{1}{r_o}) \tag{12}$$

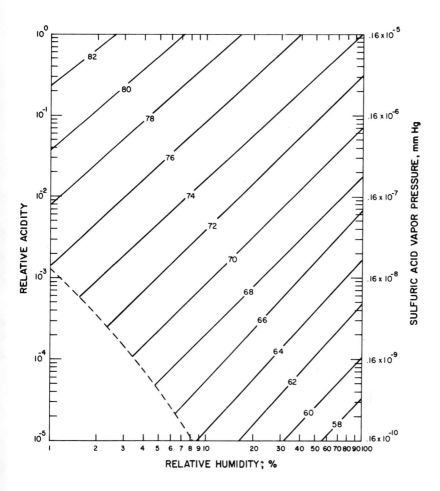

FIGURE 4. Critical composition of sulfate aerosols at 15° C
for different relative humidities and acidities. Number
represents weight percentage of H_2SO_4 in the droplet.

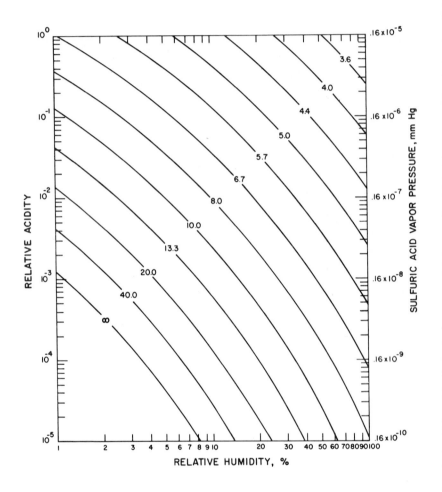

FIGURE 5. Critical size of sulfate aerosols at 15° C for different relative humidities and acidities. Number represents the critical radius in Å.

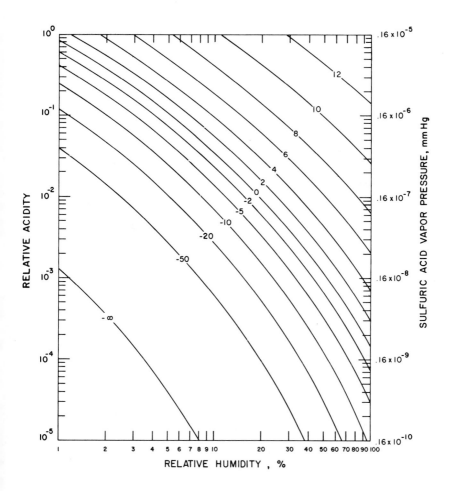

FIGURE 6. *Nucleation rate of sulfate aerosols at 15° C for different relative humidities and acidities. Integer on each curve is the power to the base 10 of the nucleation rate (no. cm^{-3} sec^{-1}).*

Eq. (12) is in the form

$$\Delta G_I = a\,r^3 + b\,r^2 + \frac{c}{r} + d \tag{13}$$

where a, b, c, and d are constants independent of r. Plots of ΔG_I against r shown in Fig. 7 demonstrate that the curve can have a single minimum, or one local minimum followed by one local maximum, or that ΔG_I is a monotonic decreasing function of r dependent on the sign and relative magnitude of the numerical values of a, b, c, and d in Eq. (13). It is obvious that these three curves correspond to three types of aerosols: stable, transitory, and unstable. The ambient conditions to form these three kinds of aerosols are shown in Fig. 8. It is assumed that $r_o^3 \ll r^3$ and ε is not a function of x. Substituting Eq. (12) into Eq. (6) yields a simultaneous equation to calculate the critical composition x* and radius r* of aerosols (7):

$$\ell n\frac{S_A}{a_A} = \frac{2}{RT}\frac{M_A\sigma}{\rho}\frac{1}{r}\left[1 + \frac{x}{\rho}\frac{d\rho}{dx} - \frac{3}{2}\frac{x}{\rho}\frac{d\rho}{dx}\right.$$

$$\left. - \frac{\varrho^2}{16\pi\sigma r^3}(1 - \frac{1}{\varepsilon})(1 + \frac{x}{\rho}\frac{d\rho}{dx})\right] \tag{14}$$

and

$$\ell n\frac{S_B}{a_B} = \frac{2}{RT}\frac{M_B\sigma}{\rho}\frac{1}{r}\left[1 - \frac{(1 - x)}{\rho}\frac{d\rho}{dx} + \frac{3}{2}\frac{(1 - x)}{\sigma}\frac{d\sigma}{dx}\right.$$

$$\left. - \frac{\varrho^2}{16\pi\sigma r^3}(1 - \frac{1}{\varepsilon})(1 - \frac{(1 - x)}{\rho}\frac{d\rho}{dx})\right] \tag{15}$$

Substituting Eqs. (14) and (15) into Eq. (12) yields a simple expression of the free energy of formation at the saddle point.

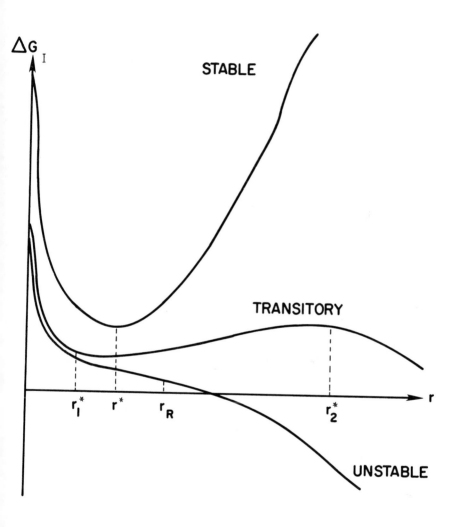

FIGURE 7. Gibbs free energy of formation ΔG_I as function of radius r for different types of aerosols.

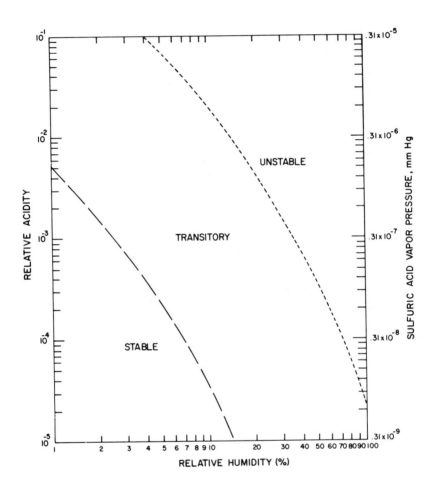

FIGURE 8. *Regions of ambient conditions on which certain types of sulfate aerosols can be formed.*

$$\Delta G_I^* = \frac{4}{3} \pi \sigma r^{*2} + \frac{Q^2}{6} (1 - \frac{1}{\varepsilon}) \left(\frac{4}{r^*} - \frac{3}{r_o} \right) \qquad (16)$$

For a stable aerosol, the nucleation rate is obviously zero; for an unstable aerosol, the nucleation rate can be taken as infinity, since no critical size can be found and the particle will grow spontaneously. What has to be considered is the nucleation rate of aerosols in the transitory region.

The change of the Gibbs free energy of formation from the point of local minimum r_1^* to the point of local maximum r_2^* for any transitory aerosols is then given by

$$\delta \Delta G_I^* = \frac{4}{3} \pi (\sigma_2 r_2^{*2} - \sigma_1 r_1^{*2})$$

$$+ \frac{2}{3} Q^2 (1 - \frac{1}{\varepsilon}) \left[\frac{1}{r_2^*} - \frac{1}{r_1^*} \right] \qquad (17)$$

where σ_1 and σ_2 are the surface tensions at compositions x_1 and x_2 which correspond to radii r_1^* and r_2^*, respectively.

Whenever the ambient conditions are given, the values of S_A and S_B can be determined. If these lie in the transitory region, two solutions of r and the corresponding values of σ can be found by solving the implicit simultaneous Eqs. (14) and (15). The values of r_1^*, r_2^*, σ_1 and σ_2 being known, the Gibbs free energy barrier $\delta \Delta G_I^*$ can be calculated by Eq. (17). Then the nucleation rate is given by

$$J_I = 4 \pi r_2^{*2} \beta_B N_C \exp \left(- \frac{\delta \Delta G}{KT} \right) \qquad (18)$$

where N_C is the number concentration of ions.

Nucleation rates for sulfate aerosols formed in the H_2SO_4-H_2O binary system and in the presence of an ion source under a variety of ambient conditions are calculated. These nucleation rates are plotted as contours in the S_B against S_A plane in Fig. 9.

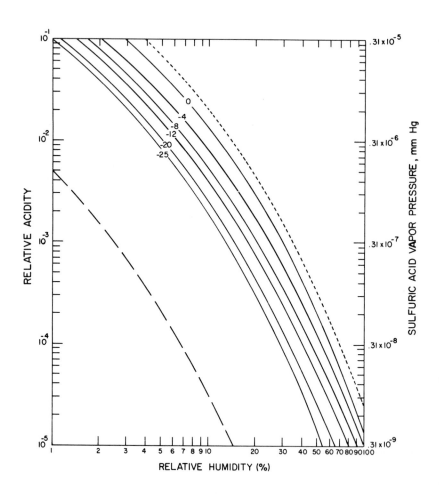

FIGURE 9. Nucleation rate of sulfate aerosols formed at the presence of an ionization source. Integer on each curve is the power to the base 10 of the nucleation rate (no. cm^{-3} sec^{-1}) divided by the number concentration of ions.

IV.   HETEROGENEOUS HETEROMOLECULAR NUCLEATION

The term heterogeneous nucleation will be used to describe
the mechanism by which binary system H_2SO_4-H_2O solution droplets
form on the surface of pre-existing solid particles. These pre-
existing solid particles are referred to as condensation nuclei
and they are assumed to be insoluble so that only physical
adsorption takes place on their surface.

The effect of a solid surface on the Gibbs free energy of
formation was considered by Fletcher (8). His arguments are
followed here, but are generalized to a binary system and
applied to heterogeneous heteromolecular nucleation to study
the formation of sulfate aerosols under different environmental
conditions.

By referring to Fig. 10 where 1 is the gas phase, 2 is the
liquid phase condensing embryo, and 3 is the substrate, the
change in Gibbs energy involved in the formation of an
embryo is given by

$$\Delta G_H = -n_A RT \ln \frac{P_A}{P_A^{sol}} - n_B RT \ln \frac{P_B}{P_B^{sol}} + \sigma_{12} S_{12}$$

$$+ (\sigma_{23} - \sigma_{13})\, S_{23} \tag{19}$$

where σ_{ij} and S_{ij} are the surface free energy per unit area and
the surface area of the interface between phase i and j.

The usual definition of contact angle θ is

$$\cos \theta = \frac{\sigma_{13} - \sigma_{23}}{\sigma_{12}} \equiv m \tag{20}$$

and the values of S_{12} and S_{23} are given by

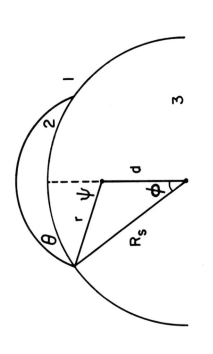

FIGURE 10. Embryo 2 on condensation nucleus 3 in nucleating phase 1.

$$S_{12} = 2\pi r^{*2} \left[1 + \frac{r^* - R_s m}{d} \right] \qquad (21)$$

$$S_{13} = 2\pi R_s^2 \left[1 - \frac{R_s - r^* m}{d} \right] \qquad (22)$$

The volume of the embryo is given by

$$V_2 = \frac{1}{3}\pi r^{*3} \left[2 + 3\frac{r^* - R_s m}{d} - \left(\frac{r^* - R_s m}{d}\right)^3 \right]$$

$$- \frac{1}{3}\pi R_s^3 \left[2 - 3\frac{R_s - r^* m}{d} + \left(\frac{R_s - r^* m}{d}\right)^3 \right] \qquad (23)$$

where $d = (R_s^2 + r^{*2} - 2rR_s m)^{\frac{1}{2}}$.

However,

$$V_2 \rho = n_A M_A + n_B M_B \qquad (25)$$

where ρ is the density of the solution and M_A and M_B are molecular weights of A and B, respectively.

Since all parts of the embryo surface must be in equilibrium with the metastable phase 1, the critical radius and composition of the embryo are those given by the solution of Eqs. (5) and (6). Following the argument given by Fletcher (8), the Gibbs free energy at the saddle point is

$$\Delta G_H^* = \frac{2}{3}\pi\sigma\, r^{*2}\, f\,(m,x) \qquad (26)$$

where

$$f(m,x) = 1 + \left(\frac{1 - mx}{g}\right)^3 + x^3 \left[2 - 3\left(\frac{x - m}{g}\right) + \left(\frac{x - m}{g}\right)^3 \right]$$

$$+ 3mx^2 \left(\frac{x - m}{g} - 1\right) \qquad (27)$$

where

$$x = \frac{R_S}{r^*} \tag{28}$$

and

$$g = (1 + x^2 - 2\ mx)^{\frac{1}{2}} \tag{29}$$

Note that for homogeneous nucleation, $x = 0$, and Eq. (26) reduced to Eq. (8).

The heterogeneous nucleation rate per unit area of the substrate surface approximated by Hidy and Brock (9) may be generalized to the case of binary systems to give

$$J_H = 2\pi r^{*2}\ \beta_B N_A^{abs}\ \exp\ (-\frac{\Delta G_H^*}{KT}) \tag{30}$$

where N_A^{abs} is the number of water molecules adsorbed per unit area of substrate.

The argument by Hamill *et al.* ("Analysis of the Nucleation of Sulfuric Acid Solution Droplets in the Stratosphere," in preparation) is followed to assume

$$N_A^{abs} = \beta_A \left[2.4 \times 10^{-16} \exp\ (\frac{10800}{RT}) \right] \tag{31}$$

where β_A is the impinging rate of water molecules

$$\beta_A = N_A \left(\frac{KT}{2\pi M_A} \right)^{\frac{1}{2}} \tag{32}$$

Since the nucleation rate J_H depends on the numerical values of R_S and θ as well as on the ambient conditions P_A and P_B, the nucleation rates J_H have been plotted as a function of R_S or θ for different values of P_A and P_B. In Fig. 11, with θ fixed at 70° and the H_2SO_4 concentration at 10^{10}, the nucleation rates J_H were plotted as a function of R_S for relative humidities of 30%, 50%, 80%, and 100%. In Fig. 12, θ was fixed at 70° and the relative humidity was fixed at 80% and the nucleation rates J_H were plotted as a function of R_S for H_2SO_4 concentration

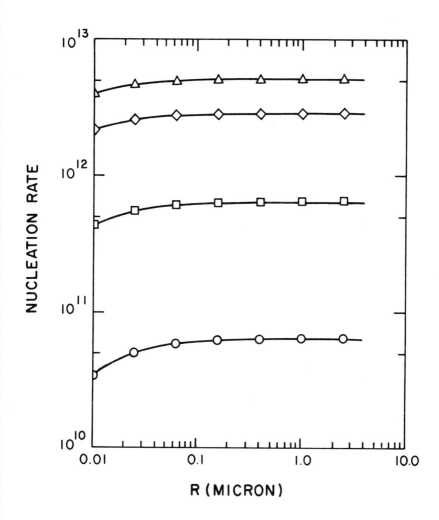

FIGURE 11.   Dependence of nucleation rate per unit area of
the substrate on radius of curvature of the surface.   Acid
concentration = 10^{10}/cc for ◯ R. H. = 30%, ▢ R. H. = 50%,
◇ R. H. = 80%, △ R. H. = 100%.

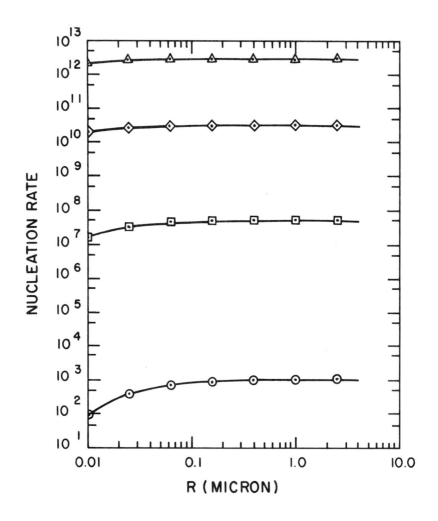

FIGURE 12.   Dependence of nucleation rate per unit area of
the substrate on radius of curvature of the surface. Relative
humidity = 80%, for ⊙ A. C. = 10^7, ▣ A. C. = 10^8, ◇ A. C. =
10^9, △ A. C. = 10^{10}.

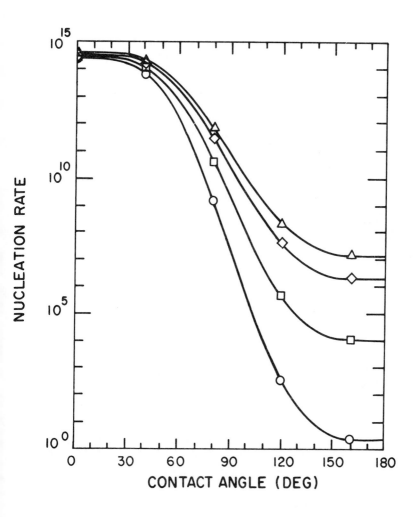

FIGURE 13. *Dependence of nucleation rate per unit area of substrate on the contract angle. Acid concentration = 10^{10}/cc for ○ R. H. = 30%, □ R. H. = 50%, ◇ R. H. = 80%, △ R. H. = 100%.*

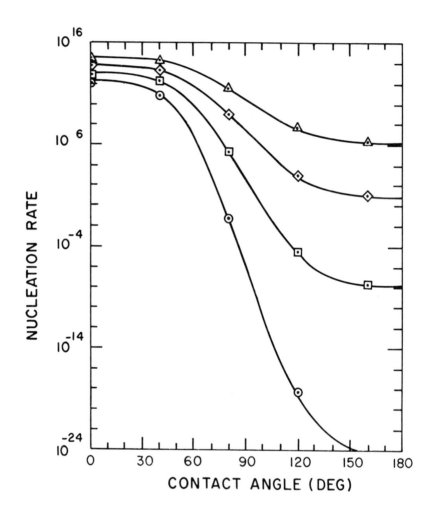

FIGURE 14. *Dependence of nucleation rate per unit area of substrate on the contact angle. Relative humidity = 80% for* ⊙ *A. C. = 10^7,* ⊡ *A. C. = 10^8,* ◇ *A. C. = 10^9,* △ *A. C. = 10^{10}.*

equal to 10^7, 10^8, 10^9, and 10^{10}. From Figs. 11 and 12, it can be seen that when R_S is larger than about 0.1 μm, the nucleation rate is almost independent of the size of the condensation nucleus; however, when R_S is less than about 0.1 μm, there is a strong dependence of nucleation rate on the radius of curvature of the underlying surface, especially when the ambient concentration of P_A or P_B is low. Generally speaking, the smaller the values of R_S, the smaller the nucleation rate J_H.

     In order to investigate the effect of the contact angle on the nucleation rate J_H, the value of R_S is fixed at 0.1 μm and J_H is plotted against θ for different ambient conditions in Figs. 13 and 14. In Fig. 13, H_2SO_4 concentration was fixed at 10^{10}, and relative humidity set equal to 30%, 50%, 80%, and 100%. In Fig. 14, relative humidity was fixed at 80% and acid concentration set equal to 10^7, 10^8, 10^9, and 10^{10}. The strong dependence of the nucleation rate on the contact angle can easily be seen from the plotted curves.

V. CONCLUDING REMARKS

     The formation of sulfate aerosols through homogeneous nucleation, ion-induced nucleation, and heterogeneous nucleation in an H_2SO_4-H_2O binary mixture under different ambient conditions has been investigated. The following conclusions can be drawn from this study:

     1. Although water vapor, sulfuric acid, and temperature are three factors controlling the formation of sulfate aerosols through homogeneous nucleation process, the concentration of water vapor is the dominant parameter determining the critical size, composition, and nucleation rate of sulfate aerosols.

2. The introduction of ions into the environment will greatly enhance the chance of formation of sulfate aerosols.

3. The decrease of the radius of the condensation nucleus will reduce the heterogeneous nucleation rate only when it is less than about 0.1μm.

4. The actual microstructure of the condensation nucleus is an important parameter to determine the nucleation rate of sulfate aerosols formed by condensing water and sulfuric acid vapors on their surface.

5. The relative importance of the homogeneous nucleation, ion-induced nucleation, and heterogeneous nucleation is determined by the following factors: the ambient concentrations of H_2O, H_2SO_4, ions and condensation nuclei, physical properties of the ions and condensation nuclei, and the temperature of the environment. Generally speaking, most of the ions will be nucleated first and the heterogeneous nucleation process is dominant in comparison with the homogeneous nucleation process.

SYMBOLS

a_A	water activity
a_B	acid activity
A	refers to H_2O
B	refers to H_2SO_4
d	$-(R_s^2 + r^2 - 2rR_sm)^{\frac{1}{2}}$
g	refers to gas phase
G	Gibbs free energy of formation
J	homogeneous nucleation rate
J_H	heterogeneous nucleation rate per unit area of substrate
J_I	ion-induced nucleation rate
K	Boltzman constant
l	refers to liquid phase
m	cosine of contact angle

M_A molecular weight of H_2O

M_B molecular weight of H_2SO_4

n_A number of moles of H_2O in embryo

n_B number of moles of H_2SO_4 in embryo

N_A number concentration of H_2O molecules

N_A^{abs} number of H_2O molecules adsorbed per unit area of
         substrate

N_B number concentration of H_2SO_4 molecule

N_C number concentration of ions

P_A ambient vapor pressure of H_2O

P_B ambient vapor pressure of H_2SO_4

P_A^o equilibrium vapor pressure over a flat surface of pure
         H_2O

P_B^o equilibrium vapor pressure over a flat surface of pure
         H_2SO_4

P_A^{sol} partial vapor pressure of H_2O over a flat surface of
         H_2SO_4-H_2O solution

Q charge of the ion core

r radius of embryo

r_o radius of ion core

$r*$ critical radius of embryo

r_1^* critical radius corresponds to local minimum of Gibbs
         free energy

r_2^* critical radius corresponds to local maximum of
         Gibbs free energy

R universal gas constant

R_s radius of substrate

S_A relative humidity

S_B relative acidity

S_B^{sol} partial vapor pressure of H_2SO_4 over a flat surface of
         H_2SO_4-H_2O solution

S_{12} surface area between the nucleating phase and condensing
         embryo

S_{13} surface area between nucleating phase and substrate

S_{23} surface area between condensing embryo and substrate

T   absolute temperature

V_2 volume of embryo

x   fraction of the mass of H_2SO_4 in solution embryo

x*   value of x at saddle point

β_A impinging rate of H_2O molecules

β_B impinging rate of H_2SO_4 molecules

$\delta\Delta G_I^*$ change of ΔG^* at r_1^* to r_2^*

ΔG change of Gibbs free energy through homogeneous
           nucleation process

ΔG^* value of ΔG at the saddle point

ΔG_I change of Gibbs free energy through ion-induced
           nucleation process

ΔG_I^* value of ΔG_I at saddle point

ΔG_H change of Gibbs free energy through heterogeneous
           nucleation process

ΔG_H^* value of ΔG_H at saddle point

ε dielectric constant of solution

θ contact angle

REFERENCES

1. Singh, J. J., and Khandelwal, Elemental Characteristics
    of Aerosols Emitted from a Coal-Fired Heating Plant.
    NASA Technical Memorandum 78749 (1978).

2. Yue, G. K., *J. Aerosol Sci. 10,* 75 (1979).

3. Nair, P. V. N., and Vohra, K. G., *J. Aerosol Sci. 6,*
    265 (1975).

4. Kiang, C. S., and Stauffer, D., *Faraday Symp. Chem. Soc.*
    7, 26 (1973).

5. Mirabel, P., and Katz, J. L., *J. Chem. Phys. 60,* 1138
    (1974).

6. Roedel, W., *J. Aerosol Sci. 10,* 375 (1979).

7. Yue, G. K., and Chan, L. Y., *J. of Colloid and Interface Sci. 68,* 501 (1979).

8. Fletcher, N. H., *J. Chem. Phys. 29,* 572 (1958).

9. Hidy, G. M., and Brock, J. R., "The Dynamics of Aerocolloidal Systems," p. 285. Pergamon Press, New York (1970).

DISCUSSION

Whitby: What is the magnitude of the nucleation rates that would be predicted from your theory for the power plant plume conditions that we have been discussing.

Yue: I think it depends on whether there is homogeneous nucleation or ion-induced nucleation process. For the homogeneous nucleation process, we did some quick estimation; if the sulfuric acid concentration is about 0.2 ppb then we found that the H_2SO_4 vapor pressure is in the range of about 0.15×10^{-6} torr. And this depends on the relative humidity. Suppose the relative humidity is about 100%, then the nucleation rate is about 10^{10}; however, if it changes, say, to 80%, then the nucleation rate is 10^8 numbers per cc per second.

Whitby: You predict a very steep change in the nucleation rates with humidity.

Yue: I think because this is purely theoretical, in the actual conditions they are competing with each other. This homogeneous nucleation process is competing with other kinds of heterogeneous nucleation and this will greatly reduce the number of aerosols formed through the homogeneous nucleation process.

Singh: I think on one of your slides you showed the nucleation rate as a function of the aerosol size and after a size of about 1.0 μm or so the nucleation rate became constant. For that size aerosols I thought that almost entirely vapor deposition on the existing aerosols would be the way they would disappear rather than through the nucleation. Does that mean that the nucleation rate becomes zero there? Does it become constant after the aerosol size has gone to about 1 micron or larger?

Yue: Yes. For different ambient conditions you have different nucleation rates here.

Singh: Take any one condition like the last one.

Yue: Take the last one. Here (see Fig. 11 in text). If you increase the size of the substrate, the nucleation rate remains the same no matter whether radius is 0.1 μm or 1.0 μm.

Singh: The condensation of the vapors on the preexisting aerosols does not present a problem for the nucleation itself?

Yue: Here the only condensation is the heterogeneous nucleation process. And theoretically we found that it is almost independent of the size of the substrate. But the unit of nucleation rate is that the number of aerosols formed per square centimeter of the substrate per second. If you increase the number concentration of preexisting aerosols, the total amount of vapors consumed through heterogeneous nucleation process will of course be increased.

ASSESSING IMPACTS OF FOSSIL FUEL ENERGY
SYSTEMS ON GEOCHEMICAL CYCLES

Robert C. Harriss

NASA-Langley Research Center
Hampton, Virginia

This paper presents the general framework of a methodology
for assessing resource impacts of energy options (ARIEO) which
is based on material accounting principles and is initially
applied to an assessment of impacts on heavy metal cycling of
competitive systems for energy production. The fluxes of mobi-
lized and embedded metals associated with each component of an
energy system for the mining of fuels and nonfuel raw materials,
upgrading of fuels and materials, production of electricity,
transmission networks, to the final disposal of all wastes are
determined and compared with natural fluxes. An ARIEO case
study is presented which suggests that mercury emissions from
U.S. fossil fuel electrical power production may exceed natural
emissions to the atmosphere by a factor of 40 during the coming
decade. The significant sources of mercury pollution from
largest to smallest are petroleum combustion, coal combustion,
iron and steel production, and cement production. The atmos-
phere will be an increasingly important pathway for the input of
mercury pollution to aquatic systems remote from the sources.

I. INTRODUCTION

A major issue facing both industrial and developing nations
is the inadequacy of existing primary energy sources and conversion systems for the long-term future. The industrial nations
face the progressive depletion of certain easily accessible,
reasonably cheap nonrenewable fossil fuel resources (1). The

83

developing nations face a dilemma of the depletion of wood,
presently a primary energy source for most low-income rural populations, and the necessity of primary sources of higher
quality energy to realize development aspirations (2). Thus,
primary energy systems which presently support most of the
world's population are in the initial phase of a major transition
to ultimately more abundant, environmentally and socially
acceptable energy sources which may be based on nuclear fusion,
solar technology, low energy lifestyles and other options.
However, the most immediate problem is how to assess energy
options available for the next 50 years, a minimum lead time for
the development of any significant large-scale shift to alternative energy systems.

Assuming that economic growth and development will continue
to be a basic objective of most societies, albeit at a reduced
rate and with more efficient use of natural resources, the
energy options for the next 10 years of the transition era can
be very generally stated as follows:

(1)   Accelerated discovery, production and consumption of
petroleum

(2)   Accelerated production and consumption of coal

(3)   Accelerated deployment and consumption of nuclear
fission power plants

(4)   Increased productivity from energy use leading to
increased material output per unit energy invested in economic
systems and, most likely

(5)   A combination of all of these options (See refs. 1 and 3
for further discussion of energy policy issues).

This paper presents the general framework of a methodology
for assessment of resource impacts of energy options (ARIEO).
The methodology is based on a materials accounting approach and
is initially being applied to an assessment of impacts of competitive systems for electrical energy production on the

geochemical cycles of selected heavy metals. This paper includes
a case study on potential mercury emissions from fossil fuel
energy systems, as a simple example of the proposed approach to
geochemical forecasting.

II. ENERGY OPTIONS AND GEOCHEMICAL CYCLES: A SYSTEMS APPROACH

A. *The System Boundaries*

The basic structure of the ARIEO methodology is given in
Fig. 1. The objective of the assessment is to quantify anthropogenic chemical flows associated with all aspects of an
electrical generation system. Each stage in a particular system
from the initial step of mining fuels to the final step of
electrical energy transmission to the consumer involves the
mobilization and/or transformation of materials.

In the ARIEO methodology anthropogenic chemical flows are
categorized into embedded flows to the system infrastructure and
emission flows to the environment. The embedded material flows
which provide the infrastructure of the system (e.g., mining
equipment, electrical generation plant, etc.) are ultimately
converted to environmental emissions because of the finite
efficient operating time span of any technological system. The
approximate operational time span for an electrical generating
plant is 30 to 50 years, after which time the infrastructure
materials are recycled or become residual wastes in the environment. For example, in the case of a nuclear generating plant
or nuclear fuel reprocessing plant, the post-operation disassembly
and disposal of spent plant materials may contribute a larger
total flow of radioactive pollution to the environment than the
cumulative flow of operational emissions. Emissions to the
environment are characterized as continuous or episodic, an
information factor critical both to the assessment of potential
biological effects and for the design of environmental monitoring

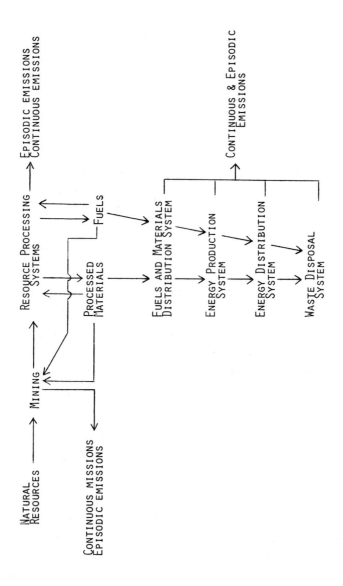

FIGURE 1. Material flow pathways associated with energy systems.

strategies. Principal episodic emissions are associated with
the manufacture of materials for the initial construction of an
energy system and the disassembly and disposal of infrastructure
components.

The ARIEO system boundaries are determined for each energy
option by the total requirements for fuel and nonfuel minerals
to build, produce, and distribute electrical power. The conceptual basis for the ARIEO methodology is similar to that used
in energy analysis (4, 5) and in the Water Energy Land Materials
and Manpower (WELMM) approach to energy assessment currently under
development at the International Institute for Applied Systems
Analysis (6).

B. *Emission Factors and Quantification*

The emission factors for pollutants released by combustion
of fuels during the direct production of electrical power are
obtained from relevant literature and by best available methods
of estimation (see, for example, refs. 7 to 13).

The calculation of embedded material flows and episodic
emissions associated with the construction and destruction of an
energy system infrastructure is a more complex process involving
several steps. First, the nonfuel material requirements for the
construction and operation of an energy option are determined.
The primary data source on material requirements for specific
energy options used in this paper is the Minerals for Energy
Production program of the U.S. Geological Survey (14). The
second step is to determine the episodic emissions associated
with the manufacture of the materials used in the construction
of an energy system. (See, for example, ref. 15.) The episodic
emissions associated with the destruction of energy system infrastructure are determined with data on the expected lifespan
of the system and information on the likely disposal options for
the material residuals.

The only remaining emissions to be quantified relate to the land disturbance produced by mining operations for both fuel and nonfuel minerals and by the development of distribution systems required for fuel transport and electricity transmission. The impacts of land use changes on geochemical cycles as a result of activities such as mining, road construction, etc. are now relatively well documented in the literature. (See, for example, refs. 6 and 16.)

The computational process for quantified emissions is illustrated by

$$
\begin{bmatrix} E_{1yz} \\ E_{2yz} \\ \cdot \\ \cdot \\ \cdot \\ E_{iyx} \end{bmatrix} = P_{yx} \begin{bmatrix} F_{1yz} \\ F_{2yz} \\ \cdot \\ \cdot \\ \cdot \\ F_{iyz} \end{bmatrix} \tag{1}
$$

where

E_{iyz} quantified emissions of type i in year y from source z

P_{yx} electrical power generation in year y from fuel type x

F_{iyz} emission factor of type i in the year y for source z.

Depending on the objective of the assessment and the electrical generation system under investigation, several approaches can be taken to evaluate environmental impacts for particular energy strategy. Emissions with similar units can be (1) summed over index iz to obtain total emissions for the life of a system x, (2) summed over index iyz to obtain total emissions for any particular year in the life of a system x, or (3) any specific emission iz related to system x can be compared with emissions from similar sources for a different system.

In some cases the calculation of summed emissions can be simplified when order-of-magnitude estimates indicate any emission E_{iz} to be small relative to other sources. For example, in the relatively simple case of mercury emissions associated with coal-fired electrical generation (a case study treated in a following section of this paper) the emission of mercury from coal cumbustion, iron and steel production, and concrete manufacture exceeds all other potential emission sources by orders of magnitude and other sources can be excluded from detailed calculations.

C. *Incorporating Temporal and Spatial Characteristics of Emissions into an Assessment*

The variations of specific emissions in time and space are important to the ultimate goal of determining exposure of organisms to potential pollutants and predicting biological effects. In the initial stage of development of the ARIEO methodology, the treatment of these variables uses the simplest possible approach, albeit a very adequate approach for the type of broad assessments being undertaken.

A general characteristic of the present application of the ARIEO methodology is that emissions are associated with a specific time y related to generation at the source. Thus, a dynamic projection of emission patterns over time is obtained by incremental calculations over set time intervals. There is no capacity in the present phase of development for handling delays between an emission source and a potential pollution impact. Buehring and Foell (17) in their more general treatment of occupational, social, and ecological impacts of electrical energy generation, have proposed using a Green's function that describes impacts at time t' due to electrical generation at time t by

$$Q(t) = E(t) \int_t^{t'} I(t,t') \, dt' \tag{2}$$

where

> $Q(t)$ quantified environmental impacts associated with electrical generation at time t
>
> $E(t)$ electrical generation at time t
>
> $I(t,t')$ impacts that occur at time t' per unit energy use at time t

The initial ARIEO approach to handling spatial considerations is to identify geographical boundaries for each calculation. Generally, national boundaries are used since the objective is to test the implications of alternative national energy strategies. However, the choice of geographical boundaries can range from national to global depending on the expected sources and transport characteristics of a particular pollutant under investigation. Later attempts to expand the methodology to an assessment of potential biological effects will require coupling with more elaborate pollutant transport models.

III. IMPACT OF PROJECTED FOSSIL FUEL ELECTRICAL GENERATION ON THE GEOCHEMICAL CYCLE OF MERCURY

The following case study on mercury emissions from projected fossil fuel electrical generation systems is presented to illustrate the basic approach used in the ARIEO methodology. This particular case study was selected for the following reasons: (1) Coal and petroleum are important transition fuels to fill the projected energy gap prior to the availability of long-term substitutes such as nuclear fusion and solar power. Considerable increases in coal and petroleum consumption are projected for the period 1975-1990 both in the United States and

on a global basis (18, 19). (2) Mercury is a ubiquitous component of coal and oil and has been previously identified as a
potential pollutant associated with fossil fuel systems (7, 8,
20, 21). (3) The high concentrations of mercury in just a few
components of fossil fuel systems (e.g., shale used for the
manufacture of Portland cement and coal used as fuel) simplifies
the calculation of summed emissions by limiting the number of
potentially signficant sources. (4) The more comprehensive
ARIEO approach provides new and extended estimates of the potential impact of proposed energy policies on the regional and
global geochemical cycle of mercury.

A highly condensed illustration of the process for calculating emissions is given in Figure 2. Because of space
limitations, this paper will only include a summary of results
on mercury emissions projected for coal and petroleum based
electrical generation in the United States for the period
1977 to 1987. The detailed calculations are available from the
author on request.

Current coal and petroleum consumption for electrical
generation were obtained from the U.S. Department of Interior
(19). A 5 percent per year growth rate in coal consumption and
a 3 percent per year growth rate in petroleum consumption were
assumed for the period 1977 to 1987. Data on projected nonfuel
minerals and materials required by the U.S. energy industry
for approximately the same time interval were taken from Albers
et al. (14). Emission factors were calculated principally from
a literature review by the United Kingdom Department of the
Environment (22) and from papers listed in the earlier sections
of this paper.

The principal sources of mercury emissions associated with
coal-fired electrical generation for the next decade are
projected as follows:

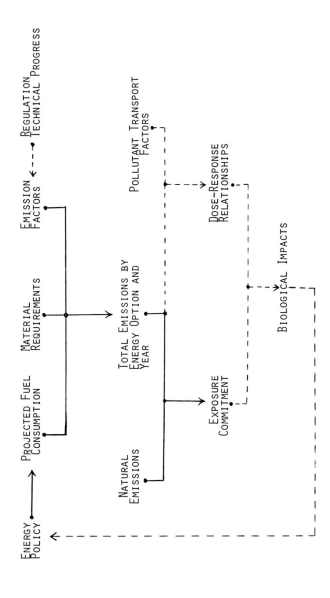

FIGURE 2. Assessment process for determining emissions from energy systems and their resource impacts (dashed lines indicate factors not discussed in the present paper).

Combustion of coal . . . . 3.0 x 10^8 g Hg

Production of cement . . . . 0.1 x 10^8 g Hg

Production of iron and steel . . . . 0.7 x 10^8 g Hg

Total (1977-1987) . . . . 3.8 x 10^8 g Hg

The principal sources of mercury emissions associated with petroleum-fired electrical generation for the next decade are as follows:

Combustion of petroleum . . . . . . 5.0 x 10^8 g Hg

Production of cement . . . . 0.1 x 10^8 g Hg

Production or iron and steel . . . . 2.0 x 10^8 g Hg

Total (1977-1987) . . . . 7.1 x 10^8 g Hg

All the emission sources listed are a result of the volatilization of mercury during high temperature processes. In the case of a coal-fired system, the cement, iron, and steel are primarily required for construction of new power plants and emissions will follow the pattern of plant construction. In the case of a petroleum-fired system, the cement, iron, and steel are primarily required for offshore exploration and development activities and emissions will follow the pattern of offshore petroleum development. Such information can be important to the design of environmental monitoring programs.

The only significant quantities of mercury embedded in the infrastructure of fossil fuel electrical generation systems are associated with the distribution of electricity and are projected to amount to 5 x 10^8 grams Hg for the period 1977 to 1987. Assuming a replacement rate of 10 percent for the decade would mean that about 5 x 10^7 grams Hg would be deposited in landfill sites. By using an emission rate determined for chlorine-production solid waste deposits at 20° C of approximately 20 µg hr^{-1} for 4 kg of solid waste (23), a preliminary estimate

of mercury vapor emissions from electrical wastes in landfills
is 2×10^4 grams Hg for the decade. In this case the loss of
mercury to aquatic pathways may predominate but more research
is required in this area.

Calculations indicate that mobilization of mercury from
land disturbance associated with mining and from the production
of nonfuel minerals and materials other than cement, iron, and
steel is less than 10^4 grams Hg per decade and can be neglected
for the purpose of national-scale assessment.

IV.   IMPLICATIONS FOR ENVIRONMENTAL MONITORING AND ASSESSMENT

The results of this ARIEO case study can be compared with
calculated natural mercury vapor fluxes from the continental
U.S. to obtain an indication of the energy-related anthropogenic
impact of the mercury cycle. The natural flux of mercury to the
atmosphere from U.S. soil can be estimated as 2.7×10^7 grams Hg
per decade, using a natural degassing rate of 0.03 μg m^{-2} hr^{-1}.
Thus, the anthropogenic input of mercury to the atmosphere over
the next decade from fossil fuel electrical power generation
alone may exceed natural inputs by a factor of approximately 40.
Since calculations by Goldberg (24) and those in this paper
demonstrate that cement, iron, and steel production are important
sources of anthropogenic mercury vapor emissions, and the use of
these materials for energy production is only a minor fraction
of their total consumption, the overall anthropogenic impact
will exceed the calculated factor of 40.

The results of this ARIEO study are in agreement with the
conclusions of a recent review and appraisal of current literature on mercury by Matheson (25). Both this study and the
literature review conclude that future environmental monitoring
for mercury should focus on the atmosphere. The ARIEO

methodology provides a guide to specific sources of anthropogenic emissions (i.e., fossil fuel plants, cement production, and iron and steel production) which can be used to design a monitoring strategy for elucidating atmospheric dispersion and deposition mechanisms for mercury.

With increasing control on the direct discharge of mercury wastes to aquatic systems, the atmosphere will become the predominant pathway for the dispersion of anthropogenic mercury emissions (25). The high volatility of mercury may enhance long distance transport. (See, for example, Refs. 26, 27.) Over the long term, the temperature dependence of the volatilization of mercury should lead to the progressive transfer of mercury to colder environments with atmospheric input being the primary source of mercury input to natural waters (23). Such a process may, at least partially, explain the unexpectedly high levels of mercury in fish in northern Canadian lakes which have no pollution sources or mineralized soils in their watersheds (D. R. Miller, Pers. Comm.).

V. ALTERNATIVE APPLICATIONS OF THE ARIEO METHODOLOGY

The idealized assessment process illustrated in Fig. 2 indicates the necessary steps to make a complete assessment where energy production can be linked to biological impacts which can in turn provide feedback to the formulation of energy policy. This paper has only presented a partial assessment of one potential pollutant; the pollutant transport models and dose-response relationships for mercury are presently inadequate to complete the process.

There are several alternative applications for the ARIEO approach which may provide useful input to energy policy analysis. For example, methodology used in this paper to calculate emissions can be computerized and operated in a simulation mode to project

the effects of variables such as changing fuel sources, improved
pollution control technologies, different fuel consumption rates
and changing nonfuel material requirements on future heavy metal
emissions. In the case of mercury the magnitude of the projected
anthropogenic impact from fossil fuel energy production suggests
that serious consideration should be given to the development of
a pollution control technology for removal of mercury from stack
gases. However, it would be critical to track mercury recovered
from stack gas; this would be automatically accomplished with the
comprehensive materials accounting approach of the ARIEO methodology.

Another alternative application of the ARIEO methodology is
the assessment of nonfuel mineral requirements associated with
any energy option. For any energy policy, nonfuel mineral
requirements can be projected and compared with available
supplies. For example, the increasing adoption of nuclear
energy for electrical power generation, particularly in countries
without internal fossil fuel resources, may produce independence
from fuel-rich countries (e.g., OPEC) but create a nonfuel
mineral supply problem for another material critical to the
nuclear fuel cycle. Nuclear energy systems require considerably
more molybdenum, nickel, tin, zirconium and other heavy metals
than do fossil fuel systems (14).

ACKNOWLEDGMENTS

This effort was initiated when the author was in residence
as a Senior Fellow at the Monitoring and Assessment Research
Center, London, during the summer of 1977. The ARIEO concept is
currently in the early stages of a second-phase investigation
with support from the Director's Research Fund, NASA Langley
Research Center.

REFERENCES

1. Lovins, A., "Soft Energy Paths." Penguin Books, New York (1977).

2. Makhijani, A., "Energy Policy for the Rural Third World." IIED Report, London (1976).

3. Hafele, W., and Sassin, W., *Energy, 1,* 147 (1976).

4. Odum, H. T., "Environment, Power and Society." Wiley, New York (1971).

5. Chapman, P., Leach, C., and Slesser, M., *Energy Policy, 2,* 120 (1974).

6. Grenon, M., and Lapillone, B., "The WELMM Approach to Energy Strategies and Options: IIASA Report RR-76-19." International Institute for Applied Systems Analysis, Laxenburg, Austria (1975).

7. Klein, D. H., Andren, A. W., and Bolton, N. E., *Water, Air, and Soil Pollution, 5,* 71 (1975).

8. Bertine, K., and Goldberg, E. D., *Science, 173,* 233 (1971).

9. Hofstader, R. A., Milner, O., and Runnels, J. (eds.), "Analysis of Petroleum for Trace Metals: Adv. Chem. Series 156." Amer. Chem. Soc., Washington, D.C. (1976).

10. Block, C., and Dams, R., *Environ. Sci. Technol. 9,* 146 (1975).

11. Andren, A. W., Klein, D., and Talmi, Y., *Environ. Sci. Technol. 9,* 856 (1975).

12. Kaakinen, J. W., Jorden, R. M., Lawasani, M., and West, R., *Environ. Sci. Technol., 9,* 862 (1975).

13. Robertson, D. E., Crecelius, E., Fruchter, J., and Ludwick, J., *Science, 196,* 1094 (1977).

14. Albers, J. P., Bawiec, W., and Rooney, L., "Demand for Nonfuel Minerals and Materials by the United States Energy Industry, 1975-90: U.S. Geological Survey Professional Paper 1006-A. "Government Printing Office, Washington, D.C. (1976).

15. U.S. Environmental Protection Agency, "Compilation of
     Air Pollutant Emission Factors." Office of Air Programs
     Report, Research Triangle Park, North Carolina (1972).

16. Turner, R. R., Harriss, R. C., and Burton, T., *in* "Mineral
     Cycling in Southeastern Ecosystems: ERDA Symposium Series
     Conf. 740513" (F. G. Howell, J. B. Gentry, and M. H. Smith,
     eds.), p. 868. U.S. Government Printing Office,
     Washington, D.C. (1975).

17. Buehring, W. A., and Foell, W. K., "Environmental Impacts
     of Electrical Generation: A System-Wide Approach: IIASA
     Report RR-76-13." International Institute for Applied
     Systems Analysis, Laxenburg, Austria (1976).

18. U.S. Federal Energy Administration, "Project Independence
     Report," 15 Vols. U.S. Government Printing Office,
     Washington, D.C. (November 1974).

19. U.S. Department of the Interior, "Energy Perspectives 2."
     U.S. Government Printing Office, Washington, D.C. (1976).

20. Billings, C. E., and Matson, W. R., *Science, 176,* 1232
     (1972).

21. Joensuu, O. I., *Science, 172,* 1027 (1971).

22. United Kingdom Department of the Environment,
     "Environmental Mercury and Man: Pollution Paper No. 10."
     HMSO, London (1976).

23. Lindberg, S., and Turner, R., *Nature, 268,* 133 (1977).

24. Goldberg, E. D., "The Health of the Oceans." UNESCO Press,
     Paris (1976).

25. Matheson, D. H., "The Atmospheric Transport of Mercury: A
     Review and Appraisal of the Current Literature." Inland
     Waters Directorate Report, Canada Centre for Inland Waters,
     Burlington, Canada (1977).

26. Klein, D. H., and Russell, P., *Environ. Sci. Technol. 7,*
     357 (1973).

27. Weiss, H. V., Kiode, M., and Goldberg, E. D., *Science, 174,*
     692 (1971).

DISCUSSION

Pack: It's not so much a question but it's an agreement with your comment about the monitoring. You mentioned going out in a structured monitoring once a month. This is precisely, I think, what one should not do. One tries to anticipate episodes, and may I point out that the NOAA Geophysical Monitoring for Climatic Change has two aspects that may be of use to this particular group. First, it was designed to measure continuously--and by continuously I mean every few seconds for some elements and never less than about once per hour 24-hours per day if the technology permitted. The second is that they have four well-equipped good logistics sites at Barrow, Alaska, Mauna Lua, Hawaii, Samoa and at the South Pole and we have, ever since I started the work, encouraged people who need data to take advantage of such logistics to try to get the kind of information that is needed for this geochemical cycling work.

Harriss: Thank you, Don. I have been talking to John Miller about doing some measurements at the NOAA stations and I think it is the next step.

Demerjian: You mentioned that in your calculations for power plants you are assuming that 60 percent of the mercury coming in with the coal would ultimately go into the gas phase. What is the basis of that? Most of the data I've seen has been 90 percent or more, even after going through scrubbers.

Harriss: The basis for that was the data that URS had obtained on scrubbers that were giving an efficiency that resulted in the 60 percent emission. I have also made calculations based on 100 percent and 90 percent emission.

Singh: Is there any metal other than mercury that might exhibit this tendency to exist longer in vapor form. Or, is mercury the only practical one?

Harriss: No, I think that there are a number of other such elements, for example, arsenic and selenium. And, as far as some of the heavy metals are concerned, zinc and cadmium may also be present in significant amounts of the vapor phase near the source.

Singh: That's rather interesting. But they don't exist in the vapor phase, normally speaking, because the vapor pressures are quite low.

Harriss: I think part of it goes back to my comment earlier this morning on their particular geochemistry in the feed material.

In the coal itself, there may be, at least for most of the
cadmium, an association of the original material with organics
which may result in ready volatilization.

MEASUREMENT OF TRACE METALS IN COAL
PLANT EFFLUENTS--A REVIEW

Jag J. Singh

NASA Langley Research Center
Hampton, Virginia

Trace metals constitute an important component of all fossil fuels--including coal. Increasing reliance on coal as the future source of energy has spurred great interest in improved techniques for the measurement of heavy metals in effluents from various types of coal plants. There are several sensitive techniques currently used for the measurement of trace metals in environmental specimens. These can be divided into four major categories: (1) Atomic Absorption/Emission Spectroscopy, (2) Nuclear Techniques, (3) X-Ray Spectroscopic Methods, and (4) Chromatography/Mass Spectrometry. Atomic absorption/emission spectroscopy is widely used in such diverse fields as biochemistry, metallurgy, and air/ water analysis. It is especially suited for aqueous sample analysis. However, the technique does not lend itself easily to multi-element analysis and is destructive of the test specimen. Nuclear spectroscopic techniques include Charged Particle Scattering, Charged Particle Activation Analysis, and Instrumental Neutron Activation Analysis (INAA). Charged particle scattering and activation techniques are usually appropriate for the deter- mination of lighter elements, most of which cannot be readily determined by other analytical methods. INAA is a widely utilized technique for fly/bottom ash elemental analysis, although certain critical elements (Be, P, S, Tl, Pb) cannot be easily determined. X-ray spectroscopic techniques include X-ray fluorescence and charged particle induced X-ray emission methods. A discussion of these last two methods will constitute the main subject of this review. Besides providing simultaneous, sensitive multi-element analyses, these techniques lend themselves more readily to depth profiling which has become increasingly important in aerosol studies. The gas/liquid chromatographic and gas chromatographic/ mass spectrometric methods are quite sensitive but rather slow, destructive of the sample, and inappropriate for airborne particulate analysis.

101

I.   INTRODUCTION

Until a few years ago, the main environmental quality monitoring effort was concentrated on the measurement of gaseous com-
ponents--such as hydrocarbons and oxides of carbon, nitrogen and
sulphur--emitted from coal-fired power plants and coal processing facilities. Recently, other effluents (such as aerosol/
fly ash, bottom ash, and sludge from stack scrubbers) have also
received increasing emphasis. Special attention has been paid
to the trace metal components of these effluents. It has been
found that most volatile compounds in coal are emitted as gases
or parts of aerosols emanating from the plant stacks. Some of
the trace elements are also left in the bottom fly ash.

In this review, special consideration will be given to the
measurement of the composition of aerosols emitted from coal
plants. These aerosols have been reported to exhibit elemental
fractionation by several authors (1, 2). Certain potentially
toxic elements have been found to be preferentially concentrated
in finer (< 2 μm) aerosols. This result is particularly significant in view of the fact that the finer aerosols are not
efficiently filtered in the nasal passages of the human
respiratory system.

There are several techniques for trace element determination
in the environmental specimens. These can be broadly grouped
into four different categories:

1.   Atomic Spectroscopic Techniques

a.   Atomic absorption/emission spectroscopy

b.   Atomic fluorescence spectroscopy

These techniques are excellent for samples in liquid form.

2. Nuclear Techniques

    a. Charged particle scattering

    b. Charged particle activation

    c. Neutron activation

    d. Photon activation

These techniques are all nondestructive and well suited for analyzing solid samples.

3. X-Ray Spectroscopic Techniques

    a. X-Ray fluorescence

    b. Charged particle induced X-ray emission

        (1) Proton induced X-ray emission and heavy ion induced X-ray fine structure

        (2) Electron excited X-ray emission

These techniques are nondestructive (or can be nondestructive) and are excellent for solid samples, although liquids as well as gases can also be analyzed.

4. Miscellaneous Techniques

    a. Gas/Liquid chromatography

    b. Ion probe mass spectrometry

    c. Photoelectron spectroscopy

Since all these techniques are standard, no detailed explanation of the experimental methods is given for any of them. Only the strong features of various techniques and their limitations will be discussed, except in the case of heavy ion induced X-ray fine structure technique which will be discussed in somewhat greater detail.

X-ray spectroscopic techniques (and, to a lesser extent, nuclear techniques) will receive major attention, not necessarily because they are among the most sensitive, but because they are suitable for nondestructive multielement aerosol analysis and also lend themselves readily to depth profiling which is becoming increasingly important in aerosol studies.

II. DISCUSSION OF VARIOUS TECHNIQUES

A. *Atomic Spectroscopic Techniques*

A brief summary of elemental detection capabilities of atomic
spectroscopic techniques will first be given for the sake of com-
pleteness of discussion. Atomic spectroscopy is widely used in
such diverse areas as geochemistry, metallurgy, and air/water
analysis. Its attractiveness stems from rather low cost, reasonable throughput rate and the flexibility of operation (flame or
nonflame atomization). The sensitivities of atomic spectroscopic
techniques for most elements of interest are summarized in
Table I (3-5). However, atomic spectroscopic techniques are not
convenient[1] for simultaneous multielement analysis and are destructive of the test specimens. They will not be discussed any
further in this review.

B. *Nuclear Techniques*

Charged particle scattering (CPS) techniques include elastic
and inelastic scattering and are more appropriate for lighter
elements which constitute most of the mass of the finer aerosols.
Figure 1 shows how the energy of the elastically scattered protons
changes with angle of scattering for various elements. It is
clear that the energy differences in the backward directions for
protons scattered from neighboring light elements are sufficiently
large to make them easily resolvable. Figure 2 shows a typical
spectrum of elastically scattered protons from aerosol sample (6)
at $\theta_{(Lab)}$ = 135°. In the inelastic scattering studies, it is

---

[1]*Some attempts at simultaneous multielement analysis using
multielement sources and multichannel approaches have recently
been reported. However, the sensitivities reported in those
studies are considerably worse than conventional AA/AF values.*

TABLE 1. *Summary of Elemental Detection Limits by Atomic Absorption/Emission Spectroscopic Technique*

Element	Flameless atomic absorption (Ref. 3)		Laser induced flame atomic fluorescence (Ref. 4)		Plasma emission spectroscopy (Ref. 5)
	$\lambda(\overset{o}{A})$	$(2\sigma), \frac{ng}{ml}$	$\lambda(\overset{o}{A})$	$(3\sigma), \frac{ng}{ml}$	$(2\sigma), \frac{ng}{ml}$
Ag	3280.68	0.005	3281	4	4
Al	3961.53	0.020	3944/3961	0.6	1
Ba	5535.55	0.150	5537	8	1
Bi	3067.72	0.100	3068	3	50
Ca	4226.73	0.050	4227	0.08	0.1
Cd	2288.02	0.003	2288	8	5
CO	3453.50	0.100	3474/3575	1000	2
Cr	3578.69	0.010	3593	1	1
Cu	3247.54	0.020	3247	1	1
Fe	3734	0.020	2967/3735	30	2
In	4511.32	0.300	4104/4511	0.2	25
K	7664.91	0.020	-	-	10
Mg	2852.13	0.004	2852	0.2	0.1
Mn	4030.76	0.010	2795	0.4	1
Mo	3798.25	0.070	3798	12	1
Na	5889.95	< 0.500	5890	0.1	1
Ni	3414.76	0.200	3524/3610	2	1
Pb	4057.83	0.050	2833/4058	13	15
Sr	4607.33	0.200	4607	0.3	1
Ti	3998.64	1.000	3999	2	2
Tl	3519.24	0.100	3776	4	-
V	4739.24	0.200	3704/4112	30	5
Zn	2138.56	0.001	-	-	2
S	1807.31	5000.000	-	-	-
Hg[a]	1849.68	0.100	-	-	-

[a]*Flameless limits are given for 100 microliters of solution*

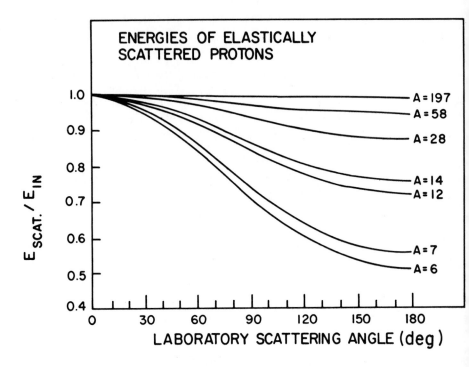

*FIGURE 1. Dependence of elastically scattered proton
energies on angle and scatterer atomic weight.*

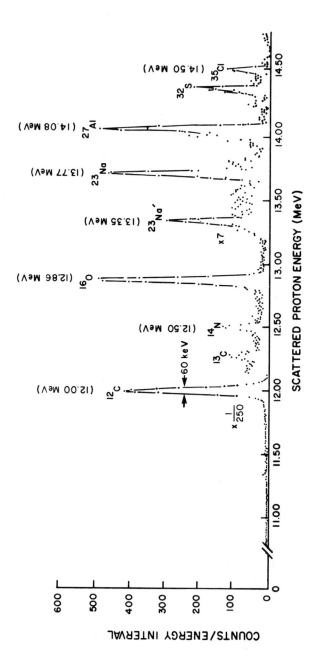

FIGURE 2. *Spectrum of 16 MeV protons scattered from an aerosol sample at* $\theta_{Lab} = 135^{O}$
$(Q = 4$ *microcoulombs), adapted from Ref. 6.*

usually the first excited state of the pollutant nucleus that is
significantly excited. Determination of the pollutant nuclei
(atoms) can be made either on the basis of inelastically scattered
proton groups or simultaneously emitted gamma rays. (For
example, see Na^{23} inelastic peak in Fig. 2.) Charged Particle
Activation Analysis (CPAA) is also appropriate for the determination of lighter elements, most of which cannot be determined
by other analytical methods, except elastic/inelastic scattering
reactions just discussed. Some typical reactions (7) for light
element detection by CPAA are listed in Table II(a). For elements
with Z > 10, CPAA is quite complementary to Instrumental Neutron
Activation Analysis, although it has not been used much except
for the elements that cannot be determined by thermal neutrons.
Some examples (7) of heavier element detection by CPAA are
summarized in Table II(b). Interferences can present a problem
in CPAA studies. However, use of several incident energies,
coupled with an appropriate choice of reactions often proves
quite useful. For example, $N^{14}(p, \alpha)C^{11}$ and $B^{11}(p, n)C^{11}$
reactions can be resolved because their relative cross sections
change from 1/2 to 1/10 as the proton energy decreases from 15 MeV
to 5 MeV. A better technique of resolving boron and nitrogen
interferences may depend on $B^{10}(d, n)C^{11}$ and $N^{14}(d, \alpha n)C^{11}$
reactions whose relative cross section ratio changes from 30 to
infinity as the deuteron energy decreases from 10 MeV to 5.9 MeV.

Instrumental neutron activation analysis (INAA) technique has
been used extensively for nondestructive analysis in several
fields--including aerosol analysis--although certain critical
elements (such as Be and Pb) cannot be easily determined.
Interferences due to threshold reactions are very rare under
normal irradiation conditions where the thermal neutron flux far
exceeds the fast neutron flux. The few reactions of concern
include: F^{20} (formed by neutron capture in F^{19} as well as (n,α)
reaction in Na^{23}) and Mg^{27} (formed by neutron capture in Mg^{26}

TABLE II. Summary of Detection Limits by Charged Particle Activation Analysis (Ref. 7)

(a) Light elements

Element	Nuclear Reaction	Sensitivity and Comments
B	(a) $B^{11}(p, n)C^{11}$ at E_p = 14.5 MeV	0.5 ppb (polycrystalline silicon matrix) 3 ppb (monocrystalline silicon matrix)
	(b) $B^{10}(d, n)C^{11}$	1 to 10 ppb
C	$C^{12}(He^3, \alpha)C^{11}$	0.3 ppm 0.01 ppm (if C/O ratio is not too unfavorable)
N	(a) $N^{14}(p, \alpha)C^{11}$	1 ppb (requires irradiation with two proton energies to separate boron and nitrogen)
	(b) $N^{14}(d, n)O^{15}$	1 to 10 ppb
O	$O^{16}(He^3, p)F^{18}$	Several ppb (in favorable matrices)
F	$F^{19}(He^3, 2p)F^{20}$	70 ppb (normalized to a beam of 100 μamps on a 100 mg/cm^2 matrix and 1.63 MeV gamma ray detection efficiency of 10 percent)

(b) Heavy elements

Element	Matrix	Reaction	Sensitivity	Interference
S	Fe Al	$S^{34}(p, x)Cl^{34m}$	0.1 ppm	$Cl^{35}(ppn)Cl^{34m}$
Ca	Si Al Mg	$Ca^{40}(He^4, p)Sc^{43}$	0.3 ppb	$K^{41}(He^4, 2n)Sc^{43}$
V	Al	$V^{51}(p, n)Cr^{51}$	36.0 ppb	–
Tl	Glass	$Tl(p, xn)Pb^{203}$	1.0 ppb	$Pb^{204}(ppn)Pb^{203}$
Pb	Pd Ta Glass	$Pb^{206}(p, n)Bi^{206}$	10.0 ppb	–
Bi	Glass	$Bi^{209}(p, 3n)Po^{207}$	1.0	–

as well as np reaction in Al^{27}). Spectral interferences resulting
from finite gamma ray detection system resolution do present
some problems. Some such interferences are: Mg^{27} (844 keV) and
Mn^{56} (846 keV); Se^{75} (121 keV) and Eu^{152} (122 keV); Hg^{203}
(279.1 keV) and Se^{75} (279.6 keV); Cu^{64} (511 keV) and annihilation
radiation.

Sometimes it may be more convenient to use fast neutron-
induced activation, particularly when cross sections for (n,p) and
(n,α) reactions of interest are accurately known (8). For example,
Si^{28}(n,p)Al^{28} reaction with 14.5 MeV neutrons at a flux of
10^9 n/cm^2 - sec easily allows silicon concentration determinations
as low as 50 ng/m^3.

Typical detection limits for several pollutant elements using
INAA technique are summarized in Table III (9, 10). Because of
the general availability of nuclear reactors and large thermal
neutron capture cross sections for many nuclei, the greater part
of activation work has been done with thermal neutrons where
sensitivities of the order of 10^{-9} g are not at all unusual.

Some elements--such as Be, C, N, O, F, Pb--which are not
highly activated with thermal neutrons can be studied with
Instrumental Photon Activation Analysis (IPAA). For beryllium,
detection of photoneutrons is used as the criterion for elemental
detection. For elements C, N, and O, the photoactivation products
decay exclusively by β^+-emission and the associated annihilation
radiation serves as the basis for their detection. For elements
heavier than oxygen, the A(γ, n/p)B reaction products are
identified by their characteristic γ-ray spectra detected with a
Ge(Li) detection system. Sometimes interferences experienced in
INAA can be avoided by using IPAA method. For example, determination of nickel may be complicated by Ni^{64}(n, γ)Ni^{65} and
Cu^{65}(n, p)Ni^{65} interferences in INAA spectrum. However, the
photonuclear reaction product Ni^{57} produced in Ni^{58}(γ, n)Ni^{57}
cannot be produced from any other element at photon bombarding
energies less than 45 MeV. IPAA technique is equally applicable

TABLE III. *Summary of Detection Limits for Several Trace*
Elements by Instrumental Neutron Activation Analysis (Refs. 9,10)

Element	Detection limit (24-hour urban sample)[a] ng/m^3	Element	Detection limit (24-hour urban sample)[a] ng/m^3
Ag	1	K	7.5
Al	8	Mg	600
Ba	40	Mn	0.6
Bi	–	Mo	–
Br	Br^{80m} (4)	Na	40
Br	Br^{82} (0.5)	Ni	20
Ca	200	Pb	–
Cd	5	S	5000
Cl	100	Si	–
CO	0.02	Sr	–
Cr	0.20	Ti	40
Cu	5	Tl	–
F	–	V	0.2
Fe	20	Zn	Zn^{65} (1.0)
In	In^{116m} (0.04)	Zn	Zn^{69m} (20.0)

[a] $\phi = 2.6 \times 10^{12}$ n/cm^2 - sec (24-hour sampling with high
volume sampler).

to biological, geochemical, and oceanographic samples. Like
other nuclear techniques, interferences occur in IPAA also.
But they can usually be avoided by appropriate choices of photon
energies and judicious choices of irradiation and cooling times
Cu, Zn, Zr, and Ag are among the elements determined in this
way in complex metal ores. Interferences between $F^{19}(\gamma, n)F^{18}$
and $Na^{23}(\gamma, \alpha n)F^{18}$ can be avoided by using a 22 MeV beam since
the $Na^{23}(\gamma, \alpha n)F^{18}$ reaction has a threshold of 23 MeV. Thus,
IPAA may be better than INAA for detecting trace elements in the
presence of large quantities of Na^{23} (as is the case for marine
aerosols). Also, Na^{23} can be studied by means of $Na^{23}(\gamma, n)Na^{22}$
as opposed to $Na^{23}(n, \gamma)Na^{24}$ which produces an overwhelming
2.76 MeV gamma ray. IPAA also has some interesting applications
in forensic studies. For example, IPAA has been used to measure
lead content of whiskey by $Pb^{204}(\gamma, n)Pb^{203}$ reaction to determine
its contraband origin!

Sometimes IPAA is based on the excitation of metastable
isomers by (γ, γ') reactions. Although (γ, γ') sensitivities are
rather low, their specificity is very good. If the irradiation
is conducted at suitably low energies ($E_\gamma < 8$ MeV), the only
activation produced in IPAA will be that due to the production
of isomeric nuclides.

Although, in general, IPAA is not as sensitive as INAA, it
can be used to measure concentrations of several elements that
are difficult (or impossible) to measure by INAA--especially
toxic elements like Ti, Ni, As, I, and Pb. The combined effect
of the electron bremsstrahlung spectrum and the excitation
function for photonuclear reactions as a function of target
atomic number, leads to a general increase in photonuclear
reaction cross section with increasing Z-value. This result
makes it possible to detect low quantities of high Z-elements in
the presence of much greater quantities of low Z-material.
Table IV summarizes limits of detectability for several elements

TABLE IV.   *Comparison of Detection Limits by IPAA and INAA Techniques for Selected Elements in Urban Aerosols (Refs. 11,12)*

Element	Detection limit, $\frac{ng}{m^3}$		Element	Detection limit, $\frac{ng}{m^3}$	
	IPAA (Ref. 11)	INAA (Ref. 12)		IPAA (Ref. 11)	INAA (Ref. 12)
As	0.2	–	Na	2	0.002
Br	30	0.005	Ni	0.05	0.25
Ca	30	2	Pb	12	–
Ce	0.4	2×10^{-5}	Sb	0.3	2×10^{-5}
Cl	0.4	0.05	Ti	0.9	–
Cr	4.5	4×10^{-5}	Zn	3	4×10^{-5}
I	0.17	–	Zr	0.2	–

1.   *It is assumed that the aerosol samples were collected from 1000 m^3 air.*

2.   *IPAA performed with bremsstrahlung from 50 μamp beam of 35 MeV electrons.*

3.   *INAA performed with neutrons at a flux of 6 x 10^{13} n/cm^2-sec.*

4.   *Gamma rays detected with 55 cm^3 Ge(Li) detector.*

in urban aerosols for IPAA and INAA techniques (11, 12). Sometimes it may be preferable to use K X-rays from (γ, n) reaction
products--particularly when $N_x(K\alpha)/N_\gamma$ is much greater than 1. A
comparison of IPAA and X-ray spectroscopic analysis following the
photon activation is given in Table V (13,14).

C. X-Ray Spectroscopic Techniques

X-ray spectroscopic techniques include X-ray fluorescence and
charged particle induced X-ray emission (both electron induced
and heavier particle induced). They can deal with samples in all
physical forms although they are not frequently used for gaseous
and liquid samples. X-ray fluorescence (XRF) can be conducted
with radioactive sources as well as tube excited X-rays. The XRF
detection limits can be improved with secondary targets in the
path of the main X-ray beam. Prominent X-ray fluorescers are:
$Ti(\alpha)$ = 4.5 keV; $Mo(K\alpha)$ = 17.5 keV; $Sm(K\alpha)$ = 40.0 keV; and
$W(K\alpha)$ = 59.3 keV. By choosing appropriate secondary targets,
most elements can be analyzed to ng/m^3 sensitivity range. Use of
pulsed beam operation, coupled with anti-coincidence guard ring
detection system for X-rays helps to improve elemental sensitivities. Sometimes, use of polarized X-rays may also be
desirable since it can minimize X-ray scattering effects. In
this context, the use of monochromatized synchrotron radiation
would be especially helpful. Desirable features of XRF technique are (1) It is rapid, (2) it provides simultaneous multielement analysis, (3) it is nondestructive and needs no special
sample preparation, (4) samples can be very small (1 mg or so),
and (5) it lends itself to automation for large sample throughput
rates.

Experimental detection limits (15) for routine trace element
concentrations in environmental samples, using tube-excited
pulsed X-ray fluorescence system, are summarized in Table VI.
The problem of elemental spectral interferences in XRF analyses

TABLE V. *Comparison of Conventional IPAA and X-Ray Spectroscopic Analysis Following Photon Activation*

Element	Reaction	$\dfrac{N_x(K\alpha)}{N_\gamma}$ (Ref. 13)	Detection Limits[a], μg	
			γ-Rays	X-Rays
Cu	$Cu^{65}(\gamma, n)Cu^{64} \xrightarrow[12.8 \text{ hrs}]{EC/\beta^+} Ni^{64}$	120	32	0.28
As	$As^{75}(\gamma, n)As^{74} \xrightarrow[17.9 \text{ d}]{EC/\beta^+} Ge^{74}$	0.6	0.7	1.1
Zr	$Zr^{90}(\gamma, n)Zr^{89} \xrightarrow[78.4 \text{ hrs}]{EC/\beta^+} Y^{89}$	1.1	0.27	0.24
Pd	$Pd^{104}(\gamma, n)Pd^{103} \xrightarrow[17.0 \text{ d}]{EC} Rh^{103}$	4.3	5.5	1.3
Sn	$Sn^{118}(\gamma, n)Sn^{117m} \xrightarrow[14.0 \text{ d}]{\gamma(M_4)} Sn^{117}$	0.18	2.2	12
Cd	$Cd^{116}(\gamma, n)Cd^{115} \xrightarrow[53.5 \text{ hrs}]{\beta^-} In^{115m} \xrightarrow[4.5 \text{ h}]{\gamma(M_4)} In^{115}$	0.7	0.7	1.0
Hg	$Hg^{198}(\gamma, n)Hg^{197m} \xrightarrow[24 \text{ hrs}]{\gamma(M_4)} Hg^{197} \xrightarrow[65 \text{ h}]{EC} Au^{197}$	2.8	0.9	0.3
Pb	$Pb^{204}(\gamma, n)Pb^{203} \xrightarrow[52.1 \text{ hrs}]{EC} Tl^{203}$	0.7	0.8	1.2

[a] Calculated assuming 6 hours decay time (from the end of irradiation) and 10^3 as minimum number of counts in 12 hours (Ref. 13,14).

TABLE VI.   Experimental Limits of Detection for Trace
Elements Using Automated Pulsed X-Ray Fluorescence System
(Ref. 15)

12-hour Samples (at 3 m^3/hr)

Secondary target	Element	Minimum Detectable limits (3σ) ng/m^3	Secondary target	Element	Minimum Detectable limits (3σ) ng/m^3
Ti	Al	40	Mo	Zn	1.06
(Elements	Si	11.78	(Elements	Ga	0.76
analyzed	P	6.58	analyzed	As	0.62
for 93.6	S	5.88	for 83.4	Se	0.52
sec)	Cl	5.32	sec)	Br	0.54
	K	1.83		Rb	0.56
	Ca	1.54		Sr	0.76
				Hg	1.20
Mo	Ti	6.26		Pb	1.82
(Elements	V	4.44			
analyzed	Cr	3.32	Sm	Cd	1.18
for 83.4	Mn	2.70	(Analysis	Sn	1.56
sec)	Fe	2.36	time =	Sb	1.62
	Ni	1.24	153 sec)	Ba	6.20
	Cu	1.30			

can be just as severe as in other analytical procedures. Some of
the interferences encountered in aerosol analyses are summarized
in Table VII. It is obvious that Ti and S suffer really bad interferences from Ba and Pb, respectively, if the latter elements are
present in significant quantities. Fortunately, this is not
always the case. Determination of K is also affected by the
presence of Cd and Sn in the specimen. The XRF analysis of air
particulate samples has to be corrected for the following effects:
(1) Matrix erfects (particle size and inter-element interference
effects), (2) particle penetration into substrate (when collected
on filter papers), and (3) bremsstrahlung background produced in
the target. A number of matrix correction procedures have been
devised which require the use of samples of known composition (16).
In the empirical approach, one obtains influence coefficients
from a multiple regression calibration involving a large number of
standards, which are then used to calculate the corrected concentrations of each element in the unknown sample. The minimum
number of calibration standards required generally equals twice
the number of elements to be analyzed. In the theoretical procedure, one corrects for matrix effects using theoretical
relations involving known values of absorption coefficients and
the fluorescence yields of the elements involved, as well as
an explicit form of the excitation radiation spectrum and determines the corrected elemental concentrations by iterative calculations. The substrate penetration correction factor is
obtained by analyzing both sides of the collection filter. The
most obvious physical cause of background in XRF spectra is the
bremsstrahlung radiation produced by the secondary electrons in
the sample. However, most of the experimentally observed background does not seem to originate in the sample. The integrated
number of background counts is 2 to 8 percent of the "high energy"
counts (incident radiation scatter peaks) depending on the
secondary target. This background appears to be the result of

TABLE VII.   Examples of Common Interferences Encountered in
XRF Aerosol Analysis

Element (a)	X-Ray line for (a) keV	Element (b)	X-Ray line for (b) keV	Interference coefficient[1] between a and b, α_{ab} (Ref.15)
K	3.314 (Kα)	Cd	3.367 (Lβ)	0.36 ± 0.05
K	3.590 (Kβ)	Sn	3.708 (Lβ)	0.24 ± 0.04
Ca	3.692 (Kα)	Sn	3.708 (Lβ)	0.23 ± 0.04
Mn	6.490 (Kβ)	Fe	6.404 (Kα)	0.017 ± 0.001
Ti	4.932 (Kβ)	Ba	4.852 (Lβ)	0.52 ± 0.03
V	4.952 (Kα)	Ba	4.852 (Lβ)	0.28 ± 0.02
S	2.308 (Kα)	Pb	2.346 (Mα)	0.50 ± 0.03

[1] C'_a = corrected concentration of element a

$$= C_b - \Sigma \; \alpha_{ab} \; C_b$$

where C_b = true concentration of b (standard)

and $\alpha_{ab} = \dfrac{\text{Apparent concentration of element a}}{\text{True concentration of element b}} = \dfrac{C_a}{C_b}$

yet unknown processes in the detector (17). Once the physical
mechanism responsible for this background is understood, it may
be possible to minimize/eliminate it.

Proton induced X-ray emission (PIXE) technique has been used
extensively for multielement characterization of aerosols (18,19).
Monoenergetic protons of energy ranging from 1 to 5 MeV, as well
as equivalent alpha particles can be used as projectiles,
although the former have been used more frequently. The analysis
is usually performed at two different proton energies for uniform
sensitivities over the entire elemental range. Elements with
atomic numbers 11 through 30 are analyzed with lower energy protons (~ 2 MeV), whereas elements heavier than zinc require
higher proton energies (~ 4 MeV). Figure 3 shows a typical
X-ray spectrum of an urban aerosol sample. Spectrum (a) was
obtained with 2 MeV protons, and 4 MeV protons were used for
spectrum (b). The results of a routine PIXE analysis of aerosol
samples collected downwind from the heating plants at Langley
Research Center are summarized in Table VIII (20). The use of
an appropriate "funny filter" (21) between the aerosol sample and
the detector also permits reasonably uniform sensitivity for
medium to heavy elements in the presence of more abundant lighter
elements. Detection limits below 1 ng/m^3 are easily achieved for
most elements. PIXE technique is rapid and lends itself easily
to pulsed automated operation with improved elemental sensitivities for routine aerosol analysis.

Goulding et al. (17) have calculated PIXE performance data
for 2 and 4 MeV protons (100 nanoamps for 200 seconds at each
energy) incident on a 5 mg/cm^2 sample containing 1 ppm by weight
of several trace elements. The detector geometrical efficiency
was assumed to be 0.3 percent. Using the 3 σ criterion for
limit of detection, the calculated detection limits for various
elements are summaried in Table IX(a). Calculations for an XRF
system for the same total counting time per sample as was allowed

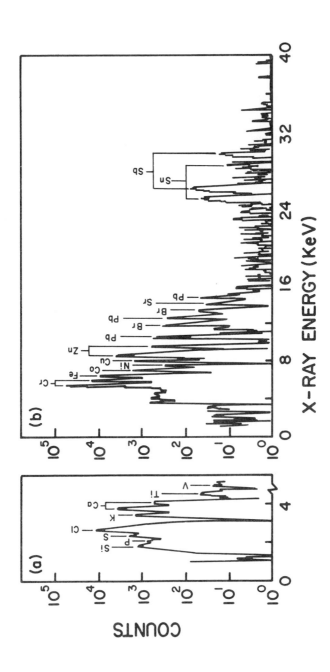

FIGURE 3. Typical X-Ray spectra from an aerosol sample. Spectrum (a) was obtained with 2 MeV protons while 4 MeV protons were used for the spectrum (b).

TABLE VIII. Elemental Concentrations Observed in Fine Aerosols (0.43 to 0.65 μm) Collected Downwind from Two Heating Plants at Langley Research Center (Ref. 20)

24-hour Samples (at 1.7 m^3/hr)

Element	Heating plant 1 (concentrations in $\frac{ng}{m^3}$)	Heating plant 2 (concentrations in $\frac{ng}{m^3}$)
V	128.11 ± 10.27	30.33 ± 10.09
Mn	< 20.61	2.70 ± 3.02
Fe	14.99 ± 1.12	40.22 ± 4.30
Ni	17.49 ± 0.54	4.43 ± 0.52
Cu	5.44 ± 0.36	23.10 ± 0.99
Zn	33.72 ± 0.55	91.06 ± 3.47
As	0.19 ± 0.44	4.43 ± 1.85
Se	0.91 ± 0.18	0.76 ± 0.15
Br	1.80 ± 0.28	13.22 ± 0.64
Rb	< 0.67	0.67 ± 0.29
Sr	< 0.69	0.75 ± 0.21
Mo	1.74 ± 0.50	1.32 ± 0.39
Ag	4.22 ± 1.52	5.64 ± 1.33
Cd	< 4.53	5.33 ± 1.55
Sn	24.66 ± 4.84	45.66 ± 5.24
Ba	< 32.58	27.20 ± 10.34
Pb	16.52 ± 1.09	53.55 ± 2.49

TABLE IX.   Calculated and Measured Elemental Detection Limits

(a) Calculated Limits of Detection for Selected
     Elements in a PIXE Analysis System Using
     2 MeV and 4 MeV Protons (Ref. 17)

| Element | Limits of detection (for 20 microcoulombs) | |
	E_p = 2 MeV	E_p = 4 MeV
S	0.24 ppm	—
Ca	0.27 ppm	—
Fe	0.05 ppm	—
Zn	0.04 ppm	—
Br	0.07 ppm	0.05 ppm
Mo	0.26 ppm	0.06 ppm
Cd	1.23 ppm	0.13 ppm
Pb(Lα)	0.24 ppm	0.21 ppm

(b) Comparison between the Calculated and Measured
     Elemental Detection Limits for an Automated
     Pulsed XRF Analysis System (Ref. 17)

Secondary target	Element	Calculated Detection Limit (ppm)	Measured Detection Limit (ppm)
Ti (100 sec)	Al	8.2	8.2
	S	2.7	2.6
	Ca	1.7	—
Mo (100 sec)	Fe	1.3	1.8
	Zn	0.6	1.1
	Br	0.4	0.5
	Pb(Lα)	0.9	1.5
Sm (200 sec)	Mo	1.0	—
	Cd	0.7	0.9

in discussing PIXE analyses, and using Ti, Mo, and Sm fluorescers
to cover roughly the same range of atomic numbers, are summarized
in Table IX(b). It is apparent that the two techniques are quite
comparable in elemental sensitivities and are indeed complementary to each other. PIXE technique is superior to XRF in the
following respects:

    1. It has microbeam capability for individual particle
analysis.

    2. It admits of depth profiling, using ion microprobes.
However, PIXE analysis is more sensitive to sample matrix than
XRF, and thus makes the latter preferable when measuring
moderately thick samples such as filter papers or larger aerosols.

    PIXE analysis is often conducted in concert with elastic
scattering, where elastically scattered particles provide useful
information about lighter elements (Z < 11) which cannot be
detected by X-ray techniques.

    Electron excited X-ray spectrometry is not as sensitive as
XRF or PIXE, mainly because of large bremsstrahlung noise produced by the primary electron beam in the sample. However, it
has excellent spatial resolution. Electron microprobe analysers
and scanning electron microscopes provide effective tools for
individual aerosol analysis.

    None of the X-ray techniques discussed so far provide any
information about the chemical forms of the elements. And yet
the chemical form is of extreme importance in toxicology. One
variant of PIXE can, however, lead to (or at least has the
potential of leading to) the chemical form of the pollutant atom.
This involves the use of heavier ions as the projectiles. The
X-ray spectra produced in such ion-atom collisions cannot be
resolved with conventional Si(Li) detectors but wavelength dispersive X-ray detectors can provide some very good results.
Recent studies (22-24) have shown that the intensity distribution
of the $K\alpha$ satellites produced in heavy ion-atom collisions is

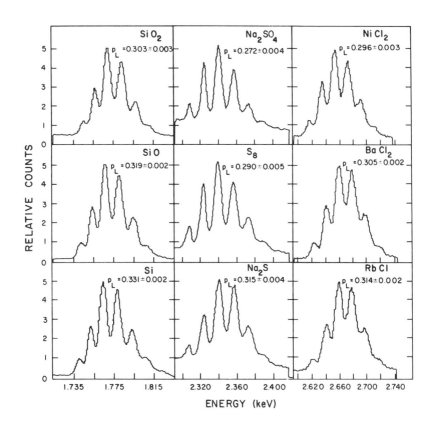

FIGURE 4.   Sample Kα satellite spectra for several Si, S, and
Cl compounds showing the variation of the relative satellite
intensities with chemical environment.   These spectra were all
taken with 32.4 - MeV oxygen ions.   (Adapted from Refs. 22 and 23)

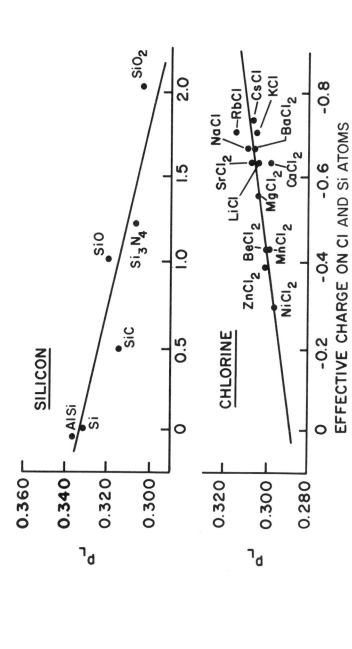

FIGURE 5. Variation of p_L with effective charge for simple compounds of silicon and chlorine. (Adapted from Refs. 22 and 23.)

quite sensitive to the projectile energy and the chemical form of the target atom. As long as $Z_{projectile}$ is less than Z_{target} the relative intensities of the $K\alpha$ satellites appear to be given by the following binary expression

$$f(n) = \text{relative intensity of } n^{th} \text{ satellite peak}$$

$$= \binom{8}{n} p_L^n (1 - p_L)^{8-n} \tag{1}$$

where p_L = average L-vacancy fraction

$$= \frac{1}{8} \sum_{n=1}^{7} nf(n) \tag{2}$$

Best results are obtained for energy per atomic mass unit of the projectile in the range 0.5 - 2.0 MeV/amu. Some typical results are shown in Figs. 4 and 5 (22,23). Figure 4 shows the $K\alpha$ satellite spectra of Si, S, and Cl atoms for several chemical forms of them under bombardment with 32.4 MeV oxygen ions. Notice the changes in p_L values for different chemical forms of the same elements. Figure 5 shows the relationship between p_L and the *effective charge* (defined as the product of oxidation number and bond ionicity) for several silicon and chlorine compounds. It is noticed that p_L decreases with increasing effective charge; this decrease suggests that all the valence electrons which happen to be localized about the target atom at the time of the collision are ionized. The definite correlation observed between L-shell vacancy fraction p_L and the effective charge on the target atom strongly support the conclusion that interatomic processes must contribute to the de-excitation of multiple ionized states following heavy ion-atom collisions. Although changes in p_L from compound to compound are rather small, it is possible that further developments in $K\alpha$-satellite spectrometry could provide a useful means for obtaining information about the chemical form of the target elements in special situations (i.e., a large variety of

compounds of third row elements) (24). Because this technique
provides information relative to the bulk conditions of the
sample, it is complementary to Electron Spectroscopy for Chemical
Analysis (ESCA) which provides information concerning the conditions at the surface only. Figure 6 illustrates how ESCA is
used to infer chemical states of elements of interest on aerosol

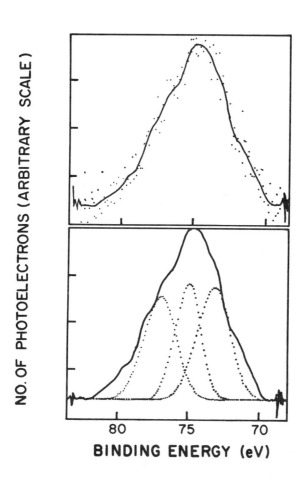

FIGURE 6. Al 2p photopeak spectrum of NASA 75-A-47 aerosols;
raw data and deconvoluted components (Ref. 25).

surfaces (25). This figure shows the spectrum of Al 2p bonding
states on surfaces of aerosols collected from space vehicle
launch rocket exhaust.

Another technique of great promise is a variation of XRF
wherein photoelectron fine structure--rather than characteristic
X-ray spectra--is measured following appropriate incident X-rays.
Extended X-ray Absorption Fine Structure (EXAFS) refers to the
oscillations in the X-ray absorption coefficient extending out to
several hundred electron volts on the high energy side of an X-ray
photoelectron edge (26,27). These oscillations are believed to
arise from the scattering of the ejected photoelectrons by atoms
surrounding the absorbing atom and are intimately related to the
chemical form of the scattering element. Like heavy ion-induced
X-ray fine structure, EXAFS can also provide chemical speciation
information about the target material. Availability of intense
synchrotron radiation sources may, in the future, make EXAFS a
practical technique for bulk chemical speciation of air pollution
samples.

D. *Miscellaneous Techniques*

Chromatographic techniques are very sensitive for gas/liquid
analysis, but they have certain inherent limitations. Gas
chromatography is restricted to gases or compounds that can be
vaporized without chemical dissociation (i.e., atomic weights
200 amu), whereas liquid chromatography requires liquid sample
preparation and consequently suffers from the danger of inadvertent
sample pollution. Furthermore, High Performance Liquid
Chromatography (HPLC) imposes (28) considerable operational
problems--such as application of very high pressures (1000 to
5000 psi) and specially designed pumps to provide a constant flow
of the order of a few milliliters per minute through the packing
column. Ion probe mass spectrometry, like chromatographic techniques, is also destructive of the sample, though it is a sensitive

technique for depth profiling of selected elements in aerosol
samples. Laser Microprobe Mass Analyzer (LAMMA) is the latest
form of an ion-probe mass spectrometer (29,30). It has excellent
sensitivity (10^{-18} to 10^{-20} g) and is essentially nondestructive.
However, LAMMA is essentially a surface instrument since it
analyzes the microplasma created by a short laser pulse which
vaporizes about 10^{-13} g of the sample surface and simultaneously
ionizes it. Photoelectron spectroscopy provides quite useful
information about elemental oxidation states in the top 20 to 25 Å
surface layer of solid specimen. However, this technique requires
sample introduction into high vacuum ($\sim 10^{-6}$ torr) and cannot
provide bulk chemical speciation information--except by a continual destructive etching with a positive argon ion stream to
expose fresh surface.

III. CONCLUDING REMARKS

Various analytical techniques used for trace element measurements in environmental specimens have been discussed. AA techniques are most suitable for aqueous solutions and can routinely
attain sensitivities of the order of ng/ml for most elements,
with few exceptions such as F, P, and S. Even though AA analysis
can be quite fast (regular flame AA ~ 6 secs/sample; graphite
furnace AA ~ 2 mins/sample) special precautions are necessary
against inadvertant sample contaminations. The various nuclear
techniques are quite sensitive, rapid, nondestructive as well as
amenable to automation. However, they do require rather expensive
nuclear sources (a neutron reactor for INAA; a betatron or linac
for IPAA; and a Van de Graaff generator or a cyclotron for CPAA).
Among the most sensitive and rapid nondestructive trace metal
detection techniques are those based on characteristic X-ray
emission from elements of interest. These techniques include
X-ray induced fluorescence and charged particle induced X-ray
emission. However, chemical speciation of the trace elements in

aerosols will require rather expensive dedicated radiation
facilities (a heavy ion accelerator for PIXE and X-ray fine
structure spectroscopy and a variable energy intense monochromatic
photon source for EXAFS).

In summary, it appears that no single technique is equally
good for analyzing all types of environmental specimens. Gas
chromatography and mass spectrometry are quite sensitive for
analyzing gaseous samples. Atomic absorption and high performance
liquid chromatography are appropriate for analyzing samples in
liquid form. Solids may be sensitively analyzed by nuclear and
X-ray spectroscopic methods, particularly when nondestructive
techniques are required. It is recommended that, whenever
feasible, two or more complementary techniques should be used
for analyzing environmental samples.

REFERENCES

1. Keyser, T. R., Natusch, D. F. S., Evans, C. A., Jr., and
   Linton, R. W., *ES & T, 12,* 768 (1978).

2. Singh, J. J., and Khandelwal, G. S., NASA Technical
   Memorandum 78749 (1978).

3. Techniques and Applications of Atomic Absorption, Publication
   #AA-322G, March 1978 (Perkin-Elmer Co., U.S.A.).

4. Weeks, S. J., Haraguchi, H., and Winefordner, J. C.,
   Anal. Chem. 50, 360 (1978).

5. Reednick, J, *Am. Lab. 11,* 53 (1979).

6. Nelson, J. W., "X-Ray Fluorescence Analysis of Environmental
   Samples," pp. 19-34. Ann Arbor Science Publishers, Inc.,
   Ann Arbor, Michigan (1977).

7. Cornelis, R., Hoste, J., Speecke, A., Vandecasteele, C.,
   Versieck, J., and Gijbels, R., "International Review of
   Science--Physical Chemistry, Series Two," Vol. 12,
   Analytical Chemistry--Part I (T. S. West, ed.), Butterworths,
   Woburn, Massachusetts (1976).

8. Pritchard, W. M., Khandelwal, G. S., and Singh, J. J. *Virginia J. of Sci. 28,* 19 (1977).

9. Heindryckx, R., and Dams, R., *Progress in Nuclear Phys. 3,* 219 (1979).

10. Dams, R., De Corte, Fe., Hertogen, J., Hoste, J., Maenhaut, W., and Adams, F, "Physical Chemistry, Series Two," Vol. 12 Analytical Chemistry--Part I (T. S. West, ed.), Butterworths, Woburn, Massachusetts (1976).

11. Natusch, D. F. S., Bauer, C. F., and Loh, A., "Air Pollution Control, Part III; Measuring and Monitoring Air Pollutants" (W. Strauss, ed.), Wiley Interscience, New York (1978).

12. Aras, N. K., Zoller, W. H., and Gordon, G. E., *Anal. Chem. 45,* 1481 (1974).

13. Segebade, C., and Weise, H. P., *J. Radioanal. Chem. 45,* 209 (1978).

14. Weise, H. P., and Segebade, C., *J. Radional. Chem. 37,* 195 (1977).

15. Jakelvic, J. M., Gatti, R. C., Goulding, F. S., Loo, B. W., and Thompson, A., Proceedings of the 4th Joint Conference on Sensing Environmental Pollutants, American Chemical Society, pp. 697-702 (1978).

16. "X-Ray Fluorescence Analysis of Environmental Samples," (T. G. Dzubay, ed.), Ann Arbor Science Publishers, Inc., Ann Arbor, Michigan (1977).

17. Goulding, F. S., and Jaklevic, J. M., *Nucl. Instr. & Meth. 142,* 232 (1977).

18. Johansson, S. A. E., and Johansson, T. B., *Nucl. Instr. & Meth. 137,* 473 (1976).

19. Walter, R. L., and Willis, R. D., *Practical Spectroscopy 2,* 123 (1978).

20. Singh, J. J., Sentell, R. J., and Khandelwal, G. S., NASA TM X-3401 (1976).

21. Cahill, T. A., "New Uses of Ion Accelerators (J. F. Ziegler, ed.), Plenum Press, New York (1975).

22. Watson, R. L., Leeper, A. K., Sonobe, B. I., Chiao, T., and Jensen, F. E., *Phys. Rev. A. 15*, 914 (1977).

23. Demarest, J. A., and Watson, R. L., *Phys. Rev. A. 17*, 1302 (1978).

24. Watson, R. L., Demarest, J. A., Langenberg, A., Jensen, F. E. White, J. R., and Bahr, C. C., *IEEE Trans. Nucl. Sci. NS-26*, 1352 (1979).

25. Dillard, J. G., Seals, R. D., and Wightman, J. P., NASA CR-3153 (1979).

26. Stern, E. A., Sayers, D. E., and Lytle, F. W., *Phys. Rev. B. 11*, 4836 (1975).

27. Lytle, F. W., Sayers, D. E., and Stern, E. A., *Phys. Rev. B. 15*, 2426 (1977).

28. Robinson, A. L., *Science, 203*, 1329 (1979)

29. Wechsung, R., Hillenkamp, F., Kaufmann, R., Nitsche, R., and Vogt, H., *Scanning Elect. Microscopy, 1*, 611 (1978); *Microscopica Acta, 52*, 281 (1978); *Mikroskopie, 34*, 47 (1978)

30. Maugh, T. H., II, *Science, 203*, 1331 (1979).

DISCUSSION

Novokov: My question deals with your ESCA slide. You showed
deconvoluting one of those complex peaks into three components.
The widths, however, of individual components are vastly
different. What is the justification of using component peaks
of so different width for the same element?

Singh: The three Al 2p bonding states in space vehicle launch
rocket exhaust are identified as Al_2O_3 . n H_2O (73.8 eV), Al_2O_3
(75.2 eV), and $AlCl_3$ (77.0 eV). The first and the third components have slightly larger half widths than the Al_2O_3 peak
because they probably include several unresolved chemically
different Al species in them. It has been postulated that
aluminum in rocket exhaust aerosol exists as a mixture of $AlCl_3$
and partially hydrolyzed $AlCl_3$, with the latter leading to a
mixture of Al_2O_3 and Al_2O_3 . n H_2O. Furthermore, the ESCA
results cannot also exclude the presence of Al_2O_3 . HCl as a
partial contributor to the central Al_2O_3 peak at 75.2 eV since
the binding energy for Al 2p in Al_2O_3 . HCl is 75.5 eV. What
figure 6 showed was the best computer fit of the experimental
peak into three resolvable Gaussian components. Obviously,
each Gaussian "component" is in itself a composite as suggested
above.

DETERMINATION OF TROPOSPHERIC AEROSOL SIZE
DISTRIBUTION FROM MULTISPECTRAL SOLAR
AUREOLE MEASUREMENT TECHNIQUES[1]

A. Deepak, M. A. Box[2] and G. P. Box[2]

Institute for Atmospheric Optics and Remote Sensing
Hampton, Virginia

*With greater utilization of coal, it becomes increasingly
important to monitor the background level of atmospheric
aerosols in order to study their impact on environment and
climate. Presented in this paper is a remote sensing technique
for the determination of columnar size distribution of tropospheric aerosols from the multispectral measurements of scattered
radiance in the solar aureole. Solar aureole is a region of
enhanced brightness close to the sun's disk, due to the predominant forward scattering by aerosols. Retrieval of aerosol
size distributions from multispectral solar aureole measurements
are made by taking account of multiple scattering effects.
Results of a solar aureole experiment will be discussed.*

[1]*This work was supported by NASA contract NAS1-15198*

[2]*Present affiliation: University of Arizona, Tucson, Arizona*

I. INTRODUCTION

   With the greater utilization of coal as a source of energy in
the next 20 years, it becomes increasingly important to monitor
the background level of atmospheric aerosols in order to study
their impact on terrestrial environment and climate. Remote
sensing techniques are, perhaps, the most practical and economical
way of monitoring atmospheric aerosols on a long-term basis.
Aerosol characteristics of interest are their size distribution
and concentration, complex refractive index, shape and spatial
distribution. Several remote measurement techniques of varying
accuracy have been developed to determine some of these characteristics. They are most often based on the multispectral measurements of one or more of the following quantities: extinction or
angular distributions of scattered intensity and polarization.
Described here is a method based on the multiwavelength measurements of forward scattered solar-aureole radiation for determining the columnar size distribution (1-6). For references to
other researchers' work on solar aureole techniques see Refs. 2
and 3.

   The problem of retrieving aerosol size distributions from
multispectral scattered or attenuated radiation is complicated
by the fact that the scattering effect is dependent on the size,
shape and complex refractive index of the particle and the
incident wavelength. It may further be complicated by the fact
that multiple scattering is also present. One, therefore, must
adopt strategies based on the physics of the problem to enable
one to extract information about the particular aerosol character-
istic(s) of interest. This is explained in the case of the
solar aureole technique described next.

II. SOLAR AUREOLE TECHNIQUE

Since 1970 solar aureole method has successfully been used to determine the altitude-integrated (or columnar) size distribution of atmospheric aerosols (1-6). Solar aureole is a region of enhanced brightness surrounding the sun's disk within about $20°$ from it, due to the predominant forward scattering by aerosols. The contribution to the solar aureole, due to aerosols, is 10^2 - 10^3 times that due to molecules. It is to take advantage of this large signal that the solar aureole technique was developed in 1970 (1). The success of the technique is due to the fact that in the forward direction, the scattered radiance is highly sensitive to the particle size distribution $n(r)$ $[cm^{-3}\ \mu m^{-1}]$, but is relatively less sensitive to the effects of particle shape, refractive index, polarization and multiple scattering. Molecular absorption effects can be neglected by working in suitably selected spectral regions. This paper describes the method and the size distribution results retrieved from the multispectral solar aureole measurement (PSAM) that were made photographically on May 6, 1977 as part of the University of Arizona's Aerosol and Radiation Experiment (UA-ARE).

A. PHOTOGRAPHIC MEASUREMENTS

The photographic system, the experimental steps and the data analysis procedures used for making the circumsolar radiance measurements are described in detail in Refs. 2 and 7. Photographs of the sun's aureole were taken with a Hasselblad camera through a series of narrow band wavelength filters with the sun occulted by a neutral density (D_n = 4.0) disk held coaxially on a stem about 1.2 meters in front of the lens. The lens focal length was 80 mm; and the film used was Kodak Plus-X (ASA 125).

Photographs of the sun's disk and the aureole were taken through three narrow band ($\Delta\lambda \approx$ 80-100 nm) spectral filters, λ 400 nm, 500 nm and 600 nm.

B. ALMUCANTAR RADIANCE

To determine the sky radiance along the almucantar, which is a scan of constant solar zenith angle, the optical density of the film at points along the conic projection of the almucantar was read with a microdensitometer, and was then converted, by the use of photometric relations, into the sky radiance data (8). The digital values of sky radiance at various angular distances from the sun along the almucantar for the three wavelengths were then used as input data in the retrieval scheme.

C. THEORY

Since the equation for radiative transfer in an aerosol medium cannot be inverted directly, in much of the earlier work the retrieval of aerosol size distribution was made tractable by assuming that multiple scattering (MS) is negligbly small compared to single scattering (SS), and could, therefore, be ignored. In the single scattering approximation, the sky radiance (L) distribution along the almucantar scan ($\theta = \psi$) is given by the relation:

$$L = L_M + L_P = \Phi_0 \sec \theta \; e^{-\tau \sec \theta} \{\tau_M P_M(\psi) + \tau_P P_P(\psi)\}/4\pi \qquad (1)$$

where Φ_0 is the incident solar flux; θ is the solar zenith angle; subscripts M and P denote molecules and particles, respectively, $\tau = \tau_M + \tau_P$, the total optical depth; ψ, the scattering angle given by the relation:

$$\cos \psi = \cos^2\theta + \cos^2\theta \cos \phi \qquad (2)$$

and $P_M = (3/4) (1 + \cos^2 \psi)$ and P_P as the molecular and particulate volume scattering phase functions (sr^{-1}), respectively. P_P and particulate volume scattering coefficient $\beta_P (km^{-1})$ are defined as follows:

$$P_P(\psi,\lambda) = \frac{1}{\beta_P 2k^2} \int_{r_1}^{r_2} (i_1 + i_2) \, n(r) \, dr \qquad (3)$$

$$\beta_P(\lambda) = \int_{r_1}^{r_2} \pi \, r^2 \, Q(x,m) \, n(r) \, dr \qquad (4)$$

where i_1 and i_2 are Mie intensity coefficients, $r_1 = 0.04$ μm and $r_2 = 13$ μm. The molecular contribution can be determined by using the molecular density profile for the particular location; so that from Eq. (1) one can then experimentally determine $P_P (\psi,\lambda)$ as a function of ψ.

This SS approach is well discussed in Refs. 1 and 2. However, it has been shown by Box and Deepak that MS contribution to the solar aureole is not insignificant compared with SS, and that the mode radius of the size distributions retrieved on the basis of the SS alone could have errors up to 10 percent for $\tau \sim 0.4$ (9). Later, it was shown (6) that one can considerably improve the accuracy of the retrievals by using a perturbation approach, which is a modification of the one suggested by Deirmendjian (1) and Sekera (11). The perturbation approach assumes that the solar aureole radiance is essentially due to SS by molecules and aerosols and MS due to molecules alone, with all aerosol events being neglected. This approach has the advantage that it permits one to retain the simplicity of the SS formulae, and, at the same time takes into consideration the effects of MS in an approximate manner. For the solar aureole region, the MS effect is taken into account by simply replacing τ_M in Eq. (1) by an effective optical depth t. The expression for t has been derived in Ref. 5; it shows that t depends on θ, τ and surface albedo A.

For the almucantar the following formula was used for t:

$$t = t_1 + \frac{At_2}{1 - At_3}$$

$$t_1 = \tau + 1.265 \ \tau^2/\mu^{1/4}$$

$$t_2 = 1.32 \ \tau\mu(1 + 0.2(\tau/\mu)^2)$$

$$t_3 = 0.9 \ \tau - 0.92 \ \tau^2 + 0.54 \ \tau^3 \tag{5}$$

Work is in progress to extend such an expression to locations other than the almucantar.

This retrieval approach was applied to a solar aureole experiment performed in Tucson on May 6, 1977. Almucantar radiance measurements were made for three wavelengths (λ = 400, 500 and 600 nm). From these measurements experimental data were obtained for $P_p(\psi)/P_p(3°)$ (Fig. 1), which were inverted by using a two-term size distribution model, each term being a Haze M model, i.e.,

$$n(r) = p_1 \ r \ [\exp \ (- \ p_2\sqrt{r}) + p_3 \ \exp \ (-p_4\sqrt{r})] \tag{6}$$

The retrievals were performed using a non-linear least squares (NLLS) inversion program. Initial estimates of the parameters p are obtained with the help of a parameterized graphical catalog of phase function plots corresponding to different size distribution models (12). The details are given in Ref. 6.

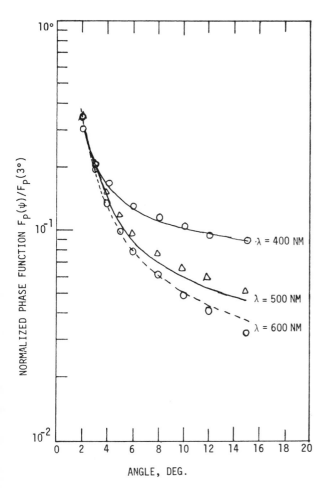

FIGURE 1. *Solid lines represent the final NLLS estimates (after 3 iterations) for the best fits to the experimental data (O, Δ) for phase function $P_p(\psi)/P_p(3^o)$ vs. scattering angle (ψ) for λ = 400, 500, and 600 nm. Aerosol refractive index is assumed to be m = 1.5 - i(0.0). Effects of multiple scattering are included.*

D. *DISCUSSION OF RESULTS*

The model used to fit the data for λ = 400, 500 and 600 nm
was a sum of two Haze M terms. Computer runs were performed to
understand the effects on retrieved results due to different
factors, such as, aerosol refractive indices, integration limit
r_1, and ignoring multiple scattering.

Three refractive indices were used: 1.45 - i(0.00), 1.50 -
i(0.00), and 1.50 - i(0.01). Essentially no difference was
found between the fits obtained for the different refractive
indices although the parameters p_i were slightly different. In
addition, the decrease of r_1 from 0.04 μm to 0.01 μm made no
difference in the final n(r) results. For scattering angles
greater than 12°, it was found that the effects of multiple
scattering need to be taken into account.

The retrieved size distribution obtained by assuming m = 1.55
- i(0) is shown in Fig. 2. Some preliminary ground truth SD data,[1]
obtained by airborne Whitby (⊛) and Royco (+) counters, have
been plotted in Fig. 2. Even though the ground-truth data is
sparse and is for May 7, 1977, the day following the day on which
solar aureole measurements were made, they are reasonably close
to the retrieved results to give us confidence in the latter.

The solar aureole technique is a simple and accurate method
for determining the average size distribution of tropospheric
aerosols. Instead of the photographic method, photoelectric
devices, such as, scanning radiometers and video cameras, can be
used to measure the angular distribution of solar aureole
radiance. Thus, multispectral radiometers used for measuring
the diffuse component of the solar àureole for solar energy
studies may also supply data that could effectively be used
in the determination of aerosol size distributions.

---

[1]*Provided by Dr. J. Reagan, U. of Arizona, Tucson, AZ*

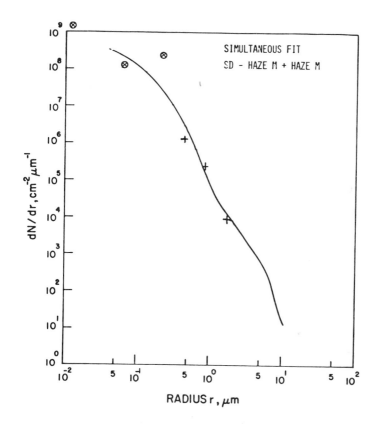

FIGURE 2. Solid curve represents the retrieved size distribution corresponding to the best fits shown in Fig. 1; and symbols represent ground truth measurements taken by Royco (+) and Whitby (⊗) counters.

REFERENCES

1. Green, A. E. S., Deepak, A., and Lipofsky, B. J., *Appl. Opt.*
    10, 1263 (1971).

2. Deepak, A., *in* "Inversion Methods in Atmospheric Remote
    Sounding" (A. Deepak, ed.), p. 297-323. Academic Press,
    New York, (1977).

3. Deepak, A., Box, G. P. , Box, M. A., and Adams, R. R.,
    Determination of Aerosol Characteristics by Photographic
    Solar Aureole Measurements, Preprint Volume: Third Conf.
    on Atmospheric Radiation, June 28-30, 1978, Davis, CA.
    (Published by the American Meteorological Society, Boston,
    Massachusetts).

4. Box, M. A. and Deepak, A. Multiple Scattering Corrections
    to the Solar Aureole, Preprint Volume: Third Conf. on
    Atmospheric Radiation, June 28-30, 1978, Davis, CA.
    (Published by the American Meteorological Society, Boston,
    Massachusetts).

5. Box, M. A. and Deepak, A., *Appl. Opt. 18,* 1376-1382 (1979).

6. Deepak, A., Box, G. P., Box, M. A., and Adams, R. R.,
    Experimental Validation of Solar Aureole Method for
    Determination of Aerosol Size Distributions, IFAORS-132-79
    (1979). (Available from IFAORS, P.O. Box P, Hampton, VA 23666.

7. Deepak, A. and Adams, R. R., Photography and Photographic
    Photometry of the Solar Aureole, IFAORS-112-78 (1978).
    (Available from IFAORS, P.O. Box P, Hampton, VA 23666.)

8. Deepak, A., *Appl. Opt. 17,* 1120-1124 (1978).

9. Box, M. A. and Deepak, A., *Appl. Opt. 17,* 3794-3796 (1978).

10. Deirmendjian, D., *Annales de Géophysique 15,* 218 (1959).

11. Sekera, Z., *Adv. Geophys. 3,* 43 (1956).

12. Deepak, A. and Box, G. P., Analytic Modeling of Aerosol
    Size Distributions, NASA-CR 159170 (1979). (Available from
    NTIS, Springfield, VA 22161.)

DISCUSSION

Stokes: What are the isophotal contours in your aureole measurements? How far down are they from the solar disk intensity? What is the relative brightness for each of those contour intervals that you showed?

Deepak: Each solar aureole photograph is about 12° x 14°. We consider the solar aureole to be a region within about 20° from the sun. Up to about 2 or 3° away from the sun we lose the radiance information due to occultation of direct sun light by a neutral density filter disk (ND x 4.0). From 3° to 17° the fall off of sky radiance is within a decade. But from the sun's disk edge to about 3°, the radiance falls off quite sharply depending upon the haziness of the sky.

Stokes: Right. Normally the approximation is $1/r^2$ when you get out beyond a degree from the limb, as in the study of forward scattering irradiance from stars, for example. I was wondering what actually is the fall off of the brightness of the sun in the isophotal contours? $1/r^2$?

Deepak: In the case of the solar aureole, the sun is about a thousand to ten thousand times brighter than the hazy sky. Close to the sun's edge, within 2°, a very sharp fall off of sky radiance occurs; from 2° outwards the rate of fall slows down. That is why I had to use a neutral density filter of ND of about 4.0, because the ordinary photographic film cannot handle this large variation in intensity of four or five orders of magnitude. The film we used had a useful dynamic range of two or at the most three orders of magnitude.

Stokes: The reason I was asking this question is that we need to have the dynamic range for the detector that we are using in our system which can perform almucantar scans. We have several mobile photometers for this purpose.

Deepak: There is a good deal of important information available from radiance measurements in the 0-3° region. I believe there are three solar telescopes, built by DOE funds that are capable of making eight spectral channel sky radiance measurements in the region from 0-3° in the almucantar. We hope to use these data combined with our multiwavelength data beyond 3° to retrieve aerosol properties.

Stokes: Ivan King measured the fall off from all bright objects from stars to the sun for what were considered clearest conditions and he found a $1/r^2$ dependence out to 25°.

Deepak: That is interesting. In another paper, we will discuss
the corrections due to forward scattering on the extinction of
direct radiation and you might get some idea as to how the fall
off in forward scattered radiation varies with aerosol size
distribution and refractive index.

SELECTED *IN SITU* AIRBORNE POLLUTANT MEASUREMENT
TECHNIQUES APPLICABLE TO THE POWER
PLANT PLUME PROBLEM

G. L. Gregory

NASA-Langley Research Center
Hampton, Virginia

The paper presents a summary of selected airborne pollutant measurement techniques that can be applied to the study of power plant plumes. Techniques discussed are aircraft in situ *techniques which are state-of-the-art, use readily available instrumentation, and have been applied in various air pollution studies. Techniques discussed range from gaseous pollutant measurements to aerosol and particulate measurements. Both* in situ *and grab sampling (requiring laboratory analyses) techniques are discussed. Although the paper is not intended to be an exhaustive review of measurement techniques, it does provide the reader a comprehensive understanding of those basic measurements which are readily available for studying the power plant plume problem. Each technique is briefly discussed in terms of theory of operation and measurement capabilities and includes suitable references for more detailed information. Where appropriate, field measurement results (not necessarily power plant plume data) are cited to illustrate the measurement technique.*

I. INTRODUCTION

Power plant plumes have been under study for a considerable period of time, dealing with both plume dynamics and plume chemistry and more recently, with interactions between dynamics and chemistry. Most researchers agree that the primary emission products are SO_2, NO_x, CO_2, hydrocarbons, and aerosols (or

   ISBN 0-12-646360-3

particulates) and that the associated environmental problem is
the ultimate fate of these emissions and/or the production of
secondary pollutants from chemical reactions among the emitted
species. However, uncertainties exist in the understanding of
the exact chemical processes occurring, the relative importance
of the chemical cycles and transformations, and the environmental impact of the chemical end products. The current energy
debate concerning the economic and environmental costs of
increased coal usage as a replacement for gas and oil has
generated an increased interest in the power plant plume and,
in particular, the coal-fired plant plume.

A power plant plume changes in species makeup and concentration as it ages because of diffusion and chemical transformation.
Current technology permits measurement of most primary emitted
species, some intermediate species critical to understanding the
chemical processes, and some end products of the chemical cycles.
Some species require sophisticated instrumental techniques and
others can be measured with readily available instruments. The
latter group of measurements are addressed in this paper. The
purpose of this paper is to summarize those readily available
measurements suitable for aircraft application that are
applicable to the power plant plume problem. The paper is not
intended to be an exhaustive review, but should provide the
reader with a basic understanding of those species that can be
measured on a routine basis. Where appropriate, suitable field
data illustrating the technique are shown and special sampling
considerations discussed. The paper also discusses some
important considerations as applied to airborne sampling in
general.

II. INSTRUMENTING AN AIRCRAFT PLATFORM

Although the measurement techniques discussed in this paper
are based on readily available instrumentation, some consideration must be given to the application of these techniques to
aircraft *in situ* measurements. An aircraft monitoring system
requires special consideration to ensure that valid airborne
measurements are being made. The transformation of a measurement system (even a simple system) from the laboratory or a
ground operation to a qualified and verified airborne sampling
system can result in numerous unanticipated problems. However,
a properly designed laboratory program to verify the airborne
system can alleviate or, at least, identify the magnitude of the
various problems. One cannot simply place the instrument aboard
the aircraft and supply outside air to the instrument and assume
a valid measurement. This section will briefly discuss problem
areas frequently encountered and sample laboratory programs
that have been conducted in the author's laboratory to identify
or alleviate these problems. For illustration purposes, the
author will focus on recently completed programs performed to
certify measurement systems flown aboard NASA-Langley's *in situ*
sampling aircraft (1). These problem areas include (but are
not limited to) five basic areas:

1. Aircraft platform selection
2. Air inlet and air sample distribution system
3. Sampling environment
4. Instrument response time
5. Onboard calibration.

A. *Aircraft Platform Selection*

The selection of the airborne platform for a field measurement program is frequently based on its availability rather than the measurement requirements. However, in cases where a choice is available, judicious selection of the airborne platform can alleviate sampling problems. For power plant plume sampling, flight altitude and flight durations are generally not an important aircraft selection criteria. All aircraft can sample at altitudes ranging to 3 kilometers and stay aloft for 3 hours without difficulty, and this is generally acceptable for most plume sampling. However, often of importance in aircraft selection are considerations of flight speed, climb rate, turbulence in the aircraft wake, and payload (instrument) capacity.

Instrument response time and flight speed determine the desired residence time of the aircraft in the plume. Flight speeds above 50 to 60 m/s can present major difficulties in obtaining concentration profiles (with the instrumentation discussed herein) in the plume, especially at locations within 10 to 15 km of the plant stack. As an example, a kilometer-wide plume sampled at 50 m/s results in an aircraft-plume residence time of only 20 seconds. Twenty seconds of sampling with a 3 to 5 second response instrument (typical of most instruments considered quick responding) is, at best, marginal for determining a cross-plume concentration profile. Although climb rate of the aircraft may not be crucial in its selection, it should be considered. For most plume sampling, aircraft climb rates of 1 to 3 m/s are most suitable for vertical profiling of the power plant plume. Climb rates above 3 m/s are frequently undesirable because of instrument response; rates below 1 m/s often result in skewed profiles through the plume as a result of aircraft wind drift.

Aircraft wake turbulence can be a major factor in aircraft
selection because minimal disturbance of the plume (physically
and chemically) is desired. Thus, generally, the hover type or
large fixed-winged aircraft, are not used in a power plant plume
measurement program. It should be noted that the wake turbulence
associated with some helicopters is comparable to fixed-wing
aircraft turbulence if a specified forward flight speed is
maintained.

Payload (instrument) capacity is an important consideration
because the selection criterion of flight speed and wake
turbulence generally results in a light, single or twin engine
craft with limited payload capacity (space and weight). In
addition to bulk payload, the center of gravity of the aircraft
and the location of the instruments onboard the aircraft must be
considered. Some measurements are sensitive to long inlet lines
and/or bends in the inlet lines. The selection of a light,
twin-engine aircraft with a usable nose baggage compartment can
provide flexibility of operation. Instruments sensitive to
inlet line parameters can be located in the nose directly in
line with the air inlet probe, and less sensitive instruments
can be located in the passenger cabin and have the air sample
ducted to them. Also, a nose baggage compartment for instrumentation alleviates potential center-of-gravity problems for
the aircraft. Reference 1 discusses several aircraft selection
considerations as they apply to a NASA sampling application.

B. *Sample Air Inlet and Air Sample Distribution System*

Once the aircraft platform is selected, the design of air
inlets, location of onboard monitoring instrumentation, and
selection of an air distribution system for the various instruments must be considered in detail. The inlet probe(s) should
be located and installed so that undisturbed free-stream (outside

of the boundary layer of the aircraft) air is sampled. Angle
of attack of the aircraft at the planned sampling flight speed
and/or isokinetic flow conditions at the probe inlet are
additional design considerations. Depending on the aircraft
and available resources, several options are generally available.
Verification of the inlet probe design through analytical
and/or experimental studies is necessary. A literature search
of aeronautical research data will in most cases provide the
analytical tools for such a verification. For example, Ref. 2
was most useful in the design of the inlet probes for the aircraft system of Ref. 1.

After selecting the inlet probe configuration and instrument
locations onboard the aircraft, the air distribution system for
each instrument can be considered. Consideration must be given
to instrument flow requirements, inlet line routing options
inside the aircraft, the environmental conditioning of the
transported airstream, and selection of materials for construction. For example, particulate and aerosol measurements require
minimal length and line-of-sight (no major bends) inlet line
configurations to maintain size distribution spectrums.
Frequently, measurement of some species requires a heated air
distribution system to minimize adsorption (loss) of the sample
as it is transported from the inlet probe to the instrument.
Other species require the entire air-distribution system to be
construction of specific materials. For example, O_3 monitoring
requires that the inlet probe and air distribution system be
constructed of Teflon. Again, it is important that the inlet
system and air distribution system design be verified. These
verification tests focus on ensuring (for example) that the
species concentration as sensed by the instrument is the same as
that which entered the inlet probe. A laboratory program is
suggested for these verifications, using actual aircraft hardware and anticipated aircraft flow rates.

Figure 1 is a schematic of a laboratory apparatus which has been used to verify the air sample distribution system for various gaseous measurement systems onboard the aircraft of Ref. 1. The systems shown within the broken lines are actual aircraft flight hardware, set up and operated as flown on the aircraft. The procedure is to introduce a constant and known gas concentration into the inlet probe (point A) and to monitor that concentration with the air distribution system (in-flight flow rates) and instrument configuration (location) planned for the aircraft sampling mission (point C, for example). Comparison of the known concentration (point A) with that at point C determines the effectiveness of the air distribution system. Concentration measurements at various points along the length of the air distribution system (point B and C, for example) can be used to assess the location of sample losses through the air distribution system. The same test sequence can be repeated for a range of gas specie concentrations, if desired. Table 1 shows typical results from these tests. In general, the experimental accuracy of a test of this type using routine laboratory procedures is about ±10 percent.

The same test apparatus, with minor modifications, can be used to evaluate chemical reactions occurring within the air distribution system that may be affecting the measurements. In this case, mixtures of species are introduced at point A and monitored at various stations (A, B, and C) along the flow path of the air distribution system. Comparison of the measurements at the various sampling stations will determine whether significant chemical reactions are occurring.

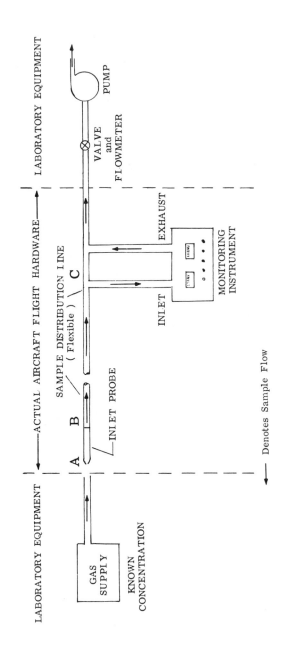

FIGURE 1. Apparatus for air sample distribution tests.

TABLE 1. Results from Air Distribution System Verification
Tests

A. *Ozone*

Concentration of O_3 at point A	$\dfrac{\text{Concentration point B}}{\text{Concentration point A}}$	$\dfrac{\text{Concentration point C}}{\text{Concentration point A}}$
95 ppb	1.06	1.00
76 ppb	1.00	0.93
41 ppb	0.97	0.91
13 ppb	1.00	0.89

B. *Nitrogen Oxide*

Concentration of NO at point A	$\dfrac{\text{Concentration point B}}{\text{Concentration point A}}$	$\dfrac{\text{Concentration point C}}{\text{Concentration point A}}$
120 ppb	1.01	1.01
58 ppb	0.97	1.00

C. *Sulfur Dioxide*

Concentration of SO_2 at point A	$\dfrac{\text{Concentration point C}}{\text{Concentration point A}}$
430 ppb	1.02
140 ppb	0.86
100 ppb	1.00
30 ppb	1.06

C. *Sampling Environment*

Because power plant plume sampling is at altitudes normally below 3 km, the ambient environment in the aircraft (presurized or unpressurized compartment) to which the instruments (electronics, etc.) are exposed presents no major problems in areas like electrical arcing, electronics temperature, etc. Instruments are satisfactorily isolated from aircraft vibration by application of normal shock mounting techniques. However, effects of reduced pressure (due to altitude) on the sensitivity of the monitoring instrumentation must be considered. Altitude effects on instrumentation sensitivity are best studied with the aid of altitude simulation chambers. However, in lieu of such a facility, laboratory equipment can be used to simulate altitude environments up to about 3 km (500 torr). Figure 2 is a schematic of one such apparatus. If concentration C_1 and flow rates Q_1 and Q_2 are known, then C_2 can be calculated. The manifold pressure P_2 is controlled by the vacuum pump and the position of valve V_1 (fully open, partially open). Q_3 must be monitored to obtain the instrument flow rate. With some care a system of the type in Fig. 2 can provide the necessary calibration data. Figure 3 illustrates typical results of altitude sensitivity studies. The data are from simulation chamber studies discussed in Ref. 3. It is recommended that where an altitude correction is to be applied to the measured data, the actual instrument be subjected to the altitude sensitivity studies rather than relying on reported literature values for a given type of instrument.

In addition to altitude effects on the instrumentation, consideration must be given to all species in the plume (even if not measured) and their potential for influencing each measurement made onboard the aircraft. For example, SO_2 (gas) is

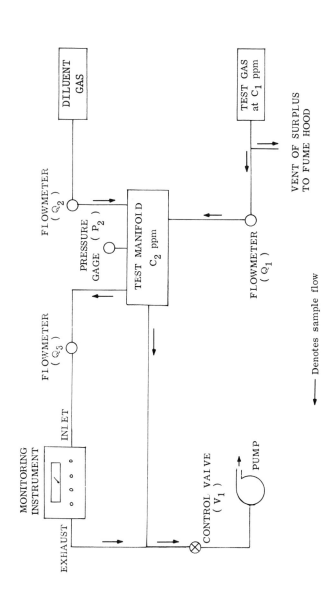

FIGURE 2. Apparatus for altitude-instrument effects test.

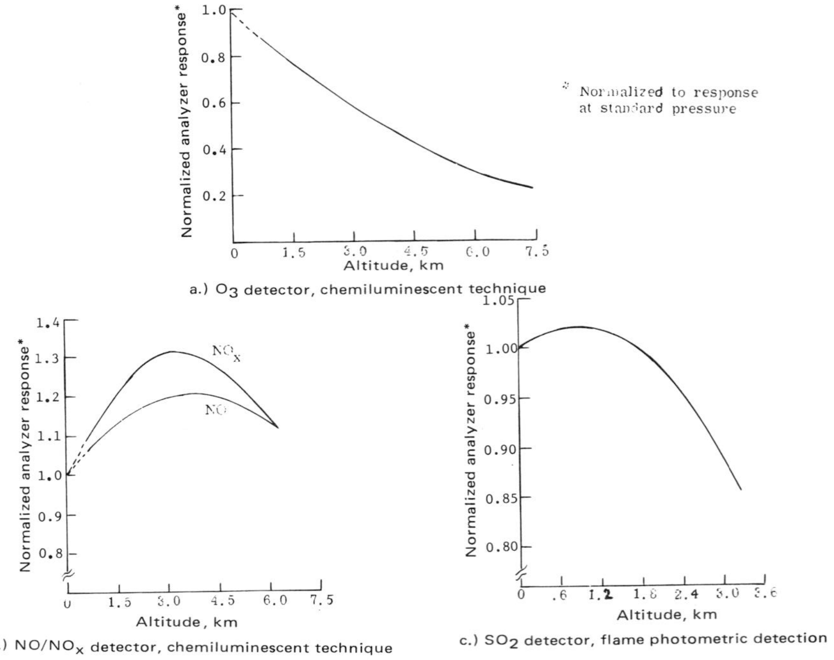
Normalized analyzer response*
1.0
0.8
0.6
0.4
0.2
0
1.5
3.0
4.5
6.0
7.5
Altitude, km
a.) O3 detector, chemiluminescent technique
* Normalized to response at standard pressure
Normalized analyzer response*
1.4
1.3
1.2
1.1
1.0
0.9
0.8
NOx
NO
0
1.5
3.0
4.5
6.0
7.5
Altitude, km
b.) NO/NOx detector, chemiluminescent technique
Normalized analyzer response*
1.05
1.00
0.95
0.90
0.85
0.80
0
.6
1.2
1.8
2.4
3.0
3.6
Altitude, km
c.) SO2 detector, flame photometric detection

FIGURE 3. Illustrative data (see Ref. 3): altitude effects on instrument sensitivity.

generally sampled with a technique that is sensitive to any
form of gaseous sulfur (like H_2S or SF_6). Thus, when using
such a technique for SO_2 measurements, the researcher must
determine that for the plume of interest, SO_2 is the dominant
form of gaseous sulfur.

D. Instrument Response Time

A primary consideration in power plant plume sampling is the
response time of the instruments onboard the aircraft in comparison to the time frequency of the event being measured.
Typically, the major consideration in the design of flight
plans for plume sampling is the response time of the instruments
to ensure sufficient residence time of the aircraft in the plume
to obtain the desired data. As discussed earlier, a 1-km-wide
plume sampled at 50 m/s provides 20 seconds of time in the
plume. Twenty seconds is sufficient for obtaining, for example,
an O_3 profile across the plume since the O_3 instrument response
is about 2 seconds to 90 percent of reading. What about a
specie like SO_2 where the instrument response time is 10
seconds? Can a cross plume profile be obtained? What will it
look like or how will it be distorted as the result of the
slow instrument response? How long must the aircraft remain in
the plume to obtain a valid plume profile? The researcher must
ask these questions and provide answers prior to planning the
sampling mission.

Again, one must turn to the laboratory (or literature if
available) to investigate these questions. For example, Fig. 4
shows the result of laboratory investigations for an SO_2 instrument (flame photometric technique, response time of about
10 seconds to 90 percent of reading) to study response time
behavior of the SO_2 measurement relative to the time frequency
of the event being measured. The procedure was to introduce

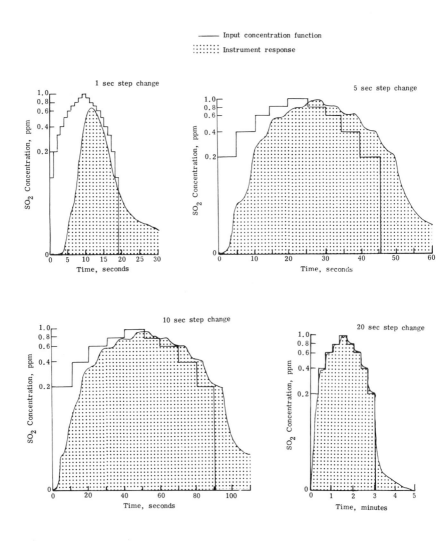

FIGURE 4. Response time data for SO_2 flame photometric detector.

known concentrations of SO_2 into the instrument and then increase
and decrease the concentration by an amount ΔC every Δt seconds.
As shown in the figure, the instrument output is then compared
with the known input concentration for assessment of the
response time problem. For the data of Fig. 4, at time zero
approximately 0.2 ppm (0.1 ppm for the 1 sec case) was introduced into the unit and increased by approximately 0.2 ppm
(0.1 ppm for the 1 sec case) every Δt seconds until 1 ppm was
reached. At this point, the process was reversed and every Δt
seconds, the input concentration was decreased approximately
0.2 ppm (0.1 ppm for 1 sec case) until 0 ppm was obtained.
Figure 4 shows the results of four test sequences: Δt = 1 sec,
Δt = 5 sec, Δt = 10 sec, and Δt = 20 sec. The reader
should note that this particular instrument output is a logarithmic function of input concentration. From data of this
type, the researcher can then assess the impact of the instrument
response on the desired measurement, and (if necessary) design
the flight program to minimize the instrument shortcomings. As
evident from Fig. 4, aircraft plume residence time of 20 seconds
with this particular instrument may or may not be acceptable
dependent upon program objectives. Where response time corrections must be applied to the data as a result of instrument
response, analytical techniques are available (for example,
Ref. 4).

E. *Onboard Calibration*

     When possible, it is recommended that onboard and in-flight
instrument calibrations be performed. Although in-flight
calibrations are often difficult, require additional instrumentation (payload penalty), and require valuable flight time, there
are no substitutes for such calibrations. Because often only
preflight and post-flight (on-the-ground) calibrations are

possible, these calibrations have to be regarded as poor substitutes for an actual in-flight calibration. On the other
hand, in-flight calibrations done poorly, under uncertain or
uncontrolled conditions, and without careful planning are worse
than no calibrations. In-flight calibration systems and
techniques will vary considerably based on aircraft configuration, inlet probe and air-distribution system design, payload
constraints, and the pollutant species. Figure 5 illustrates
a system flown, for some applications, onboard the aircraft
of Ref. 1 for calibration of selected gas specie instrumentation.
This sytem was designed by considering several key points:

1. Calibration gas is introduced into the aircraft air
distribution system as close to the inlet probe as possible.

2. Concentration of calibration gas sensed by the instrument
to be known, to be constant, and at a level to be experienced in
the actual mission.

3. Flow system used to mix the calibration gas into the
airstream not to be affected by altitude (up to 3 km) or
aircraft orientation.

4. System to be of minimal size and weight.

5. Calibration cycle to be automatic (hands-off operation)
requiring only an initiation function.

With the system of Fig. 5, the calibration gas (either pure
or a known mixture in dry air or nitrogen) is introduced into
the air sample stream immediately downstream of the inlet probe
by use of a solenoid valve. A lecture cylinder of known concentration provides the calibration gas supply. The flow
rate of gas (from lecture cylinder) into the air sample stream
is determined based on the desired concentration for calibration at the instrument, the gas concentration in the lecture
cylinder, and the air sample stream flow rate. Once determined,
this flow rate is fixed by proper setting of the supply pressure

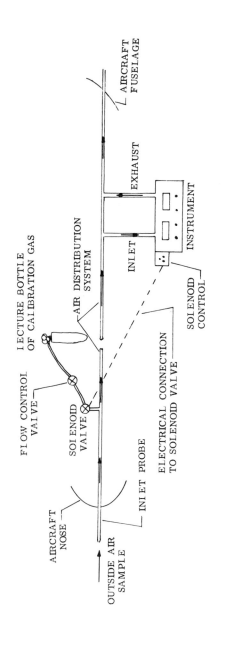

FIGURE 5. Schematic of in-flight calibration system.

and flow-control valve. These settings can be verified in the
laboratory or by in-flight testing. Generally, conditions and
settings can be found so that this flow rate is relatively
independent of altitudes up to about 3 km. Initiation of
calibration is by activation of the solenoid control (located
in the passenger cabin) which opens the solenoid (introduces
calibration gas into airstream), keeps the valve open for a
predetermined period of type (calibration cycle), and then
automatically closes the solenoid valve (restores system to
sampling configuration). The solenoid opening time is
adjustable (function switch on the solenoid control) and proper
settings are determined by laboratory and/or flight testing
of the calibration system.

Figure 6 illustrates typical calibration data obtained from
the system of Fig. 5. Both laboratory and in-flight operation
of the system are shown for an SO_2 instrument. The in-flight
data were obtained between two power plant plume cross wind
sampling passes.

III. GAS SPECIE MEASUREMENT

Of those gas species of interest in the aging power plant
plume, three will be discussed, each measurable with commercially
available instrumentation using routine aircraft sampling
techniques. These are SO_2, NO/NO_x, and O_3. Both SO_2 and NO/NO_x
are primary emissions whereas O_3 is a secondary specie introduced by ambient air entrainment and plume photochemistry.
Table II summarizes the detection characteristics for the
instruments discussed. Generally, more than one technique are
available for the given measurement. The selection of a
technique for discussion does not constitute an endorsement

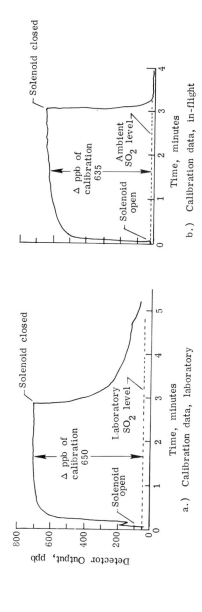

FIGURE 6. *SO$_2$ calibration data: in-flight calibration system.*

TABLE II. Detection Capabilities: Gas Species

Specie	Technique	Concentration range	Detection limit	Response to 90%
SO_2	Flame photometric	0 to 5 ppm	2 to 5 ppb	8 to 10 s
NO/NO_x	Chemiluminescent	0 to 5 ppm	2 to 5 ppb	2 to 5 s
O_3	Chemiluminescent	0 to 1 ppm	1 to 3 ppb	2 s

of that technique, but does exemplify the measurement capabilities for the specie. Often the selection of the monitoring
technique depends on the purpose of the power plant plume study
and the planned measurement scenario.

A. Sulfur Dioxide

Among the measurement techniques for SO_2 are flame photometric, infrared, colorimetric, and coulometric (electrochemical)
systems. The flame photometric technique was selected for discussion. The flame photometric detector is based on chemi-
luminescence of sulfur species produced in a hydrogen flame
(5). Sulfur containing molecules are converted to an activated
S_2 species in a hydrogen (supplied from external supply) flame
and, in turn, S_2 (activated) reverts to a lower energy state
emitting light radiation. This light is detected by a photomultiplier tube and narrowband filter system. The intensity
of the light is approximately proportional to the square of the
incoming sulfur concentration. Application of this technique
to power plant plume sampling and particularly to aircraft
operations warrants several comments.

1. The sensing technique is equally sensitive to any sulfur
gas specie in the plume including H_2S and SF_6 (if used as a
tracer). Sample preconditioning (scrubbers) is required if it
is necessary to eliminate the other sulfur species.

2. The instrument output is a voltage logarithmically proportional to SO_2 concentration (some manufacturers provide options to linearize the output). The lower detection limits specified in Table II are based on the logarithmic output.

3. Application of the technique requires a hydrogen gas cylinder (or generator) onboard the aircraft; however, gas cylinders are available which restrict (using sorbents) H_2 flow from the cylinder in the event of accidental venting of the gas cylinder (a safety consideration).

Figure 7 illustrates the type of data that can be obtained from the instrument as flown on an aircraft. The data shown were taken on June 6, 1978, downwind of the Morgantown, Maryland, power plant. Flight paths for the aircraft were approximately perpendicular to the plume transport direction at a sampling speed of $\sim$ 55 m/s. Flight altitude was approximately at the plume centerline and was estimated to be about 400 meters. Data of this type can be used to study plume growth and diffusion, SO_2 chemistry, and plume transport.

B. Nitrogen Oxides

The nitrogen oxides monitoring technique used most frequently in air quality work employs the chemiluminescent principle (5). The instrument illustrated is a dual-channel chemiluminescent unit that continually and simultaneously measures NO and NO_x. Filtered air (removes the particulates) is brought into the analyzer and separated into NO_x and NO channels. A molybdenum converter in the NO_x channel thermally converts the nitrogen oxides in the sample air to NO. The resulting NO is then mixed with ozone, and the mixture produces a chemiluminescent reaction whose emitted light is proportional to the NO_x concentration in the sample air. In the NO channel, the NO in the sample air is mixed directly with ozone. This mixture also produces a

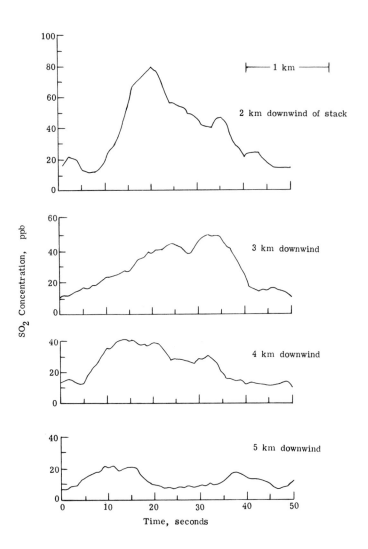

FIGURE 7.   In-plume SO₂ data: flame photometric detector.

chemiluminescent reaction whose emitted light is proportional
to the NO concentration in the sample air. For each channel,
a photomultiplier tube system is used to detect the emitted
light. A difference amplifier between the NO_x and NO channels
provides a signal proportional to the NO_2 (NO_x - NO).

Figure 8 illustrates NO/NO_x data from the chemiluminescent
detector when flown through a power plant plume. The data (6)
shown are for the Labadie, Missouri, power plant on August 4,
1974. The flight trajectory was at approximately 450 m altitude

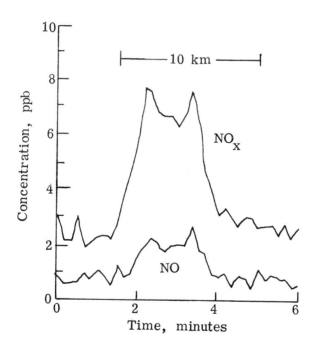

FIGURE 8. In-plume NO/NO_x data (see Ref. 6).

and 22 km downwind of the plant. The relative magnitude of
NO and NO_x concentration in the aging plume is of interest to
the photochemist and his study of plume characteristics.

C. *Ozone*

The methods commonly accepted for ozone monitoring are the
ultraviolet absorption and chemiluminescent techniques (7,8).
The chemiluminescent technique was selected for discussion. In
this technique, sample air and ethylene (C_2H_4) are drawn into a
reaction chamber where the ozone in the air reacts with the
ethylene in a chemiluminescent reaction. The emitted light is
detected with an appropriate photomultiplier tube system. This
light output is directly proportional to the ozone concentration
in the sample airstream. Two points are noted concerning aircraft operations. First, ethylene must be supplied to the
instrument from a gas cylinder flown onboard the aircraft, and
ethylene is flammable. A nonflammable gas mixture is available
as a substitute for ethylene but requires some instrument
modifications. Second, the inlet probe and air distribution
system must be made of Teflon or Teflon lined to prevent loss
of O_3 from the air sample.

Figure 9 illustrates typical O_3 data. The data shown are
from Ref. 9 and were obtained on June 22, 1974, downwind of the
Morgantown, Maryland, power plant at approximately 600 m
altitude. As shown, initially the plume shows a depletion of
O_3 as compared with ambient values (nitrogen oxides reacting
with ambient O_3 entrained by the plume) but at longer
distances the plume shows surplus O_3 as a result of plume photochemistry. Also shown for comparison are the associated plume
SO_2 concentrations. Besides providing photochemistry data on

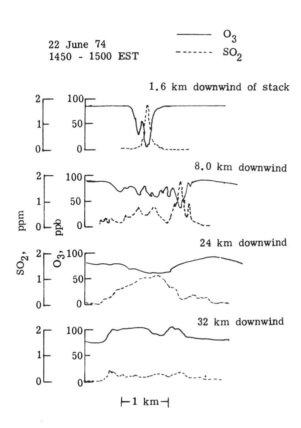

FIGURE 9.   In-plume O_3 data (see Ref. 9).

O_3 generation in the plume, ozone measurements also can define
plume growth and diffusion, transport direction, and under
certain conditions provide information on NO concentrations in
the aging plume.

IV. AEROSOL SPECIE MEASUREMENTS

Aerosol, liquid or solid, measurements in the power plant
plume onboard an aircraft are generally more difficult to
perform and interpret than the gas specie measurements. In
addition, the same instrument may be used for substantially
different purposes depending upon the requirements of the
measurement program. For example, an instrument capable of
monitoring both liquid and solid aerosols may be used to obtain
only the solid aerosols, if the inlet probe and air sample
distribution system are properly heated to help vaporize the
liquids. Although many applications of any one instrument
exist, the author has selected an application (usage) with
which he is most familiar for illustration of the technique.
In general, aerosol instrument usage tends to be less routine
when compared with the gas specie instrumentation. The
instruments discussed in the following paragraphs are commercially available, but may not in all cases be a production-line
item.

Before proceeding with the discussion of the selected
instruments, two important considerations of airborne aerosol
sampling need to be noted. The first is isokinetic sampling
conditions at the entrance to the instrument inlet probe. Only
if isokinetic flow conditions exist at the entrance of the inlet
probe, will the instrument monitor a true representative sample
of the plume aerosols. Second, care must be taken in the
design and location of an aerosol instrument onboard the aircraft to minimize modification (concentration, size distribution,

or chemical composition) of the sample as it is ducted from the entrance of the inlet probe to the instrument. For example, right angle bends in the sample flow path may result in substantial distortions of a size distribution spectrum. Likewise, multiple interactions of the aerosol sample with the walls of the air distribution system can result in composition changes of the sample.

A. Integrating Nephelometer

The integrating nephelometer is widely used for measuring the scattering coefficient of aerosols because of its rapid response, ease of operation, and the convenient form of data obtained (10). Suspended particulates in the size range of about 0.2 to 1.6 μm in diameter scatter rays of light in all directions. The amount of light scattered in any one direction depends on the size, quantity, and composition of the particles. The integrating nephelometer is designed to measure this light scattering. The air sample is drawn into a detection chamber, where it is illuminated by a pulsed flashlamp (for example, a Xenon flashlamp at about 8 pulses/second). The scattered light is detected by appropriately placed photomultiplier tubes; this signal is averaged and compared with a reference signal from another photomultiplier tube looking at the lamp. The measured parameter is the light scattering coefficient (β_{scat}) in units of length to the inverse power (m^{-1}). β_{scat} is defined as the reciprocal of the distance at which 63 percent of the light is lost from the source. β_{scat} measurements can be related to particulate mass concentration ($\mu g/m^3$) if particle size distribution and refractive index are known. If not known, then reasonable assumptions of distribution and refractive index may result in mass concentrations accurate to within a factor of 2 to 8. The reader is referred to Refs. 10 to 13 for a

discussion of the effects of these assumptions. Typically, the
air sample to the nephelometer is dehumidified to remove water
droplets which also can scatter light. Calibration is performed
by using clean filtered air and Freon-12 as reference sources.
Detection characteristics of the integrating nephelometer are
shown in Table III.

Figure 10 illustrates nephelometer data from the Morgantown,
Maryland, power plant plume, approximately 1 km downwind and at
a flight altitude of 500 m. In interpreting the nephelometer
data, one must be aware of the size distribution over which the
instrument responds, the effects of any dehumidification of the
sample in the inlet, and those assumptions required if β_{scat}
data are to be reported as mass concentration ($\mu g/m^3$).

B. Quartz Crystal Microbalance

The quartz crystal microbalance is more specialized in its
application than the nephelometer. The unit is a 10-stage
cascade impactor which measures particulate concentration
($\mu g/m^3$) as a function of time and particle diameter in the range
of 0.05 to 25 μm diameter. The instrument is discussed in some
detail in Ref. 14. Particles are separated inertially and

TABLE III. Detection Capabilities: Aerosols

Instrument	Size range	Detection limit	Response to 90%
Nephelometer	0.2 to 2 μm	$0.1 \times 10^{-4}\ m^{-1}$	0.2 s
Quartz crystal microbalance	0.05 to 25 μm	10 $\mu g/m^3$	2.0 s
Forward scattering spectrometer probe	0.5 to 45 μm	1 particle	< 1.0 s

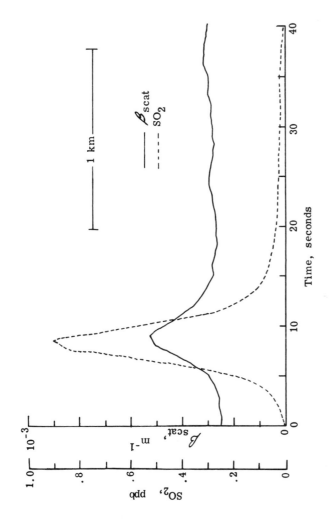

FIGURE 10. In-plume nephelometer data, 1 km downwind of power plant.

classified according to aerodynamic size. Larger particles are
collected in the first stage with each succeeding stage collecting progressively smaller particles. Each stage of the cascade
contains a microbalance consisting of two frequency-matched
quartz crystals in an oscillator circuit. One of the crystals
(the sensing crystal) serves as a collecting surface on which
the particles are impacted. It has an adhesive coating to
prevent the particles from bouncing off. As particles are
collected on the crystal's surface, the resonance frequency
decreases in proportion to the amount of mass added. This
frequency is mixed with the frequency provided by the second
crystal (reference), and thereby results in a beat frequency
which is an indication of the mass of the particulates collected.
Thus, the particles in each stage (size interval) are automatically weighed as they are collected. The signals from the
10 cascade stages are independent and are recorded as data.
Particles collected in a given stage are assigned an aerodynamic diameter equal to the 50 percent point (the diameter at
which the impaction efficiency is 50 percent for particles with
a mass density of 2). The 50-percent points are 0.05, 0.1,
0.2, 0.4, 0.8, 1.6, 3.2, 6.3, 12.5, and 25 μm diameter.

For aircraft operation, it is generally desirable to locate
the instrument as close to and within the line of sight of the
inlet probe, to design for isokinetic sampling, and to heat
the instrument inlet to reduce the relative humidity of the
sample. Deposits of moisture (especially if the plume water
content is substantially higher than ambient) on the sensing
crystal will be detected as a weight gain and if subsequently
evaporated as a weight loss resulting in potential data interpretation problems. If this dehumidification technique is
used, its effect on the representative nature of the aerosol
sample must be considered, especially if the plume includes
solid particulates with an absorbed liquid constituent.

Figure 11 illustrates typical data from the instrument. The
data shown are from Ref. 15. The data shown are not for a power
plant plume but are of a launch cloud produced during the launch
of a Titan III launch vehicle at the Kennedy Space Center,
and, as such, illustrates to the reader the potential of the
technique. A distinct advantage of the instrument is that in
addition to the real-time size distribution and concentration
information, the sensing crystals can be analyzed in the
laboratory to determine morphology and elemental composition of
the sample. For example, Ref. 15 states, in discussing the
results of these analyses for the data of Fig. 11:

> The particles from the quartz crystal microbalance
> stages, which showed positive weight gains, were examined
> with a scanning electron microscope. The particles in the
> size range of 0.4 to 1.6 µm consist of aluminum oxide
> spheres and a few irregular shaped particles containing
> sodium and chlorine. The particles in the smaller size

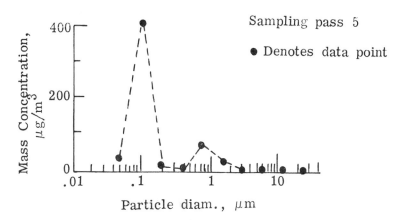

FIGURE 11. Quartz crystal microbalance data, launch
vehicle cloud.

modes (0.05- to 0.2-μm range) consist of a few single
particles and a large number of agglomerates. These
particles had a more complex makeup consisting of sodium,
aluminum, sulfur, chlorine, potassium, calcium, iron, and
zinc.

In applying the quartz crystal microbalance instrument to
power plant plume sampling, anticipated plume aerosol concentrations should be considered to ensure that the instrument
detection capabilities (Table III) and plume concentrations are
compatible. Several options are available to overcome instrument sensitivity deficiencies should this be an area of concern.

1. A single-stage instrument is available but gives no
sizing data; however, it has the advantage of collecting all
particles on a single crystal compared with diluting the sample
among 10 crystals.

2. The data reduction process can be adjusted to provide
mass concentration data on a per-pass basis through the plume
(technique used in Fig. 11) rather than mass concentration as
a function of time within a given plume sampling.

3. The aircraft flight path can be designed to provide
maximum sampling time in the plume.

C. Forward Scattering Spectrometer Probe

The forward scattering spectrometer probe measures the
number of suspended aerosols (particles/cm^3) as a function of
aerosol diameter over a size range of 0.5 to 45 μm diameter
with four selectable ranges (0.5 to 7.5, 0.5 to 15, 0.5 to 30,
and 0.5 to 45). Individual aerosol nuclei are counted and
sized when they pass through the focused portion of a laser
beam (the sampling volume). As each nuclei passes through the
sampling volume, it scatters light from the incident laser
beam. The light scattered in the near-forward direction is

directed onto a photodiode which generates a pulse. There is
one pulse for each nucleus that passes through the beam. The
magnitude of the pulse depends on the amount of light scattered
by the aerosol which is related to the size of the aerosol.
The sensing unit is flown external to the aircraft, monitoring
both liquid and solid aerosols (including water droplets); and
as such, the aerosols are not altered by the sampling process.
Figure 12 illustrates the data from the instrument as recorded
during a Titan III launch cloud sampling and serves to illustrate
the potential of the technique. The data shown are for a single
pass through the launch cloud as opposed to time information
within the cloud (15).

In application of the technique to the power plant plume,
consideration must be given to the detection capabilities
(Table III) of the technique as compared with anticipated
aerosol concentrations in the plume.

V. CONCLUDING REMARKS

The paper presents a survey of readily available measurements which can be made onboard an aircraft and which are
applicable to the power plant plume monitoring problem. The
paper focused on routine measurements frequently applied to
various air pollution problems. The paper was not intended as
an exhaustive survey, but to provide the reader with a basic
knowledge of readily available measurements, their capabilities,
and anticipated problems encountered in airborne operations.
Measurements for SO_2, NO/NO_x, O_3, and aerosols (concentration
and size distributions) in the plume can be readily obtained.
Data examples have been shown for each measurement; most
examples are from power plant plume sampling missions. Detailed

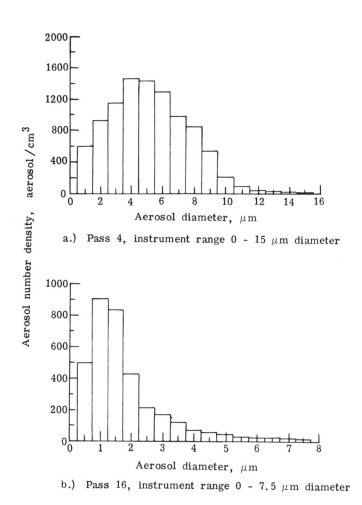

a.) Pass 4, instrument range 0 - 15 μm diameter

b.) Pass 16, instrument range 0 - 7.5 μm diameter

FIGURE 12.   Forward scattering spectrometer probe data,
launch cloud.

discussions of important considerations as applied to airborne
sampling in general are presented with emphasis on aircraft
selection, sample inlet and flow system design, the sampling
environment, instrument response time, and onboard in-flight
instrument calibration.

The discussions presented here should familiarize the
researcher with those routine measurements which are applicable
to the power plant plume and with the pitfalls of simply
placing an instrument onboard an aircraft without careful consideration and anticipating valid results. Although the
paper is not intended to define or answer the individual
researcher's problem, its purpose is to stimulate the experi-
menter to ask the appropriate questions in planning an airborne
plume sampling field program.

REFERENCES

1. Wornom, D. E., Woods, D. C., Thomas, M. E., and Tyson, R. W,
    Instrumentation of Sampling Aircraft for Measurement of
    Launch Vehicle Effluents. NASA TM X-3500 (1977).

2. Beatty, T. D., A Theoretical Method for the Analysis and
    Design of Axisymmetric Bodies. NASA CR-2498 (1975).

3. Tommerdahl, J. B., Strong, R. B., and White, J. H.,
    Validation of Airborne Air Quality Measurements. Paper
    presented at 31st Technical Conference of American Society
    for Quality Control, Philadelphia, Pennsylvania, May 16-17,
    1977.

4. Larsen, R. I., Benson, F. B., and Jutze, G. A., *J. Air
    Poll. Control Assoc. 15,* 19-22 (1965).

5. Instrumentation for Monitoring Air Quality. ASTM Special
    Technical Publication 555, Philadelphia, Pennsylvania
    (1973).

6. Oxidant Measurements in Western Power Plant Plumes. Prepared by Meteorology Research, Incorporated, for Electric Power Research Institute. EPRI Report EA-421 (1977).

7. Behl, B. A., Absolute Continuous Atmospheric Determination by Differential U. V. Absorption. Paper Presented at 65th Meeting, Air Pollution Control Association, Miami Beach, Florida (1972).

8. Warren, G. L., and Babcock, G., *Review of Scientific Instruments, 41,* 280 (1970).

9. David, D. D., Smith, G., and Klauber, G. *Science, 186,* 733-735 (1974).

10. Charlson, R. J., Ahlquist, N. C., Selridge, H., and MacCready, P. B., Jr., *J. Air Poll. Control Assoc. 19,* 12 (1969).

11. Tombach, I. Measurement of Some Optical Properties of Air Pollution with the Integrating Nephelometer. Joint Conference on Sensing of Environmental Pollutants, AIAA Paper No. 71-1101, Palo Alto, California (1971).

12. Rabinoff, R. A., and Herman, B. M., *J. Appl. Met. 12,* 184-186 (1973).

13. Hanel, G., *J. Aerosol Science, 3,* 455-460 (1972).

14. Chuan, R. L., *in* "Aerosol Measurements" (W. A. Cassatt and R. S. Maddock, eds.), pp. 137-148. ·NBS Spec. Publ. 412, U. S. Dep. Commer., Washington, D.C. (May 7, 1974).

15. Gregory, G. L., Bendura, R. J., and Woods, D. C., Launch Vehicle Effluent Measurements During the May 12, 1977, Titan III Launch at Air Force Eastern Test Range. NASA TM X-78753 (January 1979).

DISCUSSION

Whitby: This is a very good paper, and I think you have pointed out some things that are long overdue. I would like to make a couple of comments about the aerosol sampling problem because we have been with building sampling systems for a long time. Our general experience is that a single optical counter operating at a given condition isn't good for more than about a decade of size at a time. That doesn't mean that one can't use the same counter for different decades. If one looks at the combination of the variations in concentration plus the statistics due to the general r^4 drop off in aerosol number concentration, the result is that if one optimizes to measure a 0.5 μm particle, one isn't going to see usable statistics above 5 μm. If one wants to cover the whole range, one has to use several counters. If one uses only a single counter, one essentially looks at the size distribution through a one and a half to one decade knothole at the size distribution. The second point is that there are a lot of problems in getting aerosols inside of airplanes--in fact, inside of any kind of a vehicle. We found out after evaluating some of the systems in the laboratory that they were doing a lot worse than we thought they were on the basis of model calculations. Our general conclusion is that probably one can't get particles larger than about 5- or 6-microns aerodynamic diameter inside of an airplane and do it reliably. We are currently using the criteria that when the sampling efficiency has dropped to 30% the data isn't correctable any more. At that efficiency one can still correct it. However, at smaller sampling efficiencies, one is dividing by a bigger and bigger numer, so that the results become less and less trustworthy. Have you done any work on the sampling efficiency problem at all?

Gregory: First of all, I agree with you one hundred percent. It seems like that the state of the art in trying to evaluate a sampling system for particulates or aerosols is much more difficult than for the gaseous species. As a result, I think that we find ourselves--I think that we are guilty of that, too--of taking more liberties and more shortcuts in particle instrumenta-tion than we do with our gaseous species. Now I can't really address the question as to what we at NASA-Langley Research Center have done in detail in particle instrumentation as is evident from the paper. My area of expertise is the gaseous species. I can only support your comments and note that our aerosol group at Langley are aware of these problems and some of their research is directed in these areas. However, I think we are lagging behind in trying to run these types of studies on systems for particles and aerosols. All I can say here is that this is one area that we, as experimentors, must start pushing and advance it to the same state-of-the-art that exists in gaseous measurements.

Singh: I was just curious why the efficiency for gaseous detection went down as the altitude increased. Was it just the reduced pressure or was there any fundamental reason?

Gregory: With the gaseous species, it is purely a function of the instrument. Normally you would expect that because of the density effect that the sensitivity would go down, but that is not necessarily true. For example, in the case of the NO/NO_x instrument discussed in the paper, the sensitivity increased with altitude. The point I was trying to make was that, in general, you cannot always predict from theory or from logical considerations of pressure/density relationships what an instrument is going to do. So, in lieu of that, one should first try to go to the literature for information, and better yet, if possible, go into your laboratory, set up a measurement program and actually get that data on the instrument that you plan to use.

Whitby: I have one more question about the ozone sampling. Last summer the auditors during the STATE program didn't like our delivery system for getting ozone to our instrument even though we had checked it out and it seemed to be doing about as you had described. They said well, the plumbing has got to be Teflon. We had tried Teflon and we had also done a number of other things and found out that at the transport rates we were using the plumbing material didn't seem to make a difference. However, we did find out that a dirty plumbing takes out ozone. Therefore, I think that another problem that has to be worried about is whether when one gets crud on the inside of the sampling line, the ozone can react with it no matter what the tube underneath it is made of. Have you done any work on how long you can run these things in airplanes before you observe this problem?

Gregory: No, we haven't done any work in that direction. What we do is to set up a routine maintenance and cleaning program for our probes. Our particular aircraft is a chartered aircraft. So we are frequently installing and taking off equipment. And in doing this it is fairly easy for us to try to clean or replace the Teflon probes and Teflon tubing. We have not run any lifetime studies to answer that type of question. The problem that you have in this area is that it also depends on what the crud is.

Pack: Conspicuous by its absence was any mention of measuring ambient humidity, particularly in the range between about 95% and 100%. Do you have any comments?

Gregory: I have no comments directly. We purposely have deleted the temperature and humidity type measurements from the paper, based on the consideration that they were a little bit more

routine and were perhaps not directly applicable to this problem
or to this Symposium. But I think you are right. As the
humidity approaches 100%, one begins to encounter humidity
problems; however, I think that in the plume sampling programs
that we have been involved with--which have not been that many on
power plants--the plume humidity was not that high.

Pack: What I was really concerned with, among other things, was
that if you were going in and out of clouds, you know you must
be approaching that limit of relative humidity. The chemical
conversions, both for the gaseous materials as well as for
particulates, begins to behave rather differently than they do
in dry air.

Gregory: Well, that is correct and that is why when you do the
aircraft system checks, one of the parameters you consider is
relative humidity. You do look at, for instance, the loss of
species on the inlet probe and the inlet system, with the test
gas being supplied at a reasonable range of humidity. But you
are right, that is a detrimental problem which quite frequently
we tend to overlook. We tend to perform laboratory studies at
room conditions at 50% humidity; and then go fly in the cloud at
70% or 80% humidity. That can make a difference as far as the
response of the instrument and the sample losses or gains within
the sample lines, are concerned.

A COMPUTER-CONTROLLED QUADRUPOLE MASS
SPECTROMETER FOR AUTOMATED
ENVIRONMENTAL ANALYSIS

G. M. Wood
P. R. Yeager

NASA-Langley Research Center
Hampton, Virginia

A Gas Analysis and Detection System (GADS) recently developed
by the NASA Langley Research Center in collaboration with DOE,
DOD, and EPA, provides automated analysis of up to 40 constituents in an air sample at levels approaching 100 ppb. Sensitivity
of the instrument may be increased by suitable preconcentration
techniques. The GADS uses a small hyperbolic rod quadrupole
mass spectrometer, and an integrated microcomputer to provide
operational control, data acquisition, and quantitative analysis.
Automated housekeeping functions and calibration procedures are
also accomplished via computer control. Automated analysis is
accomplished by peak switching to up to 40 mass peaks selected
from the mass range 2 to 200 amu and stored in a matrix. Matrix
inversion methods are used for interference corrections and
analysis. Total mass scan and single mass monitoring are also
provided. Although the GADS is primarily intended for unattended
in situ continuous monitoring, samples may also be transported to
the laboratory for later analysis. In this paper, the GADS is
functionally described and results of laboratory tests
presented. A recently initiated program examining the GADS
capability in field monitoring is discussed.

I. INTRODUCTION

Mass spectrometry has been used routinely in the qualitative
and quantitative analysis of gases since about 1950. Its use as
in situ monitors, in other than process control applications,
has been somewhat limited for several reasons: analysis of more
than a few specific gases at a time has not been required; the
available instruments have in general been large and complex;
and operation and data analysis have required an experienced
mass spectrometrist. The Gas Analysis and Detection System
(GADS) described in this paper has been developed to alleviate
some of these limitations and to provide for long-term unattended
operation and data analysis of trace gases in air. This goal
has been accomplished by developing highly stable electronic
subsystems and by committing to computer control operational,
housekeeping, and data analysis functions normally performed
by the mass spectrometrist. Once calibrated and set up in the
laboratory, the system may then be placed in unattended operation
in the field.

II. DESCRIPTION OF SYSTEM

The analyzer subsystem (Fig. 1) consists of a quadrupole
mass spectrometer using a 2-inch-long hyperbolic rod structure,
a dual rhenium-filament closed ion source, and a continuous-
dynode electron-multiplier detector [1]. Amplified ion currents
are output through a linear current-to-frequency convertor with
three overlapping ranges of 10^{-14}, 10^{-13}, and 10^{-12} amperes/Hz,
providing a total dynamic range of 10^{8}. Ambient air is drawn
through a continuously operating heated inlet system at
approximately 15 cc/min, from which a portion is removed
through a capillary and gold foil molecular leak for introduction into the ion source. A calibration gas, e.g.,

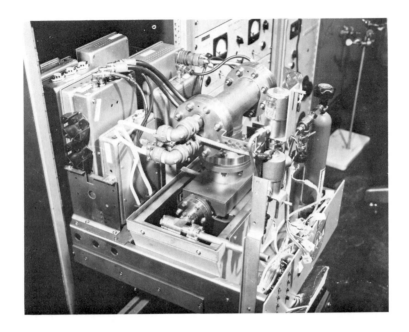

FIGURE 1. Mass spectrometer assembly.

octafluorocyclobutane (C_4F_8), is introduced from the lecture
bottle into the air sample on command via a second molecular
leak valve for mass scale calibration and correction of drifts
in instrument sensitivity. The electronic circuitry is
modularized to facilitate repair and is contained in the metal
cans located at the rear of the mass spectrometer. The ion
source, quadrupole structure, and electron multiplier are contained in a standard 10.16 cm diameter vacuum "T" as shown,
which is mounted directly to a baffled 30 sec^{-1} ion pump.

The quadrupole structure (Fig. 2) is a unique one-piece
assembly consisting of four hyperbolic surfaces separated by
ceramic insulators. The structure is formed by deposition of
tungsten from tungsten hexafluoride onto a heated mandrel to
a thickness of approximately 0.076 cm. Since the critical

FIGURE 2. Quadrupole structure--side view.

tolerances are maintained by the mandrel, it is therefore
possible to reproducibly fabricate the quadrupoles at an
estimated factor of 10 reduction in cost over methods normally
used for fabrication to hyperbolic structures. Following
deposition, the assembly is cooled to room temperature and the
mandrel removed, after which the tungsten is electrochemically
removed from the insulators, as is clearly seen in the end view
shown in Fig. 3. The structure is then mounted in the vacuum
"T" as shown in Fig. 4. Alignment of the ion source and
multiplier with the quadrupole is accomplished by a pin-and-
hole mechanism, which provides reproducible realignment whenever
removal and replacement is required.

FIGURE 3. Quadrupole structure--end view.

FIGURE 4. Quadrupole structure in vacuum "T."

The mass spectrometer is controlled by the computer subsystem shown in Fig. 5. The computer was fabricated from 7400
series MSI (TTL) logic devices, and is unique in that it uses
separate memories for data and programmed instructions. This
feature allows the computer to operate at a true rate of 3×10^{6}
instruction executions per second. The system is configured
around three busses: a 12-bit data bus, 16-bit program address
address bus, and 16-bit instruction bus. Programs are committed
to programmable-read-only-memory (PROM) and are therefore pro-
tected from accidental modification from the control panel or
loss of power. Variables stored in random-access-memory (RAM)
are retained in the event of power failure by a Power-Failure-
Detect module containing rechargeable nickle-cadmium batteries
that are activated whenever input power falls below a preselected
value. Housekeeping functions and data, described below, are
displayed on the CRT located in the computer panel and are

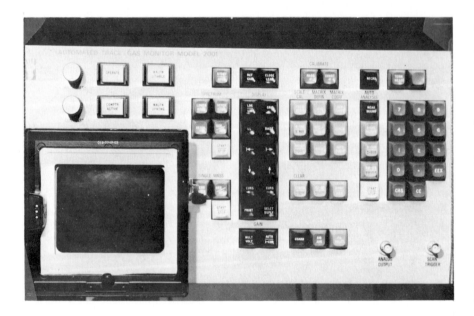

FIGURE 5. GADS computer.

simultaneously output through an I/O bus in standard ASCII format. The bus is configured to permit interfacing to peripheral recording devices and to provide for external control of the system. Any of the operating keys may be exercised by an external microprocessor or similar device via the I/O bus to provide for automated operational sequencing and data storage. Analog recording of mass spectrums may be obtained in real time through a BNC connector on the front panel. Alternatively, all or portions of a mass spectrum stored in memory may be plotted on the analog recorder on command.

The mass range of the instrument is 2 to 200 amu in 1/20 amu steps. Resolution of mass peaks is variable at both upper and lower ends of the mass range, with better than unit resolution at mass 200 achievable. The design detection limit in the absence of large adjacent peaks was 100 parts per billion (ppb); however, analysis of the major isotopes of naturally occurring atmospheric krypton (1.1 ppm) indicates that in some cases detection limits as low as 40 ppb can be obtained.

The system has three separate modes of operation. In the SCAN SPECTRUM mode a standard mass spectrum is generated between a lower and an upper amu selected from the keyboard, at a scan rate that is variable from 0.030 to 99.0 seconds/amu. Depending upon the commands given, the system will either generate one complete spectrum and halt or will continue to repeat the measurement. A typical spectrum is shown in Fig. 6, where the two large peaks are mass 18 (water) and mass 28 (nitrogen). The position of the cursor (CUR) in the alphanumeric display identifies the mass, while VAL is the peak height in hertz. Since this particular spectrum was taken with the current-to-frequency converter in the 10^{-13} amp/Hz range, the ion current after multiplication measured at the position on the mass 28 peak at which the cursor is set is 1.974×10^{-7} amperes.

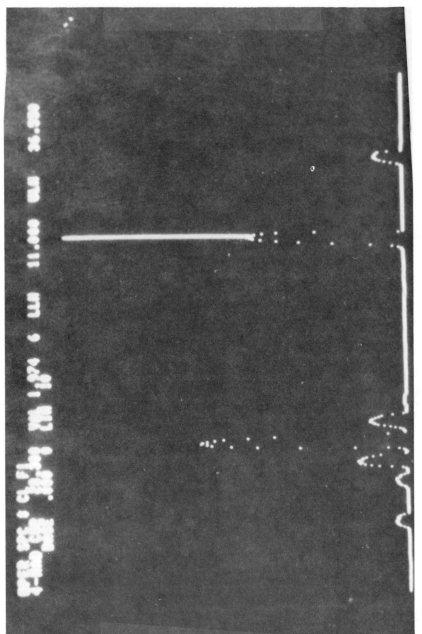

FIGURE 6. Typical mass spectrum.

In the SINGLE mode, a single mass peak is continuously
monitored to measure variations in concentration as a function
of time. The rate of measurement in this mode is variable from
0.001 to 99 seconds/measurement. Since there are 4096 data
storage locations, nonrepeated run durations may therefore be
varied from 4.1 seconds to 4.06×10^5 seconds (4.7 days). As
with the SCAN mode, repetitive measurements may also be
commanded. Data obtained in this mode appear as a continuous
line on the CRT, increasing or decreasing as a function of
changes in concentration.

In the AUTOANALYSIS mode, the computer is commanded to
search for and determine the concentration of species selected
for analysis. This command is accomplished by stepping to a
series of mass peaks indicative of the species to be analyzed,
and integrating the ion currents for preselected periods at
each peak. These ion currents are multiplied by an inverted
matrix of sensitivity coefficients to correct for interferences
resulting from contributions to a mass peak by more than one
species and to calculate concentrations of each species present.

The matrix used consists of elements denoting the instrument
sensitivity for a particular molecular (or atomic) species at
a specific mass peak. These sensitivity coefficients may be
entered manually into the matrix or, alternatively, may be
determined by the computer while introducing a suitable calibration gas at a known concentration. The matrix size may be
varied from 1 x 1 to 40 x 40, so that up to 40 species could be
measured, although 25 to 30 is probably more realistic because
of the large number of interferences that would occur. As an
example, the mass peaks displayed in Fig. 7 were selected to
measure nitrogen (14), oxygen (32), argon (40), carbon dioxide
(44), and krypton (84) and inserted into a 5 x 5 matrix. An
analysis of ambient air resulted in the data shown in Fig. 8,

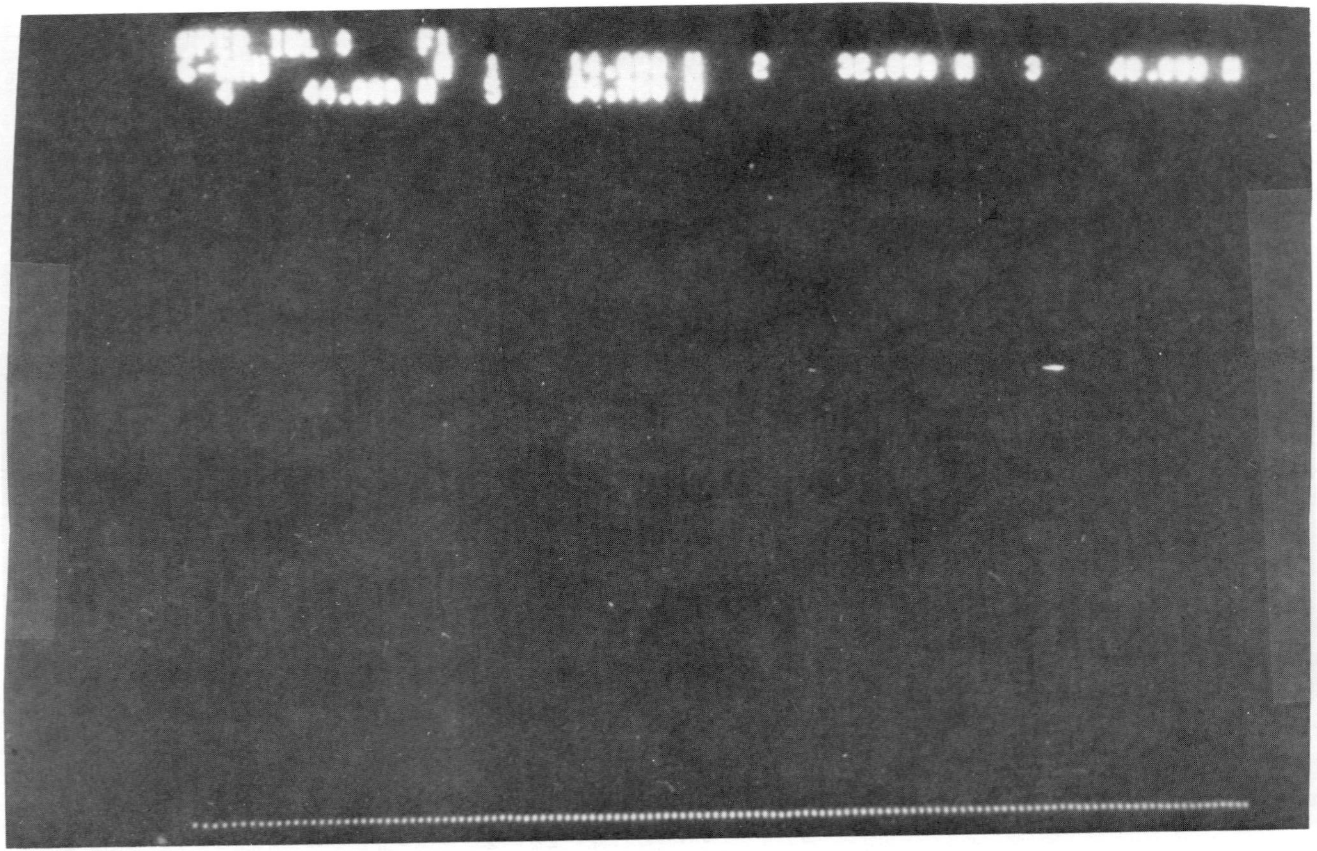

FIGURE 7. Matrix of amu values for analysis of ambient air constituents.

FIGURE 8. Calculated concentrations of ambient air constituents in fractional units.

where the concentrations in scientific notation are displayed in
fractional units. The units displayed are determined by those
used during calibration to obtain the sensitivity coefficients
stored in the matrix and may be percent, parts-per-million, and
so forth. Peak heights measured during the course of the
analysis are also retained in memory and are listed in Fig. 9.

Other functions selectable from the keyboard to assist in
the analysis include automatic subtraction of background,
normalization of background and of measured concentrations, and
corrections for drifts in instrument sensitivity and mass scale.
Signal averaging in all operational modes is also provided.
Fig. 10 shows a spectrum of naturally occurring krypton in air,
which is present at approximately 1 part per million, obtained
at a scan rate of 99 seconds per amu. Krypton has several
isotopes, the most abundant of which are at mass 82 (11.56%),
83 (11.55%), 84 (56.90%), and 86 (17.37%). These peaks are
clearly visible in the spectrum, and calculated ratios of the
peak heights are isotopically correct. Another spectrum, taken
at 1 second per amu, shows the mass 84 peak barely detectable
above background noise (Fig. 11). Figure 12 shows the same
spectrum signal averaged 54 times, in which the isotopic krypton
peaks are clearly displayed. The peak appearing at mass 85 is
not due to krypton, but to the $CClF_2$ fragment from a trace
amount of freon 114 ($C_2Cl_2F_4$) simultaneously introduced into
the sample.

The energy of the electrons used to create the ions may be
switched from an upper value of approximately 76 volts to a
preselected lower value between 15 and 35 volts. These
alternate levels may be individually selected for each of the
peaks selected in the autoanalysis mode or will cause the entire
measurement to be made at the selected level in the Spectrum
Scan or Single mass modes. The usefulness of this method is
demonstrated in Fig. 13, which shows the relative ionization

FIGURE 9. Peak intensities for the selected amu.

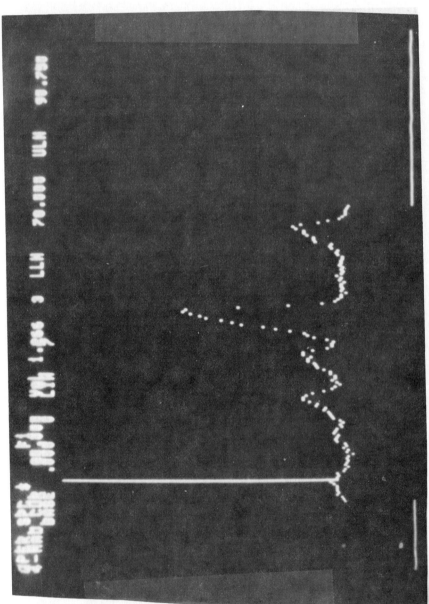

FIGURE 10. Isotopic spectrum of krypton at 1.1 ppm, 99 seconds per amu.

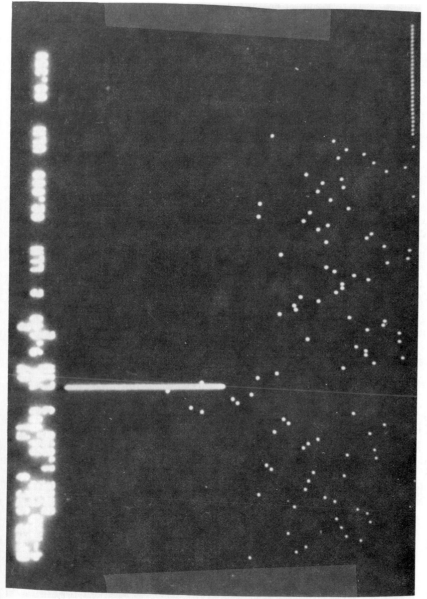

FIGURE 11. Isotopic spectrum of krypton at 1.1 ppm, 1 second per amu.

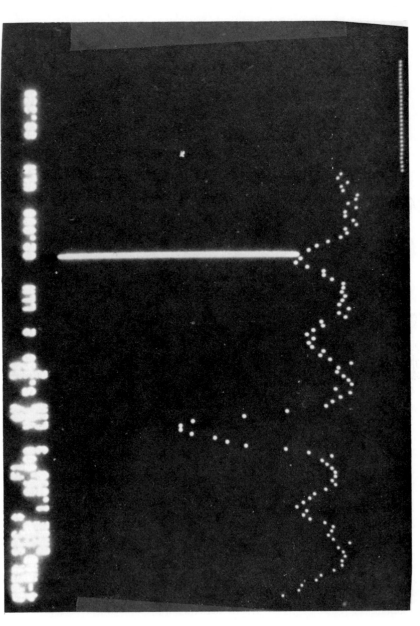

FIGURE 12. Isotopic spectrum of krypton, 1 second per amu, signal averaged 54 times.

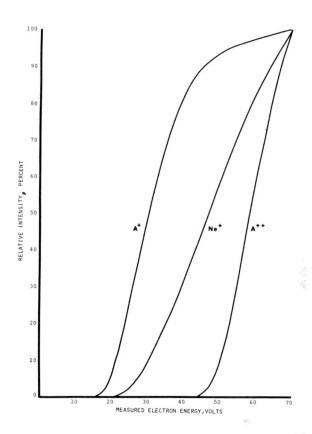

FIGURE 13. Relative ionization efficiency curves for Argon⁺, Argon⁺⁺, and Neon⁺.

efficiency curves, that is, the ratio of ion current obtained at a given ionizing electron energy to that obtained at 70 volts, for singly charged argon at mass 40, singly charged neon at mass 20, and doubly charged argon, also at mass 20 (2). The exact mass of neon is 19.9924 amu and doubly charged argon is 19.9812, requiring a resolution of 1 part in 1778 to resolve the resulting mass peaks. It is clear from the data that one cannot, at mass 20, measure a small concentration of neon in air containing 1% argon at an ionizing electron energy of 70 volts with an instrument of modest resolution. It is also

clear, however, that if the ionizing electron energy is reduced
to the neighborhood of 43 volts, the contribution to the mass
20 peak by the doubly charged argon is eliminated, while about
40% of the neon contribution remains to be measured. A
similar set of data in Fig. 14 shows relative ionization
efficiency curves obtained at mass 16 for methane (CH_4), the
NH_2 fragment from ammonia, and atomic oxygen obtained from the
dissociation of molecular oxygen and of water. From this data
it is seen that reducing the electron energy to near 20 volts
will virtually remove interferences at mass 16 for each of the

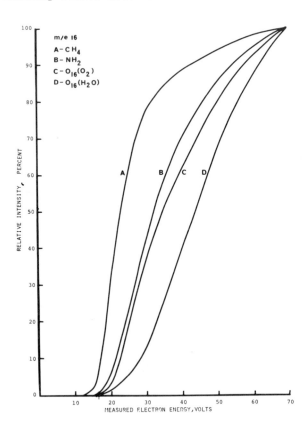

FIGURE 14. Relative ionization efficiency curves for
CH_4^+, NH_2^+, and O_{16}^+.

species other than methane. In cases where mass interferences are minimized, one can therefore detect the presence of remaining species even with an instrument of modest resolution. The data in Figs. 13 and 14 were taken with an instrument with a resolution in excess of 1 part in 2500 so that identification of each species measured at its exact mass is unambiguous (2). Tests with the GADS quadrupole mass spectrometer have demonstrated the applicability of the technique to this system.

Calibration of the GADS mass scale is accomplished by entering at least four major peaks to be found in a calibration gas into the memory, and with a small concentration of this gas mixing into the sample inlet system, commanding a mass calibration from the keyboard. The calibration gas used is typically octafluorocyclobutane (C_4F_8), perfluoro-2-butene (C_4F_8), or a high molecular weight freon such as freon-114, dichlorotetrafluoroethane, $(C_3Cl_2F_4)$. Selected peaks from the latter are shown in Fig. 15 as mass numbers 2 to 5. When commanded to perform a mass calibration, the computer will search for the selected mass ±2 amu, and when the top of the peak is located, store the exact location in the memory. The correct mass scale is then interpolated between the values entered for mass 1 and mass 6. The exact locations found for this calibration are listed in Fig. 16. Once a calibration has been performed, entry of an integer into the matrix will cause the system to go to the exact position necessary to locate the top of any peak. The system can therefore be recalibrated at any time without the necessity of reentering variables into the matrix or memory, and the appropriate corrections made. In practice, the electronics have been found to be sufficiently stable to preclude the necessity of more than weekly recalibration as long as the resulution is not changed or the controlling electronics adjusted.

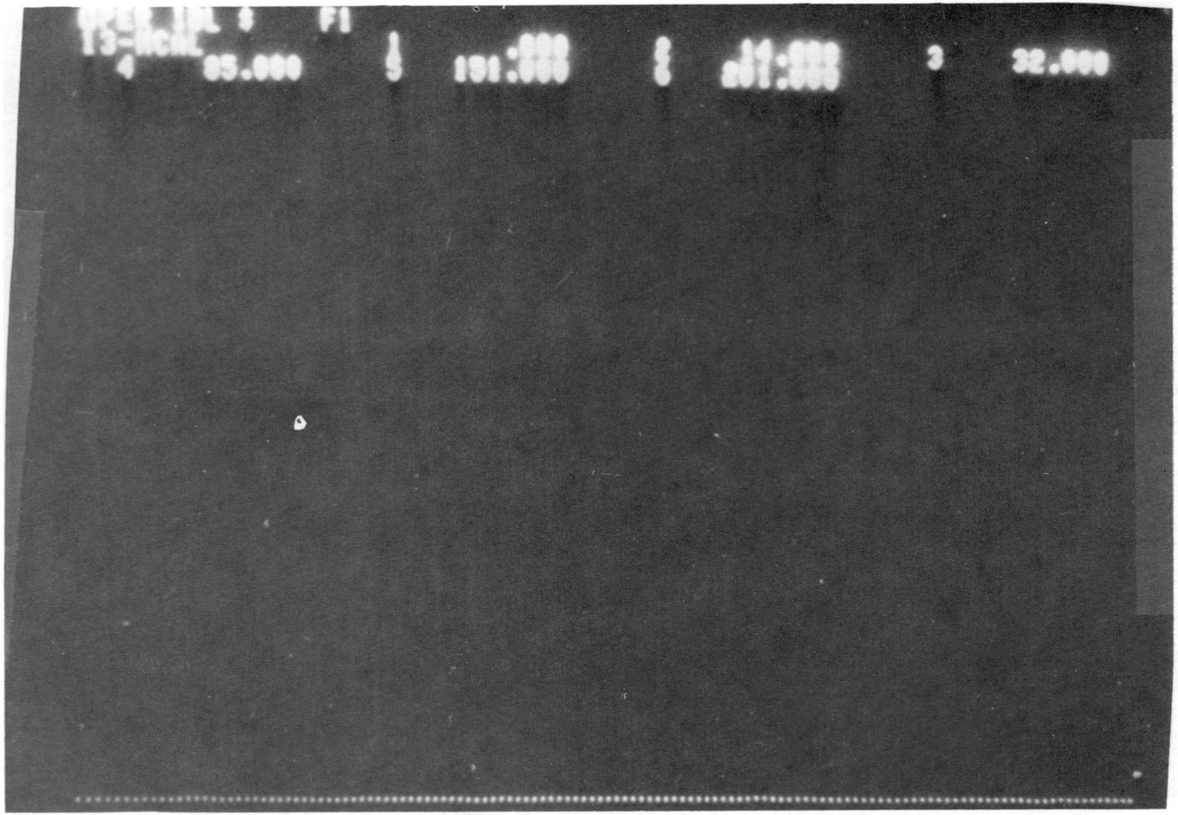

FIGURE 15. Amu selected for mass scale calibration.

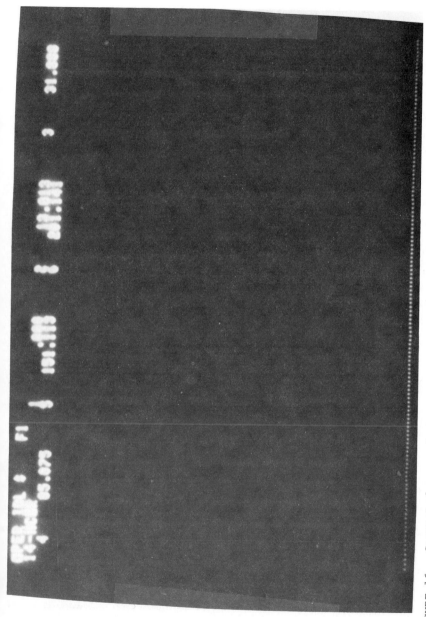

FIGURE 16. Corrected amu values.

Calibration to compensate for drifts in sensitivity is accomplished by determining the ratio of the intensity of four major peaks, which may be found in the calibration gas or may be recurring in the air sample, to the peak representative of air. When recalibration of sensitivity is commanded, these ratios are again obtained, compared with those previously stored in memory, and appropriate corrections made to the measurement data.

There are 16 information, system status, and data displays that may be observed on the computer CRT. In addition to those described are displays of dwell time for each peak in auto-analysis, of parameters used for calibration, and of operational status. The operational status display for the spectrum scan mode is shown in Fig. 17, and shows, among other parameters, that the system is to scan from mass 11 to mass 20 at 1 second per amu, and that the system is on (OPER) but not scanning (IDL). A similar display exists for the other two modes of operation. System protect functions, e.g., loss of vacuum or exceeding the range of the current-to-frequency converter, are listed in these displays if activated.

III. CONCLUDING REMARKS

The development of this system was supported by the NASA Office of Technology Utilization, DOD, DOE, and EPA, with the NASA-Langley Research Center assuming technical and contractual responsibility. The prime contractor for system development was the Analog Technology Corporation, Irwindale, California. A commerical version of the system will be produced and marketed by the Uthe Technologies Incorporated (UTI), Sunnyvale, California, during the coming year.

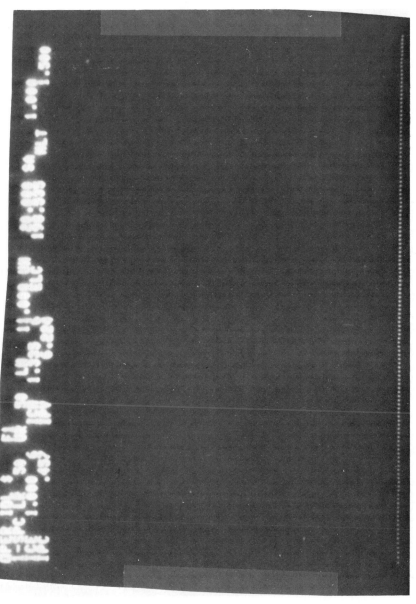

FIGURE 17. Operational status display spectrum scan mode.

REFERENCES

1. Judson, C. M., Wood, G. M., and Yeager, P. R., Paper Z-1, presented at the 24th Annual Conference on Mass Spectrometry and Allied Topics, San Diego, California, May 9-13, 1976.

2. Wood, G. M., Hughes, D. B., and Upchurch, B. T., Paper FA-6, presented at the 26th Annual Conference on Mass Spectrometry and Allied Topics, St. Louis, Missouri, May 28 - June 2, 1978.

DISCUSSION

Shaw: Can you say something about the sensitivity and accuracy, let's say, for CH_4?

Wood: We have looked at atmospheric krypton which occurs at 1.1 parts per million, and we can very readily identify the mass 84 and mass 86 isotopes. The 86 shows up at about 176 parts per billion, and it looks like the sensitivity is going to be about 40 or 50 parts per billion, except when large adjacent peaks are present. This summer we will be doing some additional work with the system in which we will be pre-concentrating the sample in order to get the detection limit down.

INFLUENCE OF SULFUR SIZE DISTRIBUTION
ON OPTICAL EXTINCTION

T. A. Cahill, J. Barone, R. A. Eldred, R. G. Flocchini,
D. J. Shadoan and T. M. Dietz

Crocker Nuclear Laboratory
University of California
Davis, California

Multiple regression analysis has been used to interpret the relationship between visibility reduction, pollutants, and meteorological conditions. One result of this study was to isolate the sulfur size mode $0.6 \ \mu m < D_p < 2 \ \mu m$ as a major factor in visibility at all sites--coastal, urban, and interior valley. However, the sulfur mode below $0.6 \ \mu m$ was only significant at a single coastal site and leads to a separation between the fine, carbonaceous sulfur containing primary particles which are not efficient light scatterers and the coarser photochemically modified sulfur particles which are effective. This result impacts on control strategies in high oxidant areas.

I.   INTRODUCTION

Mie scattering theory predicts that both particle size and complex index of refraction are necessary to calculate optical extinction. Investigation of this predicted dependency in ambient atmospheres has been hampered, however, by the few studies of visibility that include simultaneous measurements of particle size and chemical composition, as the latter factor is presumably important in the index of refraction. The situation is inherently complex because of the large number of parameters that may be implicated in visibility; thus, one is required to make either numerous detailed measurements by size and composition or to accept potentially erroneous limitations in the data set that may foreordain the conclusion. One data set that at least partially surmounts these limitations was collected in California in the period from 1973 through 1977, as it included three size fractions including a cut at $D_p \approx 0.6$ μm, and elemental analyses that routinely included sodium through lead and occasionally added hydrogen through fluorine (1). When these 24-hour values were combined with weather parameters and six gaseous pollutants on a daily basis for each 90-day season, one had a data set consisting of over 5,000 measurements from which one could extract associations with visibility (2). Results for summer 1973 have been published at four sites, two of which lie in the Los Angeles air basin. These two sites provide information regarding the influence of particle size and chemical composition, particularly sulfur species, on degradation of visibility.

II. PROCEDURES

The methodology is discussed in detail in Ref. 2. The method of multiple regressions was applied to the formula

$$b = A \ (P_1)^{b_1} \ (P_2)^{b_2} \ (P_3)^{b_3} \ \ldots \ (P_n)^{b_n} \ E$$

where P_1, P_2 ... P_n are concentrations of pollutants, relative to mean values, A, b_1, b_2, ... b_n are constants, and E is an error term. The relative values of pollutants on a 24-hour basis were compared with mean daily visibilities, synoptic weather patterns causing variations in time.over the 90-day period. Afte correcting for strongly correlated sets of elements, a limited subset of the original 60 or so parameters was found to be adequate to fit variations in visibility. The parameters, mean

values, and standard deviations are shown in Table I (taken from
Ref. 2). The beta coefficients and the fits to visibility are
shown in Table II and Fig. 1, respectively (2).

III. RESULTS

The beta coefficients in Table II indicate the relative importance of each parameter in the fit to the visibility. The two
sites in the Los Angeles air basin exhibit great similarities,
despite a separation of more than 50 km. Intermediate size sodium
is a tracer of sea salt, and both stations respond to its presence
in terms of increased visibility. Likewise, relative humidity
strongly correlates with decreased visibility at both sites, as
does NO_2 and intermediate size sulfur. Other parameters are weakly
involved or depend on local effects such as station siting. One
remarkable factor is the absence of any association between
visibility and the abundant sulfur component with diameter less
than 0.65 µm. This is all the more surprising in light of the
predicted peak in scattering efficiency around 0.5 µm. The relative association between visibility and each sulfur component is
shown in Fig. 2, and it confirms the dependencies shown in the
beta coefficients.

Resolution of this problem may be in the different natures of
the fine (<0.65 µm) and intermediate (0.65 µm to 3.6 µm) sulfur
compounds. Correlation coefficients between sulfur and the metals
vanadium and nickel are very strong in the fine fraction, and the
ratios are close to those in fuel oil. The correlations are weak
and the ratios sharply depressed for the metals in the intermediate size particles. In addition, the larger diameter sulfur
particles are only present in the Los Angeles area in summer
months, whereas the fine particles persist summer and winter.
For these and other reasons, one can associate the coarser
particles with photochemical conversion of SO_2 and H_2S to sulfate,

TABLE I. Means and Standard Deviations of Selected Variables at Four California Sites (July–September 1973)

	LOS ALAMITOS		LOS ANGELES		OAKLAND		BAKERSFIELD	
	Mean	Standard deviation	Mean	Standard deviation	Mean	Standard deviation	Mean	Standard deviation
Sodium-2 ($ng\ m^{-3}$)	239	164	224	147	462	438	89	72
Silicon-2 ($ng\ m^{-3}$)	356	141	447	403	138	78	375	140
Sulfur-2 ($ng\ m^{-3}$)	1985	1318	2109	1973	113	113	298	197
Lead-2 ($ng\ m^{-3}$)	208	168	412	288	51	26	123	74
Total-2 ($ng\ m^{-3}$)	3381	1558	3873	2448	1725	1319	1470	540
Sodium-3 ($ng\ m^{-3}$)	56	74	139	106	418	396	57	114
Sulfur-3 ($ng\ m^{-3}$)	1619	786	1517	1229	1002	561	1294	623
Potassium-3 ($ng\ m^{-3}$)	77	49	94	59	132	110	403	627
Lead-3 ($ng\ m^{-3}$)	715	684	1003	696	923	654	988	424
Total-3 ($ng\ m^{-3}$)	3083	1702	3550	2044	3892	1996	4875	2095
Humidity (%)	59.1	7.9	73.8	9.5	79.5	8.0	37.5	8.3
Wind speed (mh^{-1})	5.8	1.1	7.1	0.9	8.5	1.6	7.3	1.4
Temperature (°F)	71.2	3.6	68.3	3.2	62.0	3.4	84.0	6.2
Oxidant (ppb)	19.7	5.8	37.3	12.3	17.0	6.3	44.1	11.4
Nitrogen dioxide (ppb)	53.4	20.3	59.2	23.0	30.8	17.7	36.8	9.8
Sulfur dioxide (ppb)	22.1	10.3	*		23.2	5.8	42.8	6.1
Hydrocarbons (ppb)	*		21.5	5.0	*		*	
b_{scatt} (km^{-1})	0.435	0.214	0.448	0.201	0.197	0.055	0.192	0.068
Days	78		65		74		85	

* Indicates variable not measured.

216

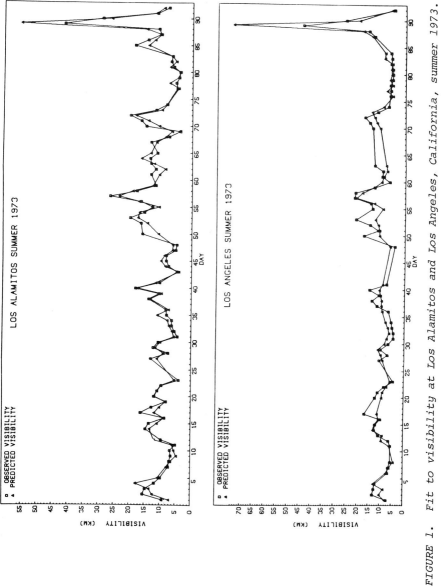

FIGURE 1. Fit to visibility at Los Alamitos and Los Angeles, California, summer 1973.

217

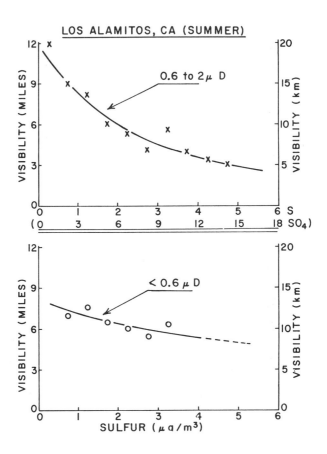

FIGURE 2. *Visibility at Los Alamitos, California, as a*
function of coarse sulfur (sulfur 2, 0.6 to about 2 μm diameter)
and fine sulfur (sulfur 3, <0.6 μm diameter) particles.
Removing the weak correlation between coarse and fine sulfur
particles in summer months removes any statistically significant
correlation between fine sulfur particles and visibility at both
Los Alamitos and Los Angeles, California, as shown in Table 2.

TABLE II. *Significant Beta Coefficients for Selected Sites, Summer 1973*

	Los Alamitos	Los Angeles	Oakland	Bakersfield
Sodium-2	-0.189	-0.172	*	*
Silicon-2	*	+0.201	*	*
Sulfur-2	+0.386	+0.283	+0.369	+0.496
Lead-2	*	*	*	*
Sodium-3	*	*	+0.198	*
Sulfur-3	*	*	+0.367	*
Potassium-3	*	*	*	+0.428
Lead-3	*	-0.369	*	-0.244
Humidity	+0.644	+0.624	*	*
Wind speed	*	*	*	*
Temperature	+0.198	*	-0.310	*
Oxidant	+0.240	*	*	+0.499
Nitrogen dioxide	+0.296	+0.282	*	+0.327
Sulfur dioxide	+0.233	**	**	**
Hydrocarbons	**	*	+0.419	*
Multiple R^2	0.848	0.730	0.631	0.645

^{2}Indicates intermediate size particles, 0.65 to 3.6 μm.

^{3}Indicates fine size particles, 0.1 to 0.65 μm.

* Indicates that the beta coefficient is not significant at the 0.1 confidence level for the t-test.

** Indicates that the variable was not measured at that site, therefore, no beta coefficient could be computed.

whereas the finer particles appear to be direct, primary emission of particles from oil combustion. The fine particles may, in fact, be much finer than 0.5 μm; thus, their ability to scatter light is greatly reduced. The relatively large size of the coarse fraction appears to be due to the presence of large amounts of water in these particles.

IV. CONCLUDING REMARKS

The implications of these results and the hypotheses are that in photochemically active atmospheres, control of gaseous sulfur emissions is more important than primary sulfur containing particles. In addition, summing up of all particles in the accumulation size mode (<2.5 μm) may obscure important dependencies vital to understanding visibility degradation; thus, some effort should be expended in generating more detailed size/composition data sets suitable for statistical studies of visibility.

REFERENCES

1. Flocchini, R. G., Cahill, T. A., Shadoan, D. J., Lange S., Eldred, R. A., Feeney, P. J., Wolfe, G., Simmeroth, D., and Suder, J., *Environ. Sci. Technol. 10,* 76 (1976).

2. Barone, J. B., Cahill, T. A., Eldred, R. A., Flocchini, R. G., Shadoan, D. J., and Dietz, T. M., *Atmos. Environ. 12,* 2213 (1978).

DISCUSSION

Perhac: Let me exercise the session chairman's prerogative
to make one or two comments. Within the electric power industry,
visibility is becoming a matter of real concern. The issue of
sulfates and health effects, believe it or not, is very quickly
getting diffused. It is turning out that more and more there
seems to be no significant relationship, at least not yet
demonstrated, between adverse health effects and sulfates in the
atmosphere. In fact, even some of the representatives of the
Federal Government, the EPA for example, will admit this. We
don't have a strong case for it, so as a result the interest
in sulfates in other aspects of air quality is becoming more and
more important. Two of these aspects are acid precipitation and
visibility, and depending on what side of the fence you are on,
you can argue one is more important than the other. I will not
get into that argument. Suffice to say--visibility degradation
is a matter of real concern to people who are in the electric
power industry, and any of you who are doing research on particle
distribution in the atmosphere are becoming more and more aware
of the emphasis on visibility impairment. So it is an important
subject, and my own personal feeling, and now I will give you an
opinion, is that probably this may be one of the most serious
issues that the industry has to face with increased coal usage
over the next 20 years. Let me qualify the last statement about
serious issues with regard to the sulfur problem in the atmosphere. There are many other problems. But with regard to the
sulfur in the atmosphere, my own personal opinion is that
visibility may be one of the more critical ones.

Singh: I am sure you are aware that according to the Mie theory,
scattering of light is a function of particle size and index of
refraction. Have you made any attempt to study the index of
refraction of these materials?

Cahill: Because the work was almost all ambient, we haven't been
able to do it. But recently, we have set up artificial systems
of dust and smoke. Agricultural smoke is a big problem in our
valley. We tried to come up with particles that we can
characterize by the index of refraction. But it's not easy in a
realistic situation. We would like to do it and I encourage
everybody here to do it. We are going to do it ourselves in a
very small way by measuring both absorption and scattering of
artifically generated aerosols in a confined volume. It is
sort of like a fog chamber, but instead of fog you work with
smoke and agricultural dust. But it's tricky, and I wish we had
more information on it.

Pack: You mentioned sort of in passing the problem of artifact formation. Since you are using filters for the very fine stage, do you anticipate any problem particularly with the oxides of nitrogen? As you know, EPA has thrown out all of their data now.

Cahill: You notice an enormous reticence out of Davis for the last 6 years on that nitrogen question. We are not going to touch it with a 10-foot pole because most of our analysis has been done in vacuum. And we know that samples are going to out gas and lose nitrogen compounds. So the answer is that until we come up with a good, stable target that will in fact take the beating we give it, we are not going to touch nitrogen. Whatever the problem with nitorgen is, we know it's ghastly. And we're not even sure whether it is formation or loss, so that's another good problem. We think it is a very important problem for nitrogen, but it does not appear significant for sulfur compounds on surface deposition filters.

Stokes: The statistical method you used of combining things that have high correlation coefficients sounds like Principal Component Analysis. Is that actually what you did?

Cahill: Close, but not quite. We backed into it. We did Principal Component Analysis manually, and the reason was because Principal Component Analysis has gotten a bad reputation by throwing together unrelated things. We are trying to be as conservative as we could, and so we backed into it by looking at the bivariant correlations, then doing the multivariant correlations, pulling out the factors and seeing the change in the multiple r^2. It is an awkward and long way, but it has a lot of redundancy to it.

Stokes: I can't agree with you more because that's really the way to do it because you have to understand the data that goes into the analysis. It has a bad reputation for cause.

Demerjian: One of your slides was rather interesting where you had a correlation in Los Angeles for good visibility versus lead. Would you like to attempt to explain what you think that might be caused from?

Cahill: As far as we can tell, at times of low wind velocity, the lead concentration reaches very high values. Lead can be plotted versus mean wind velocity and it's a very good plot. It's just a ventilation factor. We feel--as we haven't been able to prove it yet--that at times of lowest wind velocity, we were getting a drift across the local freeway which is up-wind of that site. We have seen that in other sites and have tied it down, but unfortunately the data is a little old by now and we are never going to actually prove by actually running a

profile on that freeway. But the wind factor is correct, and
the source strength is correct, so we think it's an accidental
association with wind direction.

Demerjian: So it's a source-sited problem?

Cahill: It's very difficult to put in wind direction into the
linear regression models. You have to almost regress each
quadrant separately. Again, people like to do it. You can
write an algorithm for it. But it's dangerous. Very dangerous,
indeed.

Green: I can't understand why sulfates under 0.5 µm shouldn't
correlate with visibility from the viewpoint of light scattering
theory. It seems you are just at the right wavelength and that's
where things really should be reacting.

Cahill: Right, but there is another problem. We have taken such
a low order cut at the size information, with only three size
cuts. We are very poor in size information, while being a
little rich in elemental information. We feel in fact that the
particles are quite a bit smaller than 0.5 µm, and from the
data that have been collected by other people, we feel that it's
quite a bit less. Certainly in Oakland and places like that,
0.2 µm is probably more reasonable because you have lots of
prompt, local sulfate sources. I think what happens is that the
particles are very small, in fact too small to effectively
scatter light, and I really would be awfully cautious because
the growth of particles in Los Angeles seems to be unique to
that one area of the state. We don't have enough information.
But we are going down there this summer to collect more detailed
size information. Again we have to fill out our own information
with additional size data and see if we can look at the growth
in more detail.

Whitby: I was also trying to reconcile your results with some of
ours. Last summer we had a couple of low pressure impactors
running alongside of all our other gear which included b_{scatt}
and EAA size distributions and we got excellent agreement. I
think your guess that the size is small might be part of it but
I think another factor that might be entering in here is the
thing I alluded to this morning. When sulfur is playing a role
in the aerosol growth, it tends to decrease the accumulation mode
size. If you collect a sample which is a mix of two different
growth conditions, the average modal size may not be a good index
of the actual growth conditions. It might be that one set of
conditions is generating a modal size which is fairly large
while the other one is generating a small modal size with a high
R factor. When one collects it on a filter and performs your

analysis the results are not really indicative of the true
situation. I guess this is really just a guess as to what
might be really happening.

Cahill: We have some data, that in fact indicates what you are
saying is correct. We have seen rapid growth of sulfur during
the night as humidity goes toward the dewpoint and we have seen
in seasonal cases in Central Valley where we have these valley
fogs that form because of the radiation inversion. My feeling
is that again by averaging over day and night, we are dealing
with two very different situations. We should have more diurnal
data. It was just that we couldn't get it inexpensively at the
time.

EFFECTS OF AEROSOL SIZE DISTRIBUTIONS ON THE
EXTINCTION AND ABSORPTION OF SOLAR
RADIATION[1]

P. Chylek and V. Ramaswamy

Center for Earth and Planetary Physics
Cambridge, Massachusetts

M. K. W. Ko

Atmospheric and Environmental Research, Inc.
Cambridge, Massachusetts

*Aerosol extinction and absorption was investigated as a
function of particle size distribution (geometric mean radius r_g
and geometric standard deviation σ_g in the log-normal size
distribution) for a given constant aerosol mass loading. For
size distribtuions with $r_g < 0.1$ μm the extinction and absorption
increases with increasing σ_g, whereas for size distributions
with $r_g > 0.1$ μm the extinction and absorption decreases with
increasing geometric standard deviation σ_g.*

I. INTRODUCTION

It is well known that the atmospheric extinction of a

polydispersion of aerosol particles depends in general on the

total mass loading and its size distribution. In this study, a

[1]*This research was partially supported by a grant from the
Environmental Research Laboratories, NOAA.*

method is outlined for calculating the volume extinction per unit mass loading as a function of the aerosol size distribution. The method is applied to carbonaceous soot aerosols at visible and infrared wavelengths. The results are compared with the corresponding calculation for background aerosols.

II. APPROACH

To determine the optical characteristics of a polydispersion of particles, one has to know the size distribution as well as the index of refraction and shapes of the individual particles.

For simplicity, the assumed optical properties of a randomly oriented polydispersion of irregularly shaped particles can be approximated by an "equivalent" polydispersion of spheres. It will further be assumed that the particles are homogeneous and all have the same index of refraction. The size distribution is described by the surface-area log-normal distribution of the form

$$\frac{dS}{d \ln r} = N_o \exp\left[- \frac{1}{2} \left(\frac{\ln \frac{r}{r_g}}{\ln \sigma} \right)^2 \right] \qquad (1)$$

where N_o is the normalization factor related to the total number of particles per unit volume, r_g gives the radius at which the distribution attains a maximum and σ parameterizes the width of distribution. $\frac{dS}{d \ln r}$ is related to $\frac{dN}{dr}$, the number density per unit radius, via the equation

$$\frac{dS}{d \ln r} = \pi r^3 \frac{dN}{dr} \qquad (2)$$

The volume extinction coefficient β at wavelength λ is given by

$$\beta(m,\lambda,\frac{dN}{dr}) = \int_0^\infty \pi r^2 Q_{ext}(m,r,\lambda)\frac{dN}{dr} \, dr \tag{3}$$

where m is the refractive index at wavelength λ, r is the radius, and Q_{ext} is the single particle extinction efficiency derived from Mie theory.

The mass loading of the polydispersion is given by

$$M = \rho \int_0^\infty \frac{4\pi}{3} r^3 \frac{dN}{dr} \, dr \tag{4}$$

where ρ is the bulk density of the aerosol. Experimental measurements indicate that ρ is about 2 g/cm^3.

The volume extinction per unit mass loading C is defined by

$$C(m,\frac{dN}{dr}) = \frac{\int_0^\infty \pi r^2 Q_{ext} \frac{dN}{dr} \, dr}{\rho \int_0^\infty \frac{4}{3}\pi r^3 \frac{dN}{dr} \, dr} \tag{5}$$

Note that, as expected, C is independent of N_o. Thus, once the index of refraction is determined, C can be computed as a function of r_g and σ.

The results of the measurements of the index of refraction
m = n - ki for various types of carbonaceous aerosols is shown
in Figs. 1 and 2 (1,2). The numerical study showed that the
variations of m within the measured limits have practically no
effect on the optical properties of the polydispersion. A value
of m = 1.8 - 0.6i is chosen for λ = 0.55 μm and m = 3.0 - 2.0i
for λ = 10 μm. These values are used to compute C for all
carbonaceous soot aerosols. For comparison, similar calculations are performed for background "natural" aerosol with
m = 1.525 - 0.005i and m = 1.7 - 0.3i at λ = 0.55 μm and
λ = 10 μm, respectively.

III. RESULTS AND DISCUSSION

Figs. 3 to 6 show the extinction per unit mass loading C of
carbonaceous particles and "natural" aerosols at λ = 0.55 μm
and λ = 10 μm as a function of size distribution where for
convenience ρ = 1 g/cm^3. Note that for both types of aerosol,
with larger values of r_g, the narrowest size distribution
(σ = 1.5) gives the highest extinction, whereas for smaller
values of r_g, the situation is reversed, namely, the wider size
distribution gives the higher extinction.

Another interesting feature of the result is that for a
considerable range of σ and r_g, C is almost independent of σ
and r_g. (See Figs. 4 to 6.) Therefore, one should be able to
deduce the mass loading from measurements of extinction
independent of the size distribution (3-5). Although this is
true for the "natural" aerosol only at infrared wavelengths, it
appears that the method should be applicable to soot particles at
visible as well as infrared wavelengths.

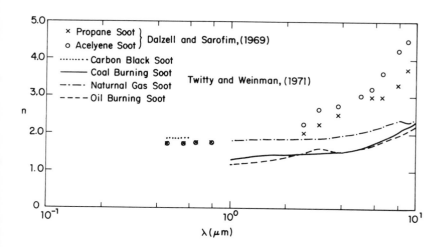

FIGURE 1.   The real part of the index of refraction for various soots.

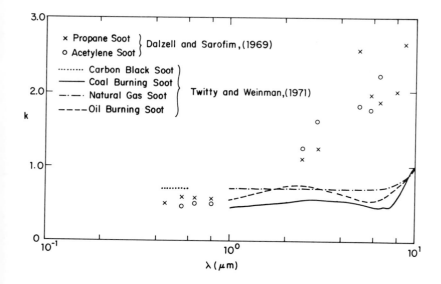

FIGURE 2.   The imaginery part of the index of refraction of various soots.

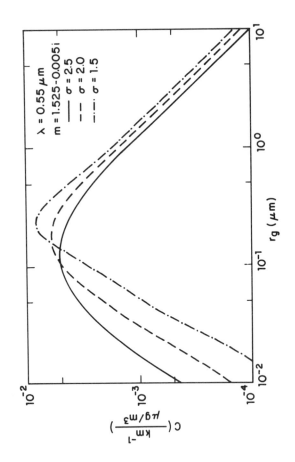

FIGURE 3. *Volume extinction per unit mass loading as a function of* r_g *and* σ *for "natural" aerosol at 0.55* μm *(m = 1.525 − 0.005i).*

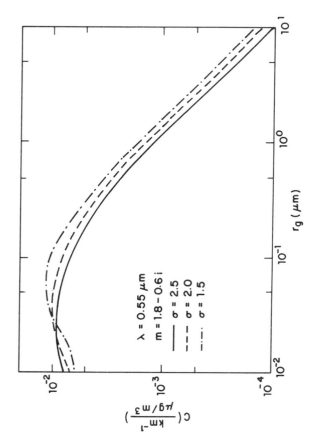

FIGURE 4. *Volume extinction per unit mass loading as a function of r_g and σ for carbonaceous aérosol at 0.55 µm ($\tilde{m} = 1.8 - 0.6i$).*

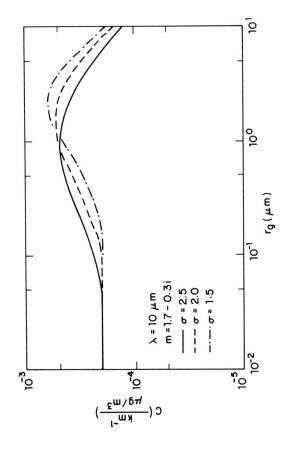

FIGURE 5. Volume extinction per unit mass loading as a
function of r_g and σ for "natural" aerosol at 10 μm
(m = 1.7 - 0.3i).

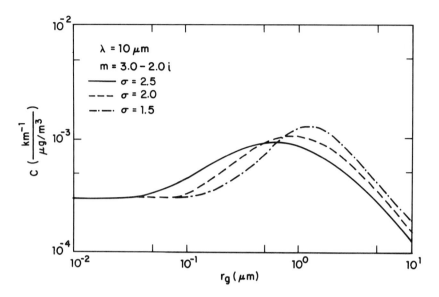

FIGURE 6.   Volume extinction per unit mass loading as a function of r$_g$ and σ for carbonaceous aerosol at 10 μm (m = 3.0 - 2.0i).

Finally, from comparisons of Figs. 3 and 4, and Figs. 5 and 6, it is noted that although the values of C for "natural" particles and soot particles essentially agree with each other for large r$_g$, there are considerable differences for small r$_g$. To better illustrate this, the visibility range R

$$R = \frac{3.9}{\beta\,(m, \lambda\frac{dN}{dr})} \tag{6}$$

was computed for the two types of aerosols with $\lambda = 0.55$ μm,
$\sigma = 2.0$ and $N_0 = 1000$ particles/cm^3. The result is plotted
in Fig. 7 and clearly illustrates the difference for
$r_g < 0.1$ μm.

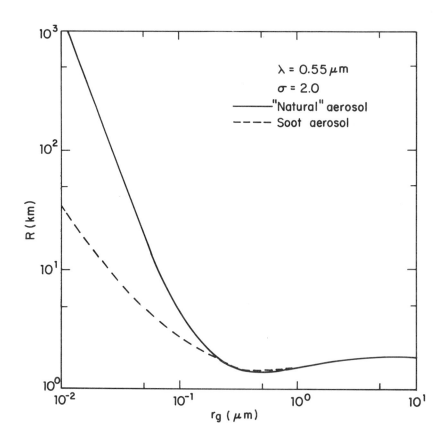

FIGURE 7.   Visibility range as a function of r_g with
$\sigma = 2.0$ for carbonaceous soots and "natural" aerosol at
0.55 μm.   The particle concentration was taken to be
10^3 particles/cm^3.

IV. SYMBOLS

$\dfrac{dS}{d \ln r}$ surface-area distribution

N_o normalization factor for log-normal distribution

r radius

r_g radius parameter for surface-area log-normal distribution

σ geometric standard deviation for log-normal distribution

$\dfrac{dN}{dr}$ number density per unit radius

m index of refraction

n real part of the index of refraction, $m = n - ki$

k imaginary part of the index of refraction, $m = n - ki$

λ wavelength

β volume extinction coefficient

Q_{ext} single particle extinction efficiency

ρ aerosol bulk density

M aerosol mass loading

C volume extinction per unit mass loading

REFERENCES

1. Dalzell, W. H. and Sarofim, A. F., *J. Heat Transfer, 91,* 100 (1969).

2. Twitty, J. T. and Weinman, J. A., *J. App. Met., 10,* 725 (1971).

3. Chylek, P., *J. Atmos. Sci., 35,* 296 (1978).

4. Chylek, P., Kiehl, J., and Ko, M., *Atmos. Environ., 13,* 169 (1979).

5. Pinnick, R. G., Jennings, S. G., Awermann, H. J., and Chylek, P., *J. Atmos. Sci. 36,* 1577 (1979).

DISCUSSION

Whitby: If the refractive index was 1.4 instead of 1.5 in what you called background aerosol, how much would that shift those curves in size?

Ko: If you shift it from 1.5 to 1.4, I don't think it would shift the curves a whole lot. Like you said, if you change the real part of refractive index, essentially it would shift them sideways. I don't think it would shift upwards at all.

Whitby: The reason I mention that is that most of these fine particles are very hygroscopic and that means that they usually have a lot of water in them. So they probably don't have a refractive index as high as 1.525. The refractive index is probably more like 1.35 to 1.4 for these aerosols, unless you get into really dry conditions.

Ko: I see. Well, another point I should mention here is this is integrated over the size distribution already, so when you shift that, that will sort of water down the effect a little. So I really don't think it would change too much.

Box: I would like to comment on precisely that point. I think probably a good point that can be made is that the anomalous diffraction approximation that Van de Hulst has popularized suggests that the most important parameter is often what he calls $\rho = 2r/\lambda(m-1)$. So if you vary the refractive index, you just vary the radius so that you keep that ratio, that coefficient constant, and that will more or less take care of it.

SCATTERING EFFECTS OF AEROSOLS
ON OPTICAL EXTINCTION MEASUREMENTS[1]

M. A. Box[2] and A. Deepak

Institute for Atmospheric Optics and Remote Sensing
Hampton, Virginia

When optical extinction measurements are performed in a scattering medium by the use of transmissometer or radiometer, some scattered radiation invariably enters the detector's finite field of view along with the direct radiation. Bouguer's (or Lambert-Beer) law, which is valid for direct radiation only, cannot be, strictly speaking, used to obtain the true optical depth τ (and true extinction coefficient β_{ext}) from transmission measurements. What one obtains are the apparent quantitites τ' and β'_{ext}. In order to obtain the true quantities, one must calculate and, if significant, correct for the scattered radiation contribution. For this purpose, one computes a correction factor R which depends on the aerosol characteristics (e.g., size distribution n(r), complex refractive $m = m' - im''$, and shape), spatial distributions of aerosols, and the geometry of experimental optics (e.g., view cone half-angle θ_D, and transmitted view cone).

---

[1]*This work was supported by NASA contract NAS1-15198.*

[2]*Present address: Institute of Atmospheric Physics,*
University of Arizona, Tucson, Arizona

I. INTRODUCTION

Determination of aerosol characteristics is being performed
by several researchers using multiwavelength measurements of the
attenuated solar radiation or laser beams traversing the atmosphere. This paper discusses some of the effects of forward
scattering due to aerosols on such optical extinction measurements.

II. OPTICAL EXTINCTION

When a light beam traverses a scattering medium, the intensity
of the light is generally reduced according to Bouguer-Beer's
law:

$$I(L) = I(0) \ e^{-\tau} \tag{1}$$

where τ, the optical thickness of the medium, is given by the
integral of the extinction coefficient along the beam path

$$\tau = \int_0^L \beta_{ex}(\ell) \ d\ell \tag{2}$$

where β_{ex}, the volume extinction coefficient (km^{-1}), is made up
of absorption and scattering components (subscripts abs and sc,
respectively):

$$\beta_{ex} = \beta_{abs} + \beta_{sc} \tag{3}$$

In an absorbing medium, where $\beta_{sc} = 0$, the experimental
determination of τ is straightforward: measure both $I(0)$ and
$I(L)$, and substitute into Eq. (1). β_{ex} (and hence β_{abs}) can then
be obtained from Eq. (2). However, in the case of scattering
media, this procedure is complicated by the fact that $I(L)$ will
almost certainly be composed of direct and scattered light.
This situation is demonstrated in Fig. 1 for the two basic

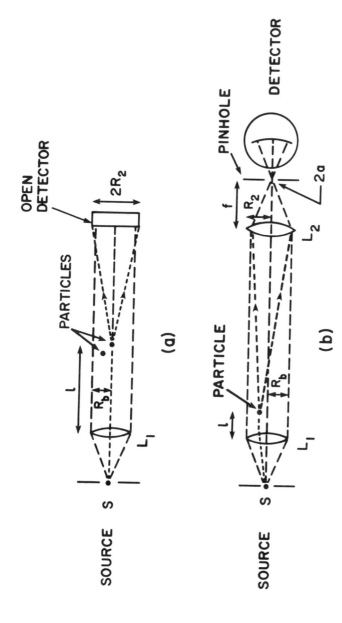

*FIGURE 1. Schematic illustrations. (a) An open detector transmissometer system.
(b) A lens-pinhole detector transmissometer system. (After Ref. 1)*

detector systems: A simple open detector, and a more sophisticated lens-pinhole detector (1-2).

In the atmosphere, and especially the lower troposphere,
natural and man-made aerosols lead to significant values of
β_{sc} at all visible wavelengths. If it is assumed that the
scattering of light by aerosol particles can be described with
sufficient accuracy by Mie theory, then the scattered light, at
least for single scattering, can be calculated.

If n(r) is the size distribution of the aerosols, then the
scattering coefficient is given by

$$\beta_{sc} = \int_0^\infty \pi r^2 \, n(r) \, Q_{sc}(kr,m) \, dr \tag{4}$$

where Q_{sc}, the scattering efficiency factor, is given in terms
of the Mie intensity functions, i_1 and i_2, by

$$Q_{sc} = (kr)^{-2} \int_0^\pi (i_1 + i_2) \sin \psi \, d\psi \tag{5}$$

where ψ is the scattering angle. Full details and definitions
can be found in Ref. 1.

III. SCATTERING CORRECTIONS

If the detection system is such as to subtend an angle of
$2\theta_d$ at the point of scattering, then the effective scattering
efficiency factor, Q'_{sc}, will be given by (1)

$$Q'_{sc} = (kr)^{-2} \int_{\theta_d}^\pi (i_1 + i_2) \sin \psi \, d\psi \tag{6}$$

Equation (1) must then be replaced by

$$I(L) = I(0) e^{-\tau'} \tag{7}$$

where

$$\tau' = \int_0^L \int_0^\infty \pi r^2 \, n(r) \, Q'_{sc} \, dr \, dl \tag{8}$$

In the case of an open detector system (Fig. 1a), θ_d will clearly depend on the location of the scattering particle along the beam path. However, for the lens-pinhole system (Fig. 1b) it is straightforward to show that

$$\tan \theta_d = a/f \tag{9}$$

where f is the focal length and 2a the pinhole diameter.

Under most practical situations, θ_d will be small, and only large particles, which scatter predominantly in the near-forward directions, will make any significant contributions to deviation from Eq. (1). For these conditions (large r, small θ_d) Q'_{sc} may be adequately approximated using Rayleigh diffraction

$$Q'_{sc} = 1 + J_0^2(kr\theta_d) + J_1^2(kr\theta_d) \tag{10}$$

where J_n is a first-kind Bessel function.

In Ref. 1, a series of plots of the correction factor $R = Q'_{sc}/Q_{sc}$ for several refractive indices and a wide range of particle sizes is presented. The accuracy of Eq. (10) is compared with the exact (Mie) result of Eq. (6). In Ref. 2 a similar series of plots of $\bar{R} = \beta'_{sc}/\beta_{sc}$ for three modified gamma size distributions, and a power law distribution, is presented. Again the accuracy of the Rayleigh approximation is also demonstrated.

IV. SOLAR RADIOMETRY

The detection system of Fig. 1(b) is typical of laboratory extinction measurements. In the case of solar extinction measurements (solar radiometry) the situation (Fig. 2) is somewhat

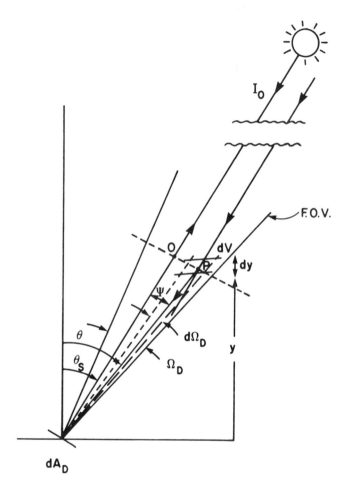

FIGURE 2. Scattering geometry for a radiometer.
(After Ref. 3).

different due to the essentially infinite breadth of the solar
beam. In Ref. 3 it is shown that in addition to the (attenuated)
direct beam (c.f., Eq. (1)), a solar radiometer will also detect
diffuse radiation, given (to a good approximation) by

$$I_{DIF} = I_o \sec \theta_s \, \ell^{-\tau \sec \theta_s}$$

$$\left[\tau_{PS}(1 - R_P) + (\tau_M + \tau_{MS}) (1 - R_M) \right] \qquad (11)$$

where τ is the total optical thickness, θ_s is the solar zenith
angle, τ_{PS} is the aerosol scattering optical thickness, τ_M is the
molecular optical thickness, τ_{MS} is an effective multiple scattering optical thickness (Refs. 4 and 5), R_P is the correction
factor for aerosols, and R_M is the molecular correction factor.

As a result, if such measurements are used to infer the
aerosol optical thickness, then the presence of scattered light
will lead to too small a value of τ_P. In Ref. 3, it is shown
that under most circumstances, the measured value τ_P' will be
related to the true value by

$$\tau_P' = \tau_P R_P - \tau_{MS}(1 - R_M) \qquad (12)$$

Except for the near UV region, the second term on the right side
of Eq. (12) is usually small.

If a series of multispectral extinction measurements is used
to derive the aerosol size distribution, then it is necessary to
make allowance for this scattered light effect. To invert a set
of multispectral extinction measurements, it is necessary to
invert the integral equation (or its matrix equivalent)

$$\tau_P(\lambda) = \int_0^\infty \pi r^2 \, N(r) \, Q(kr) \, dr \qquad (13)$$

Here $N(r)$ is the altitude-averaged size distribution.

To make allowance for scatter light, simply replace Eq. (13)
by

$$\tau_P'(\lambda) + \tau_{MS}(1 - R_M) = \int_0^\infty \pi r^2 N(r) Q'(kr) dr \tag{14}$$

where τ_P' is the measured aerosol optical thickness.

Q' in Eq. (14) depends on the acceptance angle of the radiometer, which is usually a fixed value. Thus the number of Q' values that is required is exactly the same as the original number of Q values. (See Ref. 1 for computational details.) Hence the inversion of Eq. (14) should prove no more complicated than the inversion of Eq. (13), and existing algorithms should still prove practicable, provided that the kernel values are replaced.

V. CONCLUDING REMARKS

In order to obtain accurate measurements of the extinction coefficient, from which aerosol characteristics can be determined, it is important to design the experimental system such that the correction factor R is as close to unity as possible. The design considerations are discussed in detail in Refs. 1, 2 and 3. For already existing radiometer systems, one must consider the corrections due to the forward scattered radiation entering into the view cone of the detector, as discussed in these references.

REFERENCES

1. Deepak, A., and Box, M. A., *Applied Optics, 17,* 2900 (1978).

2. Deepak, A., and Box, M. A., *Applied Optics, 17,* 3169 (1978).

3. Box, M. A., and Deepak, A., *Applied Optics, 18,* 1941 (1979).

4. Box, M. A., and Deepak, A., *Applied Optics, 18,* 1376 (1979).

5. Box, M. A., and Deepak, A., "An Approximation to Radiative
    Transfer in a Turbid Atmosphere: Almucantar Radiance
    Formulation" IFAORS-134 (1979) (available from IFAORS,
    P. O. Box P, Hampton, VA 23666).

REMOTE DETERMINATION OF AEROSOL
AND CLOUD PARAMETERS FROM SOLAR RADIATION DATA[1]

G. M. Lerfald
V. E. Derr
R. E. Cupp

Environmental Research Laboratories
National Oceanic and Atmospheric Administration
Boulder, Colorado

Radiometric measurements using the sun as a source have been
conducted at six different sites in Colorado, Montana, and
California. These data are reduced to obtain the atmospheric
optical thickness in nine wavelength bands in the range 0.3 to
10 μm, including bands chosen to measure the total amount of
precipitable water vapor and ozone in the sun-instrument path.
Also measured are the angular phase functions in the near-forward
scattering directions (solar aureole) in four wavelength bands.
Time lapse photography is used to determine general atmospheric
conditions including visible sources of aerosol pollution and to
identify cloud types. Analysis of the data yields information
on the aerosol loading, the size distribution of aerosol and
cloud particles, and the effects of aerosols and clouds on the
solar radiation budget. It is suggested that the techniques
described can be applied to definitive studies of the environmental and climatic impacts of coal utilization.

[1]The work reported on here was in part supported by the
U.S. Department of Energy, Division of Distributed Solar
Technology, Washington, D.C.

247

I.   INTRODUCTION

A program of experimental measurements and analyses using a
number of measurement techniques simultaneously has been conducted to study the effects of aerosols and clouds on solar
radiation (1). The equipment used included a large backscatter
lidar, a microwave radar, solar radiometers, aircraft-borne,
in situ particle probes, and time-lapse film cameras (sky
photography). Data have been taken with these measurement
systems at observing sites in Colorado, Montana, and California.
Usually, other supporting measurements, such as those from
balloon rawinsondes and surface meteorological instrumentation,
have also been available at the observing sites. The data
sets obtained permit unusually complete specification of many
atmospheric parameters. Analyses, including comparisons and
correlations between results obtained by the various techniques,
are currently in progress (final report in preparation).

One of the tentative conclusions emerging from these comparisons is that the solar radiometric techniques themselves
can yield valuable information on important characteristics
of atmospheric gases, aerosols and clouds. Since the solar
radiometric measurements are relatively low in cost and easy
to obtain, they are suitable for long-term monitoring programs.
It is suggested that they could be effectively applied to
studies of the environmental and climatic impacts of coal
utilization.

II.   THE ATMOSPHERIC PARAMETERS OF IMPORTANCE

Solar radiation penetrating the atmosphere is affected by
absorption and scattering by the atmospheric gases and by suspended particulate matter. The latter includes water droplets
and ice crystals composing clouds, and aerosol particles. The

physical characteristics of clouds and aerosols, and the amount
of precipitable water vapor are highly variable in both space
and time, being dependent on large scale weather patterns as
well as on local conditions. A large increase in the use of coal
as an energy source is likely to inject additional pollutants
into the atmosphere.

It is suspected that there are many subtle relationships
between the characteristics of atmospheric trace gases and aerosols and weather conditions, including the formation of clouds
and precipitation. Recent books by Pruppacher and Klett (2)
and by Twomey (3) review the state of knowledge about such known
and suspected relationships. In order to understand these
relationships better, data on such parameters as the amount of
precipitable water vapor, the aerosol loading, the size distribution of aerosol particles, and the evolution in the sizes of
cloud droplets (or ice crystals) as clouds form or disperse are
needed. The proper kinds of solar radiometric measurements are
capable of providing data on these quantities.

III. MEASUREMENT METHODS

The two primary techniques which are applied to obtaining
remote determinations of aerosol and cloud parameters are
(1) measurement of the extinction of the direct solar radiation
beam in specific wavelength bands and (2) measurement of the
angular scattering function referenced from the sun's center.
Some of the early development of these methods was done by
Deirmendjian (4) and Kerker (5). The analysis of solar radiometric data to derive parameters of aerosols or clouds is
discussed in recent papers by Herman (6), Deepak (7) and Box
and McKellar (8). To make it possible to measure optimally the
important characteristics of both clouds and aerosols, it is

desirable to have instruments which (1) measure simultaneously
in the various wavelength bands, (2) have response time in the
order of seconds, and (3) have large dynamic range capabilities.

The instruments currently in use at NOAA satisfy most but
not all of these design criteria. Table I lists the characteristics of the radiometers which have been used during the past two
years. The aureole radiometer design is described by Lerfald
(9) and the 8-channel radiometer is described and examples of
results given in reference 10.

The analysis results include determinations of total precipitable water vapor, aerosol loading estimates, the optical
thickness as a function of wavelength (in 9 wavelength bands),
and the angular scattering function. From the latter two
parameters, estimates of cloud and aerosol particle size distributions are obtained. The methods of optimizing the amount of
information derivable from the extinction and angular scatter
data are still being explored. The problem is that of applying
constraints which act to avoid spurious solutions caused by noise
in the data, and at the same time not unduly restricting the
results which represent real information on size distributions.

Time-lapse sky photography provides highly valuable qualitative data on general sky and weather conditons. The type,
geometry, and motion of clouds can be obtained from the films.
Also, local sources of pollution can often be identified and
atmospheric optical phenomena such as sun halos, rainbows, etc.
are evident. Such information results in a better insight when
interpreting the data and frequently strengthens the conclusions
which can be drawn from the analysis.

Time-lapse photography with an all-sky camera (180° field of
view) and a wide angle camera (50° field of view) mounted on the
solar tracker are normally used during measurements. Low-cost
time-lapse systems based on 8 mm motion picture cameras were
employed.

TABLE I. Design Characteristics of Radiometers

Instrument	Filter wavelength bands (μm)	Field of view (deg)	Response
Direct-Beam Extinction:			
8-Channel Solar Radiometer	0.317	1.5	Logarithmic; dynamic range is 5 orders of magnitude. (Each band is 0.01 μm wide.)
	0.333	1.5	
	0.382	1.5	
	0.501	1.5	
	0.596 (O_3 band)	1.5	
	0.875	1.5	
	0.941 (H_2O band)	1.5	
	1.06	1.5	
Infrared Solar Radiometer	8 - 12	1.5	Linear; dynamic range approximately 2 orders of magnitude.
Angular Scatter:			
Solar Aureole Radiometer	0.38 to 0.47	0.1 x 0.5 Slit, Scanned ±8 Across Sun's Disk	Logarithmic; dynamic range is 6 orders of magnitude.
	0.48 to 0.56		
	0.63 to 1.10		
	0.35 to 1.10		

TABLE II. *Potential Analysis Objectives*

Study Statistics of Aerosol Size Distributions:

   Seasonal and diurnal variations

   Variations with source of aerosols: e.g., smoke, auto
      exhaust, wind-raised dust, pollens, crop spores, etc.

   Variations with climate type

   Effect of humidity conditions

Study of Aerosol Loading and Relationship to Visibility Parameters

Study Statistics of Cloud Particle Size Distributions:

   With type of cloud

   Variations with climate type

   Variations with growth phase of clouds

Study of Precipitable Water Vapor Variations:

   Spatial structure of water vapor concentration in and near
      clouds

   Comparisons with other measurements

IV.   PROJECTED APPLICATIONS

The data derived from the solar radiometric measurements could be used to study the short and long term variations of aerosol and cloud particle characteristics and total precipitable water vapor amounts.   Knowledge of these quantities is important to a better understanding of the interactive processes which affect weather and climate.   Some possible analysis objectives are listed in Table II.

V.   CONCLUDING REMARKS

Solar radiometric techniques offer a relatively low cost method of obtaining significant data on cloud, aerosol, and water vapor parameters which play important roles in weather and climate processes.   In the context of this Symposium, it is suggested that the solar radiometric measurement techniques could represent appropriate methods to apply to study of the environmental and climatic impacts of coal utilization.

REFERENCES

1.   Lerfald, G. M., Derr, V. E., Pueschel, R. F., and Hulstrom, R. L., NOAA Technical Memorandum, ERL WPL-18 (1977).

2.   Pruppacher, H. R., and Klett, J. D., "Microphysics of Clouds and Precipitation," D. Reidel Publishing Co., Boston, Massachusetts (1978).

3.   Twomey, S., "Atmospheric Aerosols," Developments in Atmospheric Science 7, Elsevier Scientific Publishing Co., New York, (1977).

4.   Deirmendjian, D., Ann. Geophys. 15, 218 (1957).

5.   Kerker, M., "The Scattering of Light and Other Electromagnetic Radiation." Academic Press, New York (1969).

6. Herman, B. M., *in* "Inversion Methods in Atmospheric Remote Sounding" (A. Deepak, ed.), pp. 469-503. Academic Press, New York (1977).

7. Deepak, A., *in* "Inversion Methods in Atmospheric Remote Sounding" (A. Deepak, ed.), pp. 265-295. Academic Press, New York (1977).

8. Box, M. A., and McKellar, B. H. J., *Optics Lett. 3*, 91-93 (1978).

9. Lerfald, G., NOAA Technical Memorandum, ERL WPL-36 (1977).

10. Lerfald, G., and Derr, V., *Proc. Soc. Photo-Opt. Instr. Engrs. 161*, 109-113 (1978).

A SIMPLIFIED MODEL FOR THE PRODUCTION
OF SULFATE AEROSOLS

Patrick Hamill

Systems & Applied Sciences Corporation
Hampton, Virginia

Glenn K. Yue

Institute for Atmospheric Optics and
Remote Sensing (IFAORS)
Hampton, Virginia

A recently developed model for the production of sulfate
aerosol from SO_2 for a given temperature, relative humidity,
initial SO_2 concentration, and initial particle distribution is
described. The model incorporates gas phase photochemical
reactions to evaluate the rate of formation of HSO_5 radicals.
Stable $HSO_5 \cdot H_2O$ clusters are formed by binary collisions. These
clusters undergo the microphysical processes of coagulation, condensation and evaporation. Because of the Kelvin effect, small
particles can evaporate and large particles grow at the same time.
Particles can also be formed by nucleation. The model predicts
particle number density as a function of size for thirty size
categories ranging from several angstroms radius up to several
microns. Preliminary results indicate that for background conditions (SO_2 concentration = 1.2 ppb), the resultant size distribution peaks at around 0.2 to 0.4 µm radius; the peak slowly
moving to larger sizes with time. If condensation is ignored, the
size distribution is determined by coagulation and the peak is not
observed.

I. INTRODUCTION

One of the main obstacles in realizing a greater utilization
of coal as a primary energy source is the fact that coal combustion normally results in the release of sulfur oxides to the
environment. The magnitude of the problem can be appreciated by
considering that the sulfur content of coal ranges from about 0.2
to 7% by weight, and all of this sulfur is released as SO_2 or SO_3
on combustion. Consequently, burning one metric ton of 1% sulfur
coal will yield about 19 kg of SO_2.

The environmental degradation caused by sulfur oxides in the
atmosphere is a subject of much concern. Sulfur dioxide is an
important primary pollutant, and Federal regulations impose upper
limits on the allowable concentrations of this gas. However, it
is quite possible that the environmental effects of the secondary
sulfate particles formed from SO_2 may be even more serious.
Whereas SO_2 is mainly adsorbed in the mucous membranes of the
upper respiratory system, particulates can penetrate deep into the
lung. The physical and physiological processes that occur upon
inhaling a sulfuric acid mist are not well known, but it is safe
to conclude that the consequences would be deleterious to one's
health and well being. The particulates are also responsible for
degradation in visibility and they may play a role in the formation of acid rains. Furthermore, acidic particles cause much
of the damage to structures and materials in polluted areas, the
serious injury to marble statuary in Venice being a case in point.

In this paper, a model is described which simulates the formation and growth of a sulfate aerosol in air; it is assumed that
the source of sulfur in the environment can be described by a
given initial concentration of SO_2. In the following sections,
the various chemical and physical mechanisms (and the assumptions
and simplifications) which are incorporated into the model are
delineated. In this section, the general outline of the model

is described and in the final section, some model results and conclusions are presented.

The model analyzes the production of a sulfate aerosol in an
air parcel which does not interact with the environment, i.e.,
the mixing of clean or polluted air into the air parcel is not considered. Thus, the model would be applicable to air in a large
smog chamber (wall effects being negligible) or for a free parcel
under fairly stable atmospheric conditions.

In the model, it is assumed that the SO_2 is converted into
HSO_5 by photochemical reactions. A fairly standard set of chemical equations has been used to describe this process (1).

The HSO_5 molecules interact with H_2O molecules to form (presumably) stable $HSO_5 \cdot H_2O$ clusters. Through binary collisions,
these then form the smallest "particles" treated by the model,
namely, a cluster made up of two HSO_5 molecules and two H_2O molecules. Recent studies by Friend and Vasta indicate that the rate
of formation of particles under experimental conditions is approximately equal to the rate of binary collisions of HSO_5 molecules
(2). Consequently, these very small clusters are assumed to be
stable.

Since it is believed that the particles formed in sulfurous
smogs are sulfates, the assumption is made that the $HSO_5 \cdot H_2O$
clusters are transformed by an unspecified liquid phase reaction
into an $H_2SO_4 - H_2O$ binary solution. This assumption was necessary for the model calculations because the vapor pressures for
$HSO_5 \cdot H_2O$ clusters are not known and growth or evaporation properties for them could not be studied.

Since the vapor pressures for $H_2SO_4 - H_2O$ solutions are known,
the evaporation and growth by condensation of the droplets could
be calculated. As will be shown, this is a very imporant process
in the formation of the aerosol. Nucleation calculations also
require a knowledge of the pertinent vapor pressures. Consequently,

given the present stage of our knowledge of the thermodynamic
properties of the various sulfur compounds in the system, some
form of conversion from HSO_5 to H_2SO_4 has to be postulated.   In
the future, this aspect of the model will be considered more
physically.

II.   DESCRIPTION OF THE MODEL ALGORITHM

    The computer algorithm to study the time development of a
sulfate aerosol is illustrated schematically in Fig. 1. As
illustrated in the figure, the inputs to the model are the
environmental conditions, including the temperature, the relative
humidity, and the initial SO_2 concentrations. Also input are the
"process times" denoted by t_i in Fig. 1. These are the times at
which each of the various processes, namely coagulation, growth-
evaporation, nucleation, and chemistry are considered. For
example, the standard time step is 1 second; the coagulation subroutine is called every 2 seconds, the chemistry and growth subroutines every 10 seconds. The nucleation subroutines are called
every second as long as the H_2SO_4 concentration is greater than
some predetermined value.

    The main program is very simple. It consists merely of a
time counter, calling the process subroutines at the appropriate
times, and stopping the program when a predetermined time has been
reached (normally 4 hours simulated time).

    The print or output subroutine is called periodically to
allow the study of the time evolution of the system. Normally,
the results are printed out every 100 seconds (simulated time) and
selected results (such as the size distribution) are plotted every
hour.

    The present version of the model uses fixed-size time steps;
model stability requires using time steps small enough to ensure
that no significant changes occur during that time interval.

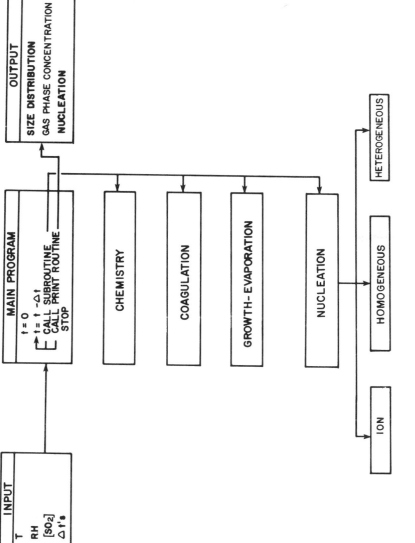

Fig. 1. Schematic of model algorithm

This requirement of taking many small time steps is, of course, not an efficient use of computer resources. Consequently, among the future modifications of the model will be a reformulation to a fully implicit scheme which will calculate time steps continuously and use the longest possible time steps commensurable with numerical stability. The various process subroutines are now considered in more detail.

III. CHEMISTRY

The model uses the set of chemical reactions presented in Table 1 (1). The main chemical process is, of course, the formation of HSO_5 from SO_2. The chemical reaction to produce HSO_3 is

$$SO_2 + OH + M \rightarrow HSO_3 + M$$

where the concentration of OH is the steady-state photochemical equilibrium concentration. The OH radical is produced by the three-step mechanism R1, R3, and R4. The OH loss mechanisms are R5, R6, and R7. It is assumed that HSO_3 is oxidized nearly instantaneously to HSO_5, and that HSO_5 is hydrated to $HSO_5 \cdot H_2O$ in a very short time interval. (This is true as long as H_2O concentrations are large.)

TABLE I. Chemical Reactions Used in the Model

$R1. \quad O_3 + h\nu \rightarrow O_2 + O(^1D)$

$R2. \quad O_3 + h\nu \rightarrow O_2 + O(^3P)$

$R3. \quad O(^1D) + M \rightarrow O(^3P) + M$

$R4. \quad O(^1D) + H_2O \rightarrow 2OH$

$R5. \quad OH + CH_4 \rightarrow CH_3 + H_2O$

$R6. \quad OH + CO \rightarrow CO_2 + H$

$R7. \quad SO_2 + OH + M \rightarrow HSO_3 + M$

$R8. \quad HSO_3 + O_2 \rightarrow HSO_5$

$R9. \quad HSO_5 + H_2O \rightarrow HSO_5 \cdot H_2O$

The $HSO_5 \cdot H_2O$ molecules collide with other $HSO_5 \cdot H_2O$ molecules to form a presumably stable cluster of two HSO_5 and two H_2O
molecules (2). These are the smallest "particles" considered by
the model. They are assumed to be liquid phase droplets of 4.15
angstroms radius that quickly transform into an $H_2SO_4 - H_2O$
solution.

IV. COAGULATION

The coagulation of aerosol particles can be described by the
well-known equation

$$\frac{dn_i}{dt} = - n_i \sum_{j=1}^{\infty} K_{ij}n_j + \frac{1}{2} \sum_{j=1}^{i} K_{kj}n_k n_j$$
$$(k=i-j)$$

where the first term on the right says that the number of particles of "size" i is decreased by collisions between "i" particles and "j" particles (where the summation over j ensures that
collisions with all other sizes are considered), and the second
term gives the increment in size "i" particles due to collisions
between smaller particles (j + k → i).

The coagulation kernel K_{ij} given by Fuchs (3) was

$$K_{ij} = 4\pi r_{ij}D_{ij} \left(\frac{r_{ij}}{r_{ij} + \triangle_{ij}} + \frac{4D_{ij}}{G_{ij} r_{ij}} \right)^{-1}$$

The coagulation algorithm used was that developed by Kritz (4).

In the model, the particle size distribution is broken up
into 30 "bins." Each bin has a volume four times larger than the
volume of the preceding bin; thus, the 30 bins cover the size
range from 4×10^{-4} μm to 2.7 μm.

The ratio of water to sulfuric acid in the particles is
determined by the environmental conditions (relative humidity and
temperature) and by the radius of the particle (Kelvin effect).
Under normal conditions (rh = 50%, T = 25°C), the weight

percentage sulfuric acid ranges from about 65% for the smaller
droplets to 42% for the larger droplets. The coagulation routine
incorporates a correction term so that the larger particles formed
by coagulation will have the correct weight percentage H_2SO_4.

V. CONDENSATION AND EVAPORATION

As mentioned earlier, the particles are assumed to be formed
of sulfuric acid and water. These droplets are continuously being
bombarded by gas phase molecules of water and sulfuric acid, and
simultaneously liquid phase water and sulfuric acid molecules are
constantly being evaporated from the droplets. Because the concentration of water molecules is so much greater than the concentration of sulfuric acid in the environment, the droplets will
quickly come into equilibrium with the water vapor. The droplets
will absorb (or lose) sulfuric acid molecules, depending on
whether the sulfuric acid partial pressure is greater (or less)
than the fugacity, i.e., the liquid phase H_2SO_4 vapor pressure.

An interesting phenomena, due to the change in vapor pressure
with radius (Kelvin effect), is that small particles can evaporate
even at the same time large particles are growing. The evaporation of the small particles is a source for gaseous H_2SO_4 which
subsequently condenses onto the large particles. Thus, small
particles "feed" the large particles. Consequently, the size distribution will peak at a value larger than the smallest radius, an
effect which is not observed if the only mechanism for the formation of large particles is coagulation.

It is important to point out that the condensation process
being considered is "heteromolecular" condensation, i.e., both
water vapor and sulfuric acid vapor are absorbed by the droplet,
and the ratio of H_2O to H_2SO_4 is determined by the condition that

the droplet maintains equilibrium with respect to water vapor. For a fuller description of this process, see Hamill (5) and Hamill, Toon, and Kiang (6).

The growth of a particle due to heteromolecular condensation can be expressed by the equation

$$\frac{dr}{dt} = \frac{\bar{v}D(P_a - P_a^{\,o})/kT}{r\chi(1 + \lambda Kn)}$$

where P_a is the partial pressure of sulfuric acid in the solution droplet. If $(P_a - P_a^{\,o}) > 0$, growth by condensation has occurred. If $(P_a - P_a^{\,o}) < 0$, evaporation has occurred.

VI. NUCLEATION

The nucleation mechanisms incorporated into the model are (a) homogeneous heteromolecular nucleation, (b) heterogeneous heteromolecular nucleation, and (c) heteromolecular nucleation onto ions.

A. *Homogeneous Heteromolecular Nucleation*

The homogeneous heteromolecular nucleation of sulfuric acid-water solution droplets has been studied theoretically by a number of investigators including Kiang and Stauffer, Mirabel and Katz, and Yue and Hamill (7-9). There are many practical and conceptual problems with the so-called classical theory of homogeneous nucleation, but in the absence of a better theoretical approach, the theory with the best available measurement for the vapor pressures of H_2SO_4 in aqueous solutions is used (10,11). The nucleation rate can be expressed as

$$J = 4\pi r^{*2} \beta_A N_B \exp(-\Delta G^*/kT)$$

where ΔG^* is the saddle point value of the Gibbs free energy, r^* is the corresponding particle radius, β_A is the impinging rate of H_2SO_4 molecules, and N_B is number density of water molecules.

B. *Heterogeneous Heteromolecular Nucleation*

The formation of sulfuric acid droplets on preexisting solid particles can be evaluated by assuming that the particles absorb water molecules onto their surface. The number of water molecules adsorbed/cm^2 surface is given by de Boer (12).

$$N^{ads} \approx \beta \left[2.4 \times 10^{-16} \exp(10800/RT) \right]$$

where β is the impinging rate of water molecules onto the surface and 10800 cal/mole is a reasonable choice for the heat of adsorption for the various clays which might be present in the atmosphere.

The rate of formation of sulfuric acid solution droplets on the surface is given by

$$J = 2\pi r^{*2} \beta_A N^{ads} \exp(-\Delta G^*/kT)$$

where ΔG^* depends on the contact angle between solution and substrate as well as on the vapor pressures.

C. *Ion Nucleation*

The nucleation onto ions is a very favorable process for the formation of new particles because the ions will tend to form small stable clusters of several water molecules. If this cluster is struck by just a few H_2SO_4 molecules, a stable solution droplet can be formed. The ion nucleation rate per ion is given by

$$J = 4\pi r^{*2} \beta_A \exp(-\delta\Delta G_i^*/kT)$$

where

$$\delta \Delta G_i^* = \Delta G_i (r^*) - \Delta G_i (r_1)$$

where r^* and r_1 are the radii of the two extrema in ΔG_i. This relation is given by

$$\Delta G_i = \Delta G_{homog} + \frac{q^2}{2} (1 - \frac{1}{\varepsilon})(\frac{1}{r} - \frac{1}{r_o})$$

The last term is the electrostatic (ionic) contribution to the free energy. For a fuller description of these nucleation processes, see paper by Yue and Hamill presented in these proceedings (13).

VII. RESULTS

The processes described were incorporated into the model framework, as described in Section II. It must be pointed out that at this stage of development, many important processes, particularly liquid phase chemical reactions, have not yet been considered. Nevertheless, it is interesting to see whether the results obtained are realistic, i.e., whether the model simulation predicts aerosol properties similar to those observed in polluted air.

Figure 2 gives our "standard model" size distribution and is a plot of the number of aerosol particles in each given size range as a function of particle radius for simulated time t = 1, 2, 3, and 4 hours. There were no particles in the system at t = 0. The main peak is at about 0.02 μm radius, after 4 hours simulated time. The motion of the peak is due to particle growth.

In Fig. 3, the total volume of aerosols is presented as a function of particle size, and in the same figure some data obtained downwind from St. Louis are presented (14). The total volume predicted by the model is less than that observed when the

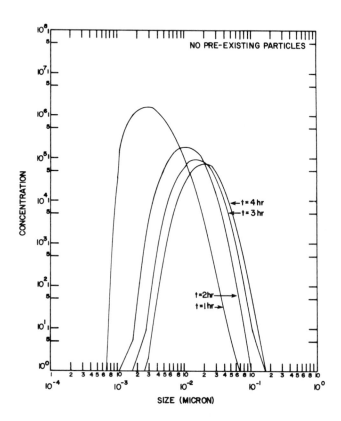

FIGURE 2. "Standard model" size distribution obtained after 1, 2, 3, and 4 hours for initial SO_2 concentration of 1.2 ppb and no preexisting particles in the system.

initial SO_2 concentration is 1.2 ppb (Fig. 3), but greater than that observed when the initial SO_2 concentration is 1.2 ppm. This condition implies that the model results are realistic.

Figure 4 gives the model predicted size distribution when an initial particle size distribution is input to the model at t = 0. This size distribution has zero particles in range r < 0.0004 µm, 500 particles in range 0.004 µm $\leq$ r $\leq$ 0.1 µm, and an $1/r^4$ fall off for r > 0.1 µm. Thus, the straight line in Fig. 4 for r $\gtrsim$ 0.25 µm represents remnants of the original size

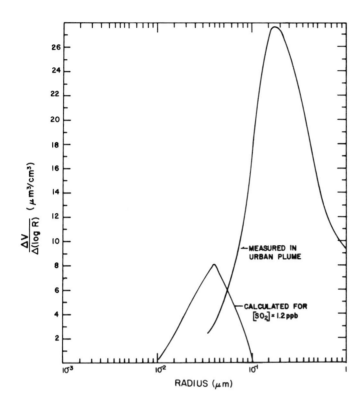

FIGURE 3. Volume distribution obtained from "standard model" and, for purposes of comparison, an experimentally determined plot of this distribution.

distribution. A comparison of Figs. 2 and 4 shows that the presence of this particular initial size distribution only affects the larger sized particles.

Figure 5 represents a run in which there was no growth by condensation; thus, the change in size distribution is due to coagulation only. Comparisons with other figures indicate that the characteristics of the size distributions are completely changed. This comparison points out the importance of incorporating the growth mechanism in the model.

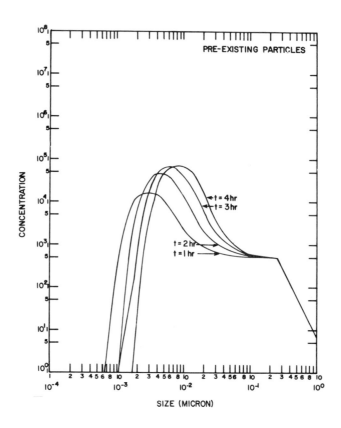

FIGURE 4. Size distribution for 1, 2, 3, and 4 hours for initial SO_2 concentration of 1.2 ppb when particles are present in the system at $t = 0$.

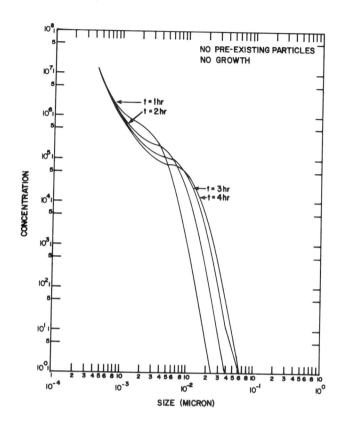

FIGURE 5. Size distribution obtained when growth by condensation is neglected.

VIII. CONCLUDING REMARKS

Although the model does not incorporate some important microphysical and chemical processes, it appears that it does simulate the formation of a sulfate aerosol rather well. Further tests and validations of the model are planned; other processes, such as mixing and sedimentation, will be included to make the model more realistic.

SYMBOLS

D_i diffusion coefficient for particle of radius r_i

$$= kT \left[\frac{1}{6\pi\nu r_i} (1 + 1.246\ Kn_i + 0.42\ Kn_i\ \exp(-0.87/Kn_i)) \right]$$

D_{ij} $= D_i + D_j$

G_i average kinetic velocity of a particle of size r_i and

       mass $m_i = (8kT/\pi m_i)^{\frac{1}{2}}$

G_{ij} $= (G_i^2 + G_j^2)^{\frac{1}{2}}$

ΔG change in Gibbs free energy

J nucleation rate

ΔJ change in nucleation rate

k Boltzmann's constant

K_{ij} coagulation kernel, $4\pi r_{ij}\ D_{ij} \left(\dfrac{r_{ij}}{r_{ij} + \delta_{ij}} + \dfrac{4D_{ij}}{G_{ij}\ r_{ij}} \right)^{-1}$

Kn Knudsen number, mean free path/particle radius

n_i number of particles in ith bin

N_B number water molecules/cm^3

N^{ads} number water molecules adsorbed on cm^2 of surface

P_a partial pressure sulfuric acid

P_a^o vapor pressure sulfuric acid over flat solution

q electronic charge (esu)

r, R particle radius

r_i radius particles in bin i

r_{ij} $= r_i + r_j$

r_o size of ionic cluster

r^* size of embryo at saddle point

t time

T temperature

$\bar{v}$ average volume of molecule in drop

V particle volume

β impinging rate H_2O molecules, $N_B(kT/2\pi m_b)^{\frac{1}{2}}$

β_A impinging rate H_2SO_4 molecules

δ_i correction factor, $\dfrac{\pi G_i}{6r_i \, 8D_i} \left[\left(2r_i + \dfrac{8D_i}{\pi G_i} \right)^3 - \left[4r_i^2 + \dfrac{64 \, D_i^2}{\pi^2 \, G_i^2} \right]^{3/2} \right] - 2r_i$

$\delta_{ij} = (\delta_i^2 + \delta_j^2)^{\frac{1}{2}}$

ε dielectric constant of sulphuric acid solution

λ correction factor, $\dfrac{1.333 + 0.71 \, Kn^{-1}}{1 + Kn^{-1}} + \dfrac{4(1 - \alpha)}{3\alpha}$

      where α is the sticking coefficient (assumed to be unity)

χ concentration of H_2SO_4 in droplet

REFERENCES

1. Davis, D. D., Ravishankara, A. R., and Fischer, S., *Geophys. Res. Lett.* 6, 113 (1979).

2. Friend, J. P., and Vasta, R., "Nucleation by Free Radicals from the Photooxidation of Sulfur Dioxide in Air," Preprint, Dept. of Chemistry, Drexel U., Philadelphia, Pennsylvania (1978).

3. Fuchs, N. A., and Sutugin, A. G., *in* "Topics in Current Aerosol Research" (G. M. Hidy and J. R. Brock, eds.), pp. 1-60. Pergamon Press, New York (1971).

4. Kritz, M., "Formation Mechanism of the Stratospheric Aerosol," Ph.D. Dissertation, Yale University (1975).

5. Hamill, P., *J. Aerosol Sci.* 6, 475 (1975).

6. Hamill, P., Toon, O. B., and Kiang, C. S., *J. Atmos. Sci.* 34, 1105 (1977).

7. Kiang, C. S., and Stauffer, D., *Faraday Sym. Chem. Soc. 7*, 26 (1973).

8. Mirabel, P., and Katz, J. L., *J. Chem. Phys. 60*, 1138 (1974).

9. Yue, G. K., and Hamill, P., *J. Aerosol Sci. 10*, 609 (1979).

10. Roedel, W., *J. Aerosol Sci.* (1979).

11. Morrison, R., and Chu, K. S., Measurements and Calculation of the Total Pressure in Equilibrium with Highly Concentrated Sulfuric Acid, in "Environmental Impact of Coal Utilization." Academic Press, New York (1979).

12. de Boer, J. H., "The Dynamical Character of Adsorption," 2nd ed., Oxford University Press, New York (1968).

13. Yue, G. K., and Hamill, P., The Formation of Sulfate Aerosols through Heteromolecular Nucleation Processes, in "Environmental Impact of Coal Utilization." Academic Press, New York (1979).

14. Komp, M. J., and Auer, A. H., Jr., *J. Appl. Meteor. 17*,1357 (1978).

DISCUSSION

Stokes: I'm a little concerned about the binning technique that
you used for your coagulation scheme. In particular, in binning
by volume, by increasing volume, aren't you in fact going to
suppress coagulation because the cross-section for collisions
in actual coagulation is going to go as r^2; and when you divide
yourself up by volume or r^3, aren't you going to skew the coagulation rate as r? It strikes me that it should be binning on
cross-sectional area rather than volume.

Hamill: I see what you are saying. There is a possibility that
it will effect somehow or another the coagulation, but this is
a well known scheme. We borrowed it from Kritz.[1] And it appears
that the results that you get from it are in good agreement with
the normal procedures in the coagulation type of calculation.

Whitby: We have a numerical model that is very similar to this
and, of course, we wanted to make the number of bins as few as
possible and so we did a bit of work and finally concluded that
an eighth of a size decade was about the minimum because if we
had any fewer than that we started to see effects on the running
of the models. So currently the largest intervals that we are
using is an eighth of a size decade.

Morrison: Do you know what effect does the equilibrium vapor
pressure of the sulfuric acid have on your results?

Hamill: That is, of course, a very important question because
everything depends on that as far as the nucleation is concerned.
Nucleation is very critically dependent on the vapor pressure of
the pure sulfuric acid. Perhaps I should say, the vapor pressure
of the solution which of course is pinned to the value of the
pure sulfuric acid. And as far as the growth is concerned, it
goes as $p_A - p_A^o$, so whether or not yours will have growth or
evaporation depends on what the vapor pressure of sulfuric acid
is.

Morrison: Is p_A the ambient partial pressure and p_A^o the
escaping tendency?

Hamill: That's right. It's the fugacity or equilibrium vapor
pressure.

---

[1]*Kritz, Mark. "Formation Mechanism of the Stratospheric
Aerosol," Ph.D. Thesis, Yale University (1975).*

Demerjian: Regarding the chemical transformation processes, I take it that the slide that you put up there was to presumably represent a clean tropospheric situation.

Hamill: Clean tropospheric in the sense that there are no pre-existing particles in there for heterogeneous reactions.

Demerjian: I was thinking more in terms of the reactions from making OH. I think you more or less indicated that the production is through singlet D processes with water. And I think that what you will find if you look a little more carefully into that, that there are other processes that create OH as well as other termination steps to be considered. And really for what you have up there you could have just as easily put in a first order process for producing OH. If you want to really understand the details of the process, then you must include NO_2 which is a major terminator of OH as well as various hydrocarbon species that ultimately produce OH. I'm a little bit concerned about your production process through the chemistry you've presented. I don't think it is very realistic.

Hamill: You're probably right. The chemical model was proposed by Doug Davis. We utilized that scheme which is, I think, generally accepted. But you are probably right. It probably is not sophisticated enough.

A STUDY OF THE INFLUENCE OF AIRBORNE
PARTICULATES ON SULFATE FORMATION

D. R. Schryer
R. S. Rogowski
W. R. Cofer III

NASA-Langley Research Center
Hampton, Virginia

A statistical analysis has been made of published measurements of sulfate, SO_2, and total suspended particulate (TSP)
concentrations. One set of data involves consecutive 3-year
running averages, over a 9-year period, of data from several
urban sites in the U.S. Northeast. A second set of data
involves daily averages of data from a single suburban site in
the Northeast taken over a 3-month period. In each case, a high,
statistically significant, correlation was found between sulfate
concentration and both SO_2 and TSP concentration, but the effect
of TSP concentration was appreciably greater than that of SO_2.
The result suggests that heterogeneous conversion of SO_2 to
sulfate on particulate surfaces may well predominate over the
competitive gas phase photo-oxidation. Further analysis
involving a larger data set, as well as consideration of
additional parameters, is in progress. Also in progress is a
laboratory study of the interaction of SO_2 with carbonaceous
surfaces with emphasis on the potential synergistic effect of
other atmospheric trace gases--such as NO_2, NH_3, and O_3--on
this interaction.

                    ISBN 0-12-646360-3

I. INTRODUCTION

This paper will report some results of two phases of an
on-going study of the role of particulates in the formation of
sulfates from SO_2. One phase involves statistical analysis of
published field measurements of sulfate, SO_2, and particulate
concentrations as a technique for evaluating the relative
importance of heterogeneous and homogeneous processes for sulfate
formation in the atmosphere. The other phase is a laboratory
investigation of factors affecting the chemisorption of SO_2 on
carbonaceous surfaces with particular emphasis on the effect of
other atmospheric gases, such as NO_2, on the amount of such
chemisorption.

II. BACKGROUND

The formation in the atmosphere of sulfate aerosols from SO_2
can occur by several alternative processes. However, at present,
the predominant process is assumed to involve homogeneous gas-
to-particle conversion. Certainly, most atmospheric models
which consider SO_2 oxidation assume gas phase kinetics
exclusively.

The currently favored gas phase reaction scheme involves
attack of SO_2 by photolytically generated radicals, such as OH
and HO_2, followed by a complex series of reactions which
culminate in formation of H_2SO_4 and various of its derivatives
(1,2). The molecules thus formed are then assumed to nucleate
and grow--frequently on various particulate surfaces present
in the atmosphere--and result in the complex, sulfate-bearing
aerosol particles commonly observed. It is important to note
that in the homogeneous reaction scheme just outlined, the
conversion of SO_2 to sulfate occurs entirely in the gas phase,

ambient particulates serving only as sites for condensation and growth of molecules formed in the gas phase. The rate of formation of sulfate in such a process should be first order with respect to the concentration of SO_2 and independent of the concentration of ambient particulates.

Although gas phase formation of sulfate may well be important, it has not been shown conclusively to predominate over all alternative processes. In fact, several investigators have presented evidence that other sulfate formation mechanisms are at least competitive with the gas phase process. (See references 3 to 12). One such class of alternative mechanisms involves the heterogeneous oxidation of SO_2 on catalytic particulate surfaces including soot and various metal oxides (3-8). The formation of sulfate by such heterogeneous processes should be a function of the catalytic surface area available in the ambient atmosphere. Unfortunately, this quantity is rarely known, but it may be approximated by assuming it to be proportional to the concentration of total suspended particulates [TSP]. Because of observed saturation effects, the heterogeneous oxidation of SO_2 may be effectively zero order in SO_2 concentration (3,13).

The role of other atmospheric gases, including NO_2, in the gas phase oxidation of SO_2 has been well studied, but the effect of such gases on particle-catalyzed heterogeneous oxidation has largely been neglected (1,2,8,14,15).

III. ANALYSIS OF PUBLISHED FIELD MEASUREMENTS

The debate as to the significance of alternative sulfate formation mechanisms has generally been based on theoretical calculations and/or extrapolation of laboratory data (2-4, 9-12). The presently reported study constitutes an attempt to evaluate the relative contributions of homogeneous and heterogeneous sulfate formation processes in the atmosphere by means of statistical analysis of field measurement data. It is assumed in this study that over a given finite measurement period, a steady state or quasi-equilibrium exists. This assumption implies that at a given time, the amount of atmospheric sulfate attributable to homogeneous, gas phase oxidation of SO_2 is proportional to the concentration of SO_2 whereas that attributable to heterogeneous catalytic oxidation is proportional to the concentration of particulates. Numerous other factors such as the concentration of reactive radicals, the relative humidity, the concentrations of other trace gases, etc., may well effect the yield of sulfate; but much of the influence of such factors should be averaged out if properly averaged data are considered.

A. Urban Northeast Data

Table I presents a set of sulfate and SO_2 concentrations originally presented by Altshuller (16). The values presented are averaged both spatially and temporally. They are 3-year running averages of measurements from 32 urban sites in the U.S. Northeast. Note that over the 9-year interval involved, the concentration of SO_2 was reduced by 55% whereas that of sulfate declined by only 15%. Thus, the concentration of sulfate in the U.S. urban Northeast clearly was not simply proportional to the concentration of SO_2 during the time interval considered. This is an example of the well-known "urban sulfate anomaly" which Altshuller and others have attempted to explain by various hypotheses.

TABLE I. *Three-Year Running Average Concentrations of*
Sulfate, Sulfur Dioxide, and Total Suspended Particulates for
Urban Sites in the U.S. Northeast

Years	$[SO_4^{-2}]^a$ $(\mu g/m^3)$	$[SO_2]^a$ $(\mu g/m^3)$	$[TSP]^b$ $(\mu g/m^3)$
1963-65	18.4	147	116
1964-66	18.2	146	115
1965-67	17.7	132	110
1966-68	17.2	119	104
1967-69	16.9	100	97
1968-70	16.6	74	95
1969-71	15.7	66	93

[a]Data from Altshuller, A. P., J. Air Pollut. Control Assoc.
26, 318 (1976).

[b]Data from EPA-450/1-73-001-Vol. 1.

Linear regression of the sulfate-SO_2 data from Table I
reveals that concentrations of the two species are not unrelated
but, rather, that they are highly correlated by the linear
equation

$$[SO_4^{-2}] = 0.028 \ (\pm \ 0.003, \ 1\sigma) \ [SO_2] + 14.1 \ (\pm \ 0.4, \ 1\sigma)$$

with a correlation coefficient of 0.97. One interpretation of
this result is that a fraction of the measured atmospheric
sulfate--that represented by the term involving $[SO_2]$--was
indeed generated by gas phase oxidation of SO_2, but that a
significant additional fraction, represented by the additive
intercept, was generated by alternative processes which are
effectively zero order with respect to SO_2. In fact, the
intercept should not be thought of as truly constant, rather
as merely independent of $[SO_2]$.

To evaluate the contribution of suspended particulates to
the measured sulfate values in Table 1, a linear regression of
sulfate-TSP data was performed. Average TSP data for urban
Northeastern sites were extracted from Ref. 17 and are presented
in Table I as 3-year running averages for consistency. Again,
statistical analysis reveals a high degree of linear correlation
between the variables.

$$[SO_4^{-2}] = 0.096 \ (\pm \ 0.011, \ 1\sigma) \ [TSP] + 7.2 \ (\pm 1.2, \ 1\sigma)$$

The correlation coefficient is again 0.97. The term involving
[TSP] is interpreted as representing sulfate generated by
particle-catalyzed heterogeneous oxidation of SO_2 and the
intercept includes sulfate from all other processes including
homogeneous gas-phase oxidation.

The fractions of the measured sulfate due to homogeneous
and heterogeneous processes, as determined by this analysis,
are presented in Table II. Note that in all cases, although
the homogeneous fraction is significant, the heterogeneous
fraction predominates. This result clearly indicates that the
concentration of atmospheric particulates is of importance in
determining the amount of sulfate formed and strongly suggests
that particulate concentration can, at least in certain cases,
be even more important in this regard than the concentration of
SO_2. The results further suggest that heterogeneous sulfate
formation is strongly competitive with the gas phase process
and cannot realistically be ignored. It should be noted that
the partition of atmospheric sulfate into a fraction which is
proportional to the atmospheric SO_2 concentration and a significantly large fraction which is not proportional to SO_2 provides
a simple explanation of the much debated urban sulfate anomaly.

TABLE II. Results of Linear Regressions of Data from Table 1.

$$[SO_4^{-2}] = 0.028 \ (\pm \ 0.003, \ 1\sigma) \ [SO_2] + 14.1 \ (\pm \ 0.4, \ 1\sigma)$$

$$R = 0.97$$

$$[SO_4^{-2}] = 0.096 \ (\pm \ 0.011, \ 1\sigma) \ [TSP] + 7.2 \ (\pm \ 1.2, \ 1\sigma)$$

$$R = 0.97$$

Years	$\dfrac{0.028 \ [SO_2]}{[SO_4^{-2}]}$	$\dfrac{0.096 \ [TSP]}{[SO_4^{-2}]}$
1963-65	0.22	0.61
1964-66	0.22	0.61
1965-67	0.21	0.60
1966-68	0.19	0.58
1967-69	0.16	0.55
1978-70	0.12	0.55
1969-71	0.12	0.57

It would, of course, have been preferable to have analyzed the sulfate data by means of a multiple regression in $[SO_2]$ and $[TSP]$ rather than by separate linear regressions. In fact, a multiple regression was performed and the result again indicated the predominance of the TSP term but, because of the limited number of data points, the result was not statistically significant.

B. *Suffolk County Data*

Multiple regression of $[SO_4^{-2}]$ with respect to $[SO_2]$ and $[TSP]$ was successfully performed on a second set of field measurement data reported by Meyers and Ziegler (18). These

data represent 14 sets of 1-day averages of $[SO_4^{-2}]$, $[SO_2]$, and
[TSP], as well as several other parameters. The data were all
taken in Suffolk County, New York, in the winter of 1975. The
values of $[SO_4^{-2}]$, $[SO_2]$, and [TSP] are presented in Table III.
In contrast to the broadly averaged data presented in Table I,
the data in Table III represent a much narrower scope both
spatially and temporarily; thus, the effects of parameters other
than $[SO_2]$ and [TSP] should be less effectively averaged out.
Nevertheless, a statistically valid regression has been
achieved.

In their study, Meyers and Ziegler performed individual
linear regressions of the sulfate concentrations with various of
the measured parameters including $[SO_2]$ and [TSP]. They found
the best correlation with [TSP], although they also found a
moderately good correlation with $[SO_2]$. The multiple regression
of their sulfate data in $[SO_2]$ and [TSP] yields the equation:

$$[SO_4^{-2}] = 0.052 \ (\pm \ 0.033, \ 1\sigma) \ [SO_2] + 0.16 \ (\pm \ 0.05, \ 1\sigma) \ [TSP]$$

A small negative intercept was also obtained which was statistically indistinguishable from zero and which therefore has been
dropped. Note that the coefficient of [TSP] is larger than
that of $[SO_2]$ and has a smaller relative standard error. This
result is consistent with the results obtained by the original
investigators. The fractions of the measured sulfate proportional to $[SO_2]$ and to [TSP]--and, therefore, by implication to
homogeneous and heterogeneous processes, respectively--are
presented in Table IV. It can be seen that the fraction proportional to [TSP] ranges from 0.51 to 0.80 with an average of
0.68. Again, as with the data from Altshuller, there is a clear
indication of the importance of atmospheric particulates in
determining the yield of sulfates and a strong implication of
the importance of heterogeneous processes.

TABLE III. Concentrations of Sulfate, Sulfur Dioxide, and
Total Suspended Particulates for Suffolk County, New York, in
Winter of 1975[a]

$[SO_4^{-2}]$ ($\mu g/m^3$)	$[SO_2]$ ($\mu g/m^3$)	$[TSP]$ ($\mu g/m^3$)
3.26	24.4	13.03
1.92	15.5	13.61
2.77	15.2	12.91
8.22	45.1	36.29
5.51	45.6	31.02
7.78	57.1	35.44
6.53	51.9	35.12
5.43	60.0	23.73
9.58	57.6	19.79
7.52	42.4	35.29
4.69	18.6	19.11
6.28	35.1	24.37
10.31	31.6	42.47
2.79	22.0	17.81

[a]Data from Meyers, R. E., and Ziegler, E. N., Environ. Sci.
Technol. 12, 302 (1978).

TABLE IV.   Results of Multiple Regression of Data from
Table III.

$[SO_4^{-2}] = 0.052$ $(\pm\ 0.033,\ 1\sigma)$ $[SO_2] + 0.16$ $(\pm\ 0.05,\ 1\sigma)$ $[TSP]$

R = 0.83

$\dfrac{0.052[SO_2]}{0.052[SO_2]\ +\ 0.16[TSP]}$	$\dfrac{0.16[TSP]}{0.052[SO_2]\ +\ 0.16[TSP]}$
0.38	0.62
0.27	0.73
0.28	0.72
0.29	0.71
0.33	0.67
0.35	0.65
0.33	0.67
0.46	0.54
0.49	0.51
0.28	0.72
0.24	0.76
0.32	0.68
0.20	0.80
0.29	0.71

C. *Discussion*

The foregoing analyses are admittedly somewhat simplistic and involve certain simplifying assumptions. Nevertheless, they do provide an approximate measure of the relative importance of SO_2 and particulate concentrations, as well as of homogeneous and heterogeneous processes, based on actual field measurements rather than on theoretical calculations or laboratory data. They also, as previously noted, provide a simple explanation of the urban sulfate anomaly.

At present, the analyses are being extended to a larger data base including data from the Regional Air Pollution Study (RAPS). In doing so, the effects of other parameters will be considered in addition to $[SO_2]$ and $[TSP]$. This work, at present, is just getting underway and no results from it are available yet.

IV. GRAVIMETRIC SO_2 CHEMISORPTION STUDY

A laboratory study of the effect of various factors on the chemisorption of SO_2 on carbonaceous materials is currently in progress at Langley Research Center. Novakov and coworkers have demonstrated that carbon surfaces can be catalytic in the heterogeneous oxidation of SO_2 to sulfates (3). Other investigators have demonstrated that chemisorption of SO_2 on carbon surfaces leads to such catalytic oxidation (5).

A. *Experimental Procedure*

A carbonaceous sample of known Brunnauer-Emmett-Teller (BET) surface area per unit mass is placed on the pan of a continuously recording microbalance. A known concentration of a given test gas, or combination of test gases, in an inert carrier gas is then flowed over the sample at a measured flow rate for a measured length of time. If any sorption occurs, it is indicated by a recorded weight gain. The sample is then exposed to a measured flow of dry inert gas to desorb any physically adsorbed test gas. The flow of inert gas is continued until no additional desorption is detected. The net weight gain of test gas over the complete sorption/desorption cycle is taken as the amount of the gas which is chemisorbed. Test gases which exhibit net weight gains are run at various concentrations and for various exposure times. Particular emphasis is being placed in this study on the effect of other atmospheric gases on SO_2 sorption and, thus, test gases are run in various combinations as well as individually.

B. *Results and Discussion*

Table V summarizes the results of a preliminary series of test runs made with samples of a commercially available carbon black. The tests indicated no detectable weight gain over the complete sorption/desorption cycle except when air, water vapor, and SO_2 were all three present in the test gas. When this criterion was met, the carbon black samples not only exhibited detectable weight gains, they also gave positive tests for acidity and for sulfate when dispersed in distilled water. The effects of SO_2 concentration and of exposure time on the amount of chemisorption are presently being investigated.

TABLE V. Results of Preliminary Gravimetric Sorption Runs on Commercial Grade Carbon Black

Test gas	Net weight gain?
Dry air	No
Humidified air	No
SO_2 in dry N_2	No
SO_2 in humidified N_2	No
SO_2 in dry air	No
SO_2 in humidified air	Yes

After the preliminary tests summarized in Table V, a series
of tests was initiated in which NO_2 was added to the test
stream. Although this series of tests is currently in progress,
some interesting initial results have been obtained which are
presented in Table VI.

Despite some variability in the concentrations of SO_2 and
NO_2 employed in these early runs, the large increases in weight
retention associated with the SO_2-NO_2 combination are quite
clear. For example, the weight gains for the SO_2-NO_2 runs are

TABLE VI. Data for SO_2-NO_2 Test Runs[a]

SO_2 (ppm)	NO_2 (ppm)	Net weight gain per unit sample weight	Exposure time (hr)
16	0	0.036	16
23	0	0.034	16
23	0	0.037	16
0	34	0.087	16
0	49	0.113	16
16	15	0.469	16
16	15	0.746	16
19	10	0.568	16
19	10	0.591	16
41	10	0.635	16
79	10	0.599	16
79	10	1.521	40

[a]Carrier gas for all runs was 70% air--30% N_2 at 70%
relative humidity.

in every case more than an order of magnitude greater than those
of runs involving SO_2 with no added NO_2. Furthermore, the
SO_2-NO_2 run weight gains are in every case more than a factor
of 3 greater than the largest sum of the weight gains for
individual SO_2 and NO_2 runs.

It is interesting to note that, despite some scatter in the
weight gain data, the gains for the SO_2-NO_2 runs appear to be
independent of the concentration of SO_2. On the other hand,
comparison of the weight gain for the one 40-hour run with
those for the 16-hour runs indicates that the weight gain is
time dependent for the time intervals involved. Both of these
observations will be investigated further.

V. CONCLUDING REMARKS

The results of the statistical analyses of published field
measurement data presented in this paper indicate that atmospheric particulates play an important role in the oxidation
of SO_2 to sulfates. The results further suggest that heterogeneous processes of sulfate formation are strongly competitive
with the alternative homogeneous gas phase process and cannot
realistically be ignored.

The observation that the presence of NO_2 greatly enhances
the chemisorption of SO_2 on a carbonaceous surface, although
preliminary, suggests that the degree of interaction of SO_2
with such surfaces may be much greater than previous laboratory
studies have indicated. This effect tends, therefore, to lend
support to the prominent role of particulates which is
suggested by the field data analyses.

REFERENCES

1. Davis, D. D., and Klauber, G., *Int. J. Chem. Kinet.,*
   Symp. n1, 543 (1975).
2. Calvert, J. G., Su, F., Bottenheim, J. W., and Strausz,
   O. P., *Atmos. Environ. 12,* 197 (1978).
3. Novakov, T., Chang, S. G., and Harker, A. B., *Science, 186,*
   259 (1974).
4. Chang, S. G., and Novakov, T., *in* "Man's Impact on the
   Troposphere--Lectures in Tropospheric Chemistry" (J. S.
   Levine and D. R. Schryer, eds.), NASA RP 1022, 349 (1978).
5. Barbaray, B., Contour, J. P., and Mouvier, G., *Atmos.*
   Environ. 11, 351 (1977).
6. Liberti, A., Brocco, D., and Possanzini, M., *Atmos.*
   Environ. 12, 255 (1978).
7. Haury, G., Jordan, F., and Hofmann, C., *Atmos. Environ.*
   12, 281 (1978).
8. Barbaray, B., Contour, J. P., and Mouvier, G., *Environ.*
   Sci. Technol. 12, 1294 (1978).
9. Erickson, R. E., Yates, L. M., Clark, R. L., and McEwen, D.,
   Atmos. Environ. 11, 813 (1977).
10. Larson, T. V., and Harrison, H., *Atmos. Environ. 11,*
    1133 (1977).
11. Beilke, S., and Gravenhorst, G., *Atmos. Environ. 12,*
    231 (1978).
12. Hegg, D. A., and Hobbs, P. V., *Atmos. Environ. 12,* 241
    (1978).
13. Altshuller, A., *Environ. Sci. Technol. 7,* 709 (1973).
14. Schroeder, W. H., and Urone, P., *Environ. Sci. Technol.*
    12, 545 (1978).
15. Farlow, N. H., Snetsinger, K. G., Hayes, D. M., and Lem,
    H. Y., *J. Geophys. Res. 83,* 6207 (1978).

16. Altshuller, A., *J. Air Pollut. Control Assoc. 26*, 318 (1976).

17. "The National Air Monitoring Program: Air Quality and Emissions Trends Annual Report, Volume 1," EPA-450/1-73-001A-Vol. 1 (1973).

18. Meyers, R. E., and Ziegler, E. N., *Environ. Sci. Technol. 12*, 302 (1978).

DISCUSSION

Unidentified Speaker: Do you have a mechanism that will account for the synergism?

Schryer: No. As a matter of fact, this observation of the synergistic effect is very recent. We have, for example, just recently completed a long duration run with SO_2 and NO_2 in humidified air. We wanted to find out at what time we would achieve saturation and we found that at about 120 hours we finally achieved saturation. There was no additional weight gain after that period of time, but we have not yet had time to analyze these results. At this point, I think you could argue that some of the reactions that occur heterogeneously are vey analagous to those that occur homogeneously.

Pack: Your statistical analyses were both for areas where TSP is modestly high and sulfate is definitely high in the eastern U.S. Will the RAPS data, whose concentrations I can't recall off-hand, provide a test of this relationship when the SO_4 value is relatively low? What I am thinking of is low SO_2, modest TSP, and then the consequential sulfate. Have you done any testing at all in that kind of a mode?

Schryer: As I say, we have just started looking at the RAPS data. I would comment that Sandberg *et al.* did a similar linear regression of sulfate vs. SO_2 concentrations for the San Francisco Bay area where these concentrations were relatively low.[1] No attempt was made to explicitly consider the effect of particulates, but a relatively large intercept was obtained. We may assume that the intercept represents all sulfate that was not formed homogeneously. This was in California where there was quite a bit of sunlight and one would have expected the homogeneous mechanism to dominate, but it did not do so. The term in SO_2 was much smaller than the intercept.

Whitby: I guess I really question this whole approach. Well, one has to be very careful. In the first place, it's quite well known that the sulfate in most of the country is in the fine particle mode, whereas most of the TSP mass is in the coarse particles. Most of the sampling stations during the period in which you have used data, have been located in cities. Many of them were on rooftops and places where they served as pretty good

---

[1]*Sandberg, J. S., et al., J. Air Pollut. Control Assoc.* *26, 559 (1976).*

dust collectors. The reason TSP has gone down is quite well
known. Altshuller stated this in his paper. TSP is down because
we have cleaned up the places dumping out the rocks. The reason
the SO_2 has gone down is because we cleaned up the local sources
and built tall stacks which put it over our heads. Furthermore,
if one looks at SO_2 to sulfate conversion, the time constant is
in the hours range or more. In fact, in a few hours one has only
accounted for 10% to 15% of the total conversion that occurs. In
other words, it takes days for substantial conversion to occur.
The coarse particles which, therefore, dominated the historical TSP
data base are long on the ground before there is any time for
the coarse particles which dominated the mass that you used in
your correlation to have reacted with anything. So if you put
this all together, it is very difficult to see how chemistry
could possibly have anything to do with your correlations. So I
guess I wonder how one could really draw the conclusion that you
draw.

Schryer: First, I would have been surprised if there had not
been a challenge to this because I realize it goes against the
currently prevailing theory. Second, I would note that
Altshuler's explanation of the urban-sulfate anomaly in terms of
long range transport of sulfate has been challenged by the EPA's
own Science Advisory Board.[2] The concept that we have significantly dispersed the SO_2 sources out into the rural areas and
that the sulfate we are seeing in the cities is largely the
result of nonurban generated SO_2 is certainly attractive. It
enables one to explain away the inconvenient fact that measured
sulfate levels are not simply proportional to ambient SO_2 concentrations without abandoning the prevailing assumption that
sulfate formation occurs predominantly by a homogeneous mechanism.
However, it is still just a hypothesis, as is the alternative
explanation I have offered here. Further study is needed to
determine which explanation corresponds best to reality.

---

[2]*"Scientific and Technical Issues Relating to Sulfates."*
Report by an ad hoc panel of the Science Advisory Board-
Executive Committee (N. Nelson, Chairman, U.S. Envir. Prot.
Agency (1975).

THE VAPOR PRESSURES AND LATENT HEAT
OF AQUEOUS SULFURIC ACID AT
PHASE EQUILIBRIUM[1]

Kwo-Sun Chu
R. Morrison

Department of Physics
Talladega College
Talladega, Alabama

Based on the hydrodynamic equations, Knudsen's measurement
of effusion current, and on simple phenomenological considerations, a self-contained theory is developed for calculating the
partial and total pressures of aqueous acid at quasi-static
thermodynamic equilibrium with its vapor. The equilibrium
temperatures and pressures are in the range which is pertinent
to aerosol nucleation. The experimental determinations of the
total effused mass rates at steady states are carried out from
-10° C to 70° C. The acid weight concentration is found to
maintain a value of 76.3% once hydrodynamic equilibrium is
reached. The results are compared with those of Roedel's
recent measurements by using radioactive sulfur isotopes as a
tracer technique.

---

[1]This work is supported by NASA under Grant No. 2217 and
partially assisted by RCSE (The Resource Center for Science and
Engineering).

I. INTRODUCTION

It is well established that sulfate aerosol plays an
important role in climate and atmospheric chemistry. In the
stratosphere the sulfate aerosol reduces visibility, leads to
acid rain, and has other meteorological effects. It is now
accepted that the sulfate aerosol is largely created *in situ*
by gas-to-particle conversion. The classical theory to describe
homogeneous nucleation and the modifications of it known to
the authors depend sensitively on the vapor pressure in equilibrium with the bulk liquid (1,2). Recent models of the stratosphere aerosol layer also depend on this quantity (3).

Experimental and theoretical information available was
reviewed by Verhoff and Banchero, and Banchero and Verhoff
(4,5). Measurements of the total pressure (the same as the
partial pressure of water) exist for concentrations less than
70%, and measurements at higher concentrations exist for
temperatures about 100° C. To be useful for modeling of
atmospheric pressures, measurements and accurate theories are
necessary for concentrations near 100% and for temperatures
between -60° C and 60° C. Until recently, most authors have
used the calculation of Gmitro and Vermeulen to estimate the
equilibrium vapor pressures (6). However, Ref. 6 concluded
that the extrapolation of the total pressure to atmospheric
conditions was unreliable, and the predicted partial pressures
could be off by a factor of 10 or worse, because of the
extreme sensitivity of the calculation to small errors in the
thermodynamic parameters and because of the lack of information
about the sulfuric acid partial pressure.

Very recently, Roedel (7) has reported a measurement of the
rate of diffusion of sulfur atoms in the equilibrium vapor
above 96% acid at 23° C. Then, using ideal gas assumptions,

this value was translated into a total pressure determination.
The value reported by Roedel was more than an order of magnitude
less than that predicted by Gmitro and Vermeulen.(7).

II. THEORY

Consider a Knudsen effusion capsule which contains aqueous
sulfuric acid at thermodynamic equilibrium with its vapor. The
equilibrium temperature and pressure are in the domain within
which the vapor behaves as an ideal gas mixture. The size of
the orifice of the capsule is negligible compared with the volume
of the vapor. The measurement of the effusing current of the
molecules through the orifice I_o at steady state is related to
the vapor pressure P by the Knudsen equation (8)

$$P = 17144 \ (I_o/ak)\sqrt{T/M} \tag{1}$$

In this relation, P is in torrs, $I_\cap$ is in grams per second,
a is the opening area of the small circular orifice in cm^2,
T is the equilibrium temperature in °K, and M is the effective
molecular weight in grams and refers to the vapor species. The
"clausing factor" k is determined by the ratio of the length of
the orifice to its radius.

In order to determine M, it is necessary to obtain the
partial pressures P_i for H_2SO_4, H_2O, and SO_3. Application of
the hydrodynamic equations will prove this:

$$\left. \begin{array}{c} \dfrac{d}{dt} \iiint \rho \ dV = - \iint \hat{n} \cdot \rho \vec{u} \ dS \\[3mm] \ell(\dfrac{\partial}{\partial t} + \vec{u} \cdot \nabla)\vec{u} = \dfrac{\rho}{m}\vec{F} - \nabla \cdot \overset{\leftrightarrow}{P} \end{array} \right\} \tag{2}$$

where ρ is the mass density, $\vec{u}(\vec{r},t) = \langle\vec{v}\rangle$ is the thermal average
of the microscopic velocity or the velocity of a collective
mode. $\vec{F}$ is the external force and $\overset{\leftrightarrow}{P}$ is the pressure tensor.

For the case of a steady irrotational flow under these considerations,

$$
F = -\nabla \phi
$$

$$
\frac{\partial \vec{u}}{\partial t} = 0
$$

$$
\nabla T = 0 \tag{3}
$$

$$
\nabla x \vec{u} = 0
$$

$$
\frac{\partial \rho}{\partial t} = 0
$$

where T is the equilibrium temperature. It follows that

$$
\rho = \rho_o \exp \left[- \frac{1}{K_B T} (\frac{1}{2} mu^2 + \phi) \right] \tag{4}
$$

where ρ_o is the static equilibrium mass density.

The gravitational potential, ϕ in Eq. (4), can be neglected in this case except in the cases of lower temperatures and larger sizes of system containers.

By using the simple kinetic theory of ideal gas, one finds the hydrodynamic velocities at the orifice for each component in the gas mixture

$$
u_{i0} = \frac{4k\sqrt{RT}}{3\sqrt{2\pi M_i}} \tag{5}
$$

The continuity equation combined with Eq. (5) leads to the approximation of u_{iz} at distance z above the liquid surface with area A, which changes negligibly with time.

$$
u_{iz} = \left[\frac{RT}{M_i} \pm \frac{RT}{M_i} \left(1 - \frac{32a^2 k^2}{9\pi (A - \pi z^2)^2} e^{\frac{-8k}{9\pi}} \right)^{\frac{1}{2}} \right]^{\frac{1}{2}} \tag{6}
$$

The density ρ_{iz} and the ith component effusing current thus are

$$
\left.\begin{array}{l}
\rho_{iz} = \dfrac{2ak\sqrt{2RT}}{3\sqrt{\pi M_i}\,(A - \pi z^2)^2\, u_{iz}}\, e^{\dfrac{-4k^2}{9\pi}}\, \rho_{oi} \\[4ex]
I_i = \dfrac{2k\sqrt{2RT}}{3\sqrt{\pi M_i}}\, e^{\dfrac{-4k^2}{9\pi}}\, \rho_{oi}
\end{array}\right\} \tag{7}
$$

where ρ_{oi} is the ith component static equilibrium density.
Equations (5), (7), and the ideal gas law give the following
relations:

$$
\left.\begin{array}{l}
\dfrac{P_{io}}{P_{ko}} = \dfrac{P_{oi}}{P_{ok}} = \dfrac{n_i}{n_k} \\[3ex]
P_{io} = e^{\dfrac{-4k^2}{9\pi}}\, P_{oi} \\[3ex]
I_o = \sum_i I_i
\end{array}\right\} \tag{8}
$$

where n_i is the ith component number density.

Now the findings of the ratio appearing in Eq. (8) will be
explained. Denote the numbers of the H_2SO_4 and H_2O molecules
evaporated from the liquid phase by Δn_1 and Δn_2, respectively,
in the time interval Δt. The increments of the numbers of
H_2SO_4, H_2O, and SO_3 are denoted by $\Delta n_1'$, $\Delta n_2'$, and $\Delta n_1'$. Because
of the number conservation, the following relations at
equilibrium result:

$$
\left.\begin{array}{l}
\dfrac{n_1}{n_2} = \left| \dfrac{\Delta n_1}{\Delta n_2} \right| \\[3ex]
\dfrac{\Delta n_1}{\Delta n_2} = \dfrac{(1 - \rho)c}{1 + \rho c}
\end{array}\right.
$$

$$\frac{n_3}{n_2} = \frac{\Delta n_3}{\Delta n_2} = \frac{c\gamma}{1 + c\gamma}$$

$$\frac{n_1'}{n_2'} = \frac{\Delta n_1'}{\Delta n_2'} = c$$

(9)

where C is the acid weight concentration, n_1, n_2, and n_3 are, respectively, the number concentrations of H_2SO_4, H_2O, and SO_3 in vapor phase while n_1' and n_2' are those in the liquid phase. The parameter γ in Eq. (9) indicates the dissociation and association rate of the following reactions at chemical equilibrium.

$$H_2SO_4 \underset{\leftarrow}{\overset{\rightarrow}{}} SO_3 + H_2O$$

The positive γ represents dissociation and the negative represents association. From the law of mass conservation in a hydrodynamic flow,

$$\frac{M_1}{M_2}c = \frac{I_1 + (1 + M_2/M_3)I_3}{I_2 - (M_2/M_3)I_3}$$

(10)

Substitution of Eq. (7) into Eq. (10) leads to

$$\gamma = \frac{13.20}{12.14c - 1.06}$$

(11)

Again, from the simple kinetic theory of ideal gas or by using Eqs. (1) and (7), one obtains the expression for the effective molecular weight of the vapor in terms of the molecular weights of the component species

$$\frac{\sqrt{M_1}P_{10} + \sqrt{M_2}P_{20} + \sqrt{M_3}P_{30}}{P_{10} + P_{20} + P_{30}} = \sqrt{M}$$

(12)

Substituting Eqs. (8), (9), and (11) into Eq. (12) yields

$$(120.18 - 12.14\sqrt{M})c^2 + (2.31 - 2.12\sqrt{M})c -$$

$$(4.49 - 1.06\sqrt{M}) = 0$$

(13)

Since the relative number concentration of the liquid has to be real and single-valued, the discriminant of Eq. (13) has to vanish. This gives the effective molecular weight and the acid concentration (wt.).

$$M = 76.73 \text{ grams} \\ c = 76.32\% \qquad\Bigg\} \tag{14}$$

With the knowledge of M, therefore, the total static equilibrium vapor pressure can be calculated by Eqs. (1) and (7) with measured values of $a = 2.58 \times 10^{-3}$, and $K = 0.7184$.

$$P_o = \frac{17144 \ I_o}{ak} \sqrt{\frac{T}{M}} \ e^{\frac{4k^2}{9\pi}} \tag{15}$$

By utilizing Eqs. (9), (11), (14), and (15), the static equilibrium partial pressures are calculated with

$$P_{o1} = 0.55 \ P_o$$

$$P_{o2} = 0.08 \ P_o$$

$$P_{o3} = 0.37 \ P_o \tag{16}$$

III. THE EXPERIMENT

The technique reported here is a standard one for the absolute measurement of low vapor pressures (8). A sample cell containing the liquid and its vapor at a temperature T is suspended from a sensitive microbalance. A hole whose area is small compared with the surface area of the liquid allows a small mass current I_o to effuse from the vapor. The slope of the mass-against-time curve then gives the mass current. If the effusion hole is sufficiently small to ensure that the perturbation of the equilibrium pressure is negligible, and if

the effective molecular weight of the effusing species M is
known, then the equilibrium vapor pressure P is given by the
Knudsen equation (8). Reagent grade sulfuric acid was distilled
under dry nitrogen using the techniques of Kunzler (9).
According to this reference, sulfuric acid has a constant boiling
concentration of 98.5%. The acid was injected into a pyrex
bubble, with outer diameter, 0.5 cm., wall thickness, 0.02 cm,
and effusion hole with circular cross-section area of 0.00258 cm.
The bubble was suspended from a microbalance (Cahn model RG-UHV),
which was ultra-high-vacuum compatible. The voltage from the
microbalance was recorded by a chart recorder (Houston
Instruments omniscribe). Figure 1 diagrams the experiment.
The temperature of the sample cell was determined by a cryoshroud, which was optically dense, with the exception of a
small hole for the suspension wire. The shroud also has an
open structure so that after one or two bounces, molecules can
escape into the vacuum system.

The vacuum system was ion pumped but unbaked. The residual
vacuum pressure was approximately 3×10^{-8} torr, of which more than
70% was due to hydrogen. The microbalance temperature was held
constant at $\pm 0.1°$ C for various temperatures ranging from $30°$ C
to $40°$ C. The temperature controller sensed the temperature
of a thermister mounted firmly on the beam of the microbalance.
Within the range mentioned, the vacuum system temperature did
not affect the data. Thermal isolation of the vacuum system
was provided by 15 cm of fiberglass insulation surrounding it.
Vibration isolation was accomplished through various means:

1. The vacuum system, which weighs more than 150 lb., was
mounted in a sandbox (10 cm of sand) on concrete blocks.

2. All electrical connections traveled to the vacuum system
by flexible light cables that traveled first to the sand in the
box.

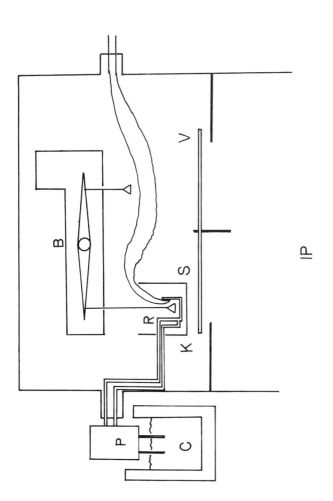

(a) Details of the temperature determination of the sample

FIGURE 1. Experimental setup. The balance B, the cryoshroud S, the acid K, the thermocouple T, the pump P, and the coolant reservoir C are shown. A valve V shown here was removed in this experiment and replaced by a wire screen. The ion pumps IP are below.

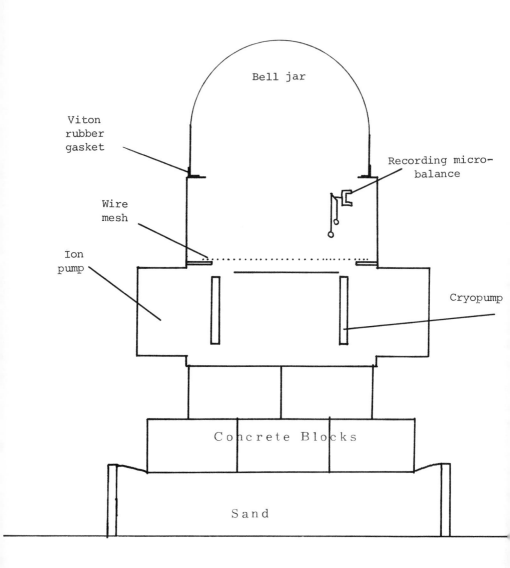

(b) The vacuum system used in the experiment

FIGURE 1--concluded

3. The coolant flow rate to the cryoshroud was minimized to prevent turbulence.

4. The experiment was performed on the stone floor in a basement, and traffic in the neighborhood of the experiment was discouraged.

5. Windows in the room were closed and prevented thermal transients.

It is not possible to determine the acid concentration directly during the measurement since it must change during vacuum exposure. In practice, the 98.5% acid reached quasi-static equilibrium within 2 days of pumpdown, as evidenced by the constant effusion current. The experiment was also attempted with reagent-grade acid (96% to 98%). In this case the effusion currents were much higher, and more than a week elapsed before the effusion current stabilized. From this it is concluded that the concentration is closer to 98.5% than it is to 97%.

The experiment took place during a 6-month period, using a single sample. Data at a given temperature consisted of the slope of the mass-against-time curve. The temperature of the vacuum system or the temperature of the cryoshroud coolant could also be monitored simultaneously. The procedure following a change in the sample temperature was as follows. The chart ran continuously. After the effusion current stabilized, all the data were analyzed with the exception of those times that were obviously faulty (such as cases of great noise in the signal, known thermal transients, or flooding in the laboratory). The slopes were averaged by using a weighting of the time elapsed in each measurement. For the lowest pressure, time periods of integral numbers of days were used to minimize diurnal variations. It was found that the data could be reproduced after 6 months to within the weekly standard deviation. Also trends in the data after thermal stability was reached were not seen. From this it was concluded that the concentration of the acid did not vary significantly from its constant-boiling value during the experiment.

Because of the small mass changes involved, touching the sample to determine its temperature would distort the data. Because of the need to exhaust the effusate to the vacuum pumps, it is also impossible to completely close the cryoshroud around the sample. Therefore the sample is, in general, not at the cryoshroud temperature, and a direct determination of its temperature is difficult. The sample temperature was determined by replacing the hangdown wires with 0.04 cm-diameter thermo-couple wires, and allowing the sample cell to touch the thermocouple junction. The copper wire of the copper-constant thermocouple was more than 30 cm long, and the calculation included the heat transferred from the vacuum system to the sample via the thermocouple wire. The wires did not touch the cryoshroud. The smoothed results of the experiment and the results of the theory are shown in Table 1. The temperature .uncertainty is estimated at ± 3° for temperatures less than 10°, at ± 1.5° for temperatures less than 20°, and greater than 30°, and ± 1/2° for temperatures between 20° and 30°.

IV. DISCUSSION

The theory presented here is self-contained and requires only the values of the effusion current and the ideal gas law. The specific nature of the acid then enters the theory only through the effusion current. The discrepancy between the predicted concentration 76% and the azeotrope concentration 98.5% used here is not surprising because of the restricted nature of the theory. The predicted values of the partial pressures must also be viewed in this light. A more fundamental theory, involving the inter-molecular potentials, has been started.

TABLE 1. Results of Vapor Pressures

T K	$I_0(g/sec)$	P(torr)	P_1(torr)	P_2(torr)	P_3(torr)
273	1.0E-11	1.5E-7	8.2E-8	1.2E-8	5.6E-8
283	6.6E-11	9.3E-7	5.1E-7	7.4E-8	3.4E-7
293	4.0E-10	5.8E-6	3.2E-6	4.6E-7	2.1E-6
303	2.4E-9	3.6E-5	2.0E-5	2.9E-6	1.3E-5
313	1.5E-8	2.6E-4	1.4E-4	2.1E-5	9.6E-5
323	9.0E-8	1.5E-3	8.2E-4	1.2E-4	5.6E-4
333	5.6E-7	9.0E-3	4.9E-3	7.2E-4	3.3E-3

A comparison between the pressures and the theory of
reference indicates that their predictions are high by more than
an order of magnitude. The experiment pressures are within a
factor of 2 of the values quoted by Roedel at 23° C, when the
pressures are adjusted for the different concentrations used by
Roedel. Considering the current uncertainties, this result
is considered to be in reasonable agreement.

Further experiments to refine the total-pressure measurements
and to measure the partial pressures are in progress. A
molecular-beam, mass-spectrometer apparatus will be used.

The "Third Law" graph of ln P against 1/T indicates that,
over the temperature range studied here, the effective latent
heat of vaporization is a constant of 1.3×10^5 joules/mole
(10).

Utilization of the developed theory to calculate the
vaporization coefficients and critical data will be published
in the near future.

REFERENCES

1. Volmer, M., and Weber, A., Z. *Physik, Chem. (Leipzig)*, *119*, 277 (1925).

2. Becker, R., and Doring, W., *in* "Kinetic Theory of Liquids" (J. Frenkel, ed.), p. 719. Oxford University Press, London (1946).

3. Turco, R. P., Hamill, P., Toone, D. B., Whitten, R. C., and Kiang, C. S., *J. Atmos. Sci. 36*, 699 (1979).

4. Verhoff, F. H., and Banchero, J. T., *Am. Ind. Chem. Eng. J. 18*, 1265 (1965).

5. Banchero, J. T., and Verhoff, F. H., *J. Inst. Fuel, 81*, 76 (1975).

6. Gmitro, J. I., and Vermeulen, T., *Amer. Ind. Chem. Eng. J. 10*, 740 (1964).

7. Roedel, W., Measurement of Sulfuric Acid Vapor Pressure: Implications for Aerosol Formation by Heteromolecular Nucleation. Inst. fur Unwelfph, der Un. Heidelberg. D-6900 Heidelberg, Fed. Repub. Germany (1979).

8. Margrave, J. L., *in* "Physico-Chemical Measurements at High Temperatures" (J. O'M. Bockris, J. D. White, and J. D. MacKenzie, eds.), Ch. 10. Butterworths Sci. Pub., Ltd., London (1959).

9. Kunzler, J. E., *Analyt. Chem., 25*, 93 (1953).

10. Lewis, G. N., and Randall, M. (Revised by K. S. Pitzer and L. Brewer) *in* "Thermodynamics," 2nd ed., Ch. 15. McGraw-Hill, New York (1961).

DISCUSSION

Ko: It seems to me that maybe the real way to measure the composition and also the temperature of the cell would be to perform
some kind of high resolution spectroscopy, so that you can
calculate the temperature from the infrared spectrum and also the
composition.

Morrison: The possibility of doing far infrared on it is
interesting. Do you mean IR spectroscopy through the liquid
phase?

Ko: Both liquid and vapor phases. You can put the vapor in a
white cell and try to measure the absorption.

Morrison: That would work, yes. That is a good idea.

Ko: And the temperature of the liquid, you can determine from
the radiation of the liquid.

Morrison: Yes. That is a possibility and we should be looking
into doing that. I didn't realize the problem until too late.
I am building right now a molecular beam device with an inhouse
mass spectrometer to actually do mass spectrometer determination
of the partial pressures. That is well along the way.

Hamill: First, I would like to make a comment, namely, that
Roedel's latest measurement is about a factor of 2 larger than
yours which is just what you predicted. Second, I would like
to ask you a question. Are you planning to formulate a set of
tables like those of Gmitro and Vermeulen, that we can use for
determining sulfuric acid vapor pressure?

Morrison: That is what the theory is for. I wish I had time to
share more about that with you.

Yue: You know that the temperature of stratosphere is around
-55°C. If you can get vapor pressure of sulfuric acid at low
temperature then it would be extremely helpful for modelers to
investigate the possibility of forming sulfuric acid aerosols
in the stratosphere. Are you going to extend your experiment to
much lower temperatures or can you estimate the values? What
is the difference between your result and that of Gmitro and
Vermeuler at low temperature?

Morrison: My refrigerator is capable of going to -80°; however,
I am not willing to wait two centuries to accumulate a microgram
of mass loss. So the answer is no, but the calculations will go
down to those temperatures, and if you want to believe them, you
are welcome to.

CS_2 AND COS IN ATMOSPHERIC SULFUR BUDGET[1]

N. D. Sze
M. K. W. Ko

Atmospheric and Environmental Research, Inc.
Cambridge, Massachusetts

The stratospheric sulfur layer may play an important role in the Earth's radiation budget. It is usually assumed that SO_4 is formed in the stratosphere by oxidation of SO_2 and removed by sedimentation. The supply of SO_2 is thought to be maintained by upward diffusion from the troposphere and by photolysis of COS in the stratosphere. In order to assess the environmental impact of sulfur emission from anthropogenic activities, it is necessary to understand the natural atmospheric sulfur cycle. By using an atmospheric model, the background concentrations of SO_2, COS and CS_2 were computed and the contribution of each species toward the sulfate concentration in the remote tropo-sphere and stratosphere is discussed.

[1] This work is partially supported by the Air Force Geophysics Laboratory under contract F19628-78-C-0215 and by the Manufacturing Chemists Association.

I. INTRODUCTION

The persistence and pervasion of the stratospheric sulfate
(Junge) layer during periods of low volcanic activities have
puzzled many aeronomers (1,2). Junge suggested that SO_2 is
probably the gaseous precursor for the sulfate aerosol layer (3).
The origin of stratospheric SO_2 is, however, uncertain. The gas
may originate from the troposphere through upward diffusion
processes according to Junge, while Crutzen argued that photolysis
of COS may provide the dominant source for stratospheric SO_2
(3,4). More recently, Sze and Ko argued that CS_2 may be an
important precursor for atmospheric COS, and oxidation of CS_2
could also provide a comparable source for stratospheric
sulfate (5,6). This paper discusses the roles of CS_2 and COS
in atmospheric sulfur budget with emphasis on recent atmospheric
and laboratory data.

II. SULFUR CHEMISTRY

The troposphere contains appreciable concentration of COS
of about 0.5 ppb (7-9). Recent measurements by Inn *et al.*
indicated that COS is also present in the stratosphere with
concentration of about 15 ppb around 31 km (10).
    Crutzen estimated that photolysis of COS

$$COS + h\nu \rightarrow CO + S \tag{1}$$

could provide a stratospheric sulfur source of about 5×10^4 ton
(S) per year (4). He further argued that COS might have a
relatively long tropospheric lifetime (> 10 years) and cautioned
that future industrial release of COS (especially in the area
of fuel conversion technology) might perturb the atmospheric
COS budget, with a consequent impact on the Junge layer.

A new development in COS chemistry concerns the reaction of COS with OH,

$$COS + OH \rightarrow CO_2 + HS \qquad\qquad (2a)$$

$$\rightarrow CO + SOH \qquad\qquad (2b)$$

Kurylo reported a rate constant k_2 at room temperature of 6×10^{-14} cm^3 s^{-1}, significantly faster than the upper limit ($\approx 7 \times 10^{-15}$ cm^3 s^{-1}) set by Atkinson et $al.$ (11,12). A simple 1-D calculation suggested that reaction (2) may remove as much as 3×10^6 ton (S) of COS per year, a removal rate almost 10^2 times larger than that due to reaction (1) (5). This discovery motivated a search for atmospheric sources for COS and led to the suggestion that oxidation of CS_2 might provide a significant COS source (5,13).

Carbon disulfide (CS_2) was first detected by Lovelock in seawater samples (14). More recent measurements by Sandalls and Penkett showed that CS_2 may also have tropospheric concentration ranging from 0.07 to 0.37 ppb (9).

The important atmospheric reactions involving CS_2 molecule include the following:

$$CS_2 + h\nu \rightarrow CS + S \qquad (\lambda < 2200\ \overset{o}{A}) \qquad\qquad (3a)$$

$$\rightarrow CS_2^* \qquad (\lambda > 2200\ \overset{o}{A}) \qquad\qquad (3b)$$

$$CS_2 + OH \rightarrow \begin{bmatrix} S-C-S \\ | \\ OH \end{bmatrix}^* \rightarrow COS + SH \qquad\qquad (4a)$$

$$\rightarrow CS + SOH \qquad\qquad (4b)$$

and

$$CS_2 + O \rightarrow CS + SO \qquad\qquad (5a)$$

$$\rightarrow COS + S \qquad\qquad (5b)$$

$$\rightarrow CO + S_2 \qquad\qquad (5c)$$

The rate constant for reaction (4) has been measured by Kurylo who reported a value $k = 1.9 \times 10^{-13}$ cm^3 s^{-1} (11). The reaction products for reaction (4) have not yet been determined, although Kurylo argued that the more probable branch is reaction (4a) (15). By assuming that COS is mainly formed by reaction (4a) and removed by reaction (2), a first-order balance in COS budget would require the relation

$$\frac{(CS_2)}{(COS)} \simeq \frac{k_2}{k_{4a}} \tag{6}$$

be approximately satisfied in the lower troposphere (5). Existing atmospheric data of COS and CS_2 and laboratory data of k_2 and k_4 apparently lend some support to the hypothesis that CS_2 may be a precursor of atmospheric COS.

It is likely that sulfur radicals (SH, CS, and S) produced by reactions (1 to 5) would be oxidized readily in the atmosphere to form SO_2. Removal of SO_2 may be accomplished by either heterogeneous processes or by homogeneous gas phase reactions such as (16,17)

$$OH + SO_2 + M \rightarrow HSO_3 + M \tag{7}$$

$$HO_2 + SO_2 \rightarrow OH + SO_3 \tag{8}$$

$$CH_3O_2 + SO_2 \rightarrow CH_3O + SO_3 \tag{9}$$

Reaction (7) has been studied by several groups and probably represents the major sink for SO_2 (18-20). Reaction (8) appears to be unimportant based on the upper limit $k_8 < 2 \times 10^{-17}$ cm^3 s^{-1} set by Burrows et al. (21). Reaction (9) could be important; however, there is considerable uncertainty regarding the concentration of atmospheric CH_3O_2. The key reaction rates for sulfur chemistry are summarized in Table I.

TABLE I. Reaction Rates

Reaction	Rate constant[a]	Reference
$COS + h\nu \rightarrow CO + S$	$J_1^\dagger$	(22)
$COS + OH \rightarrow CO_2 + HS$	$k_2 = 6(-14)$	(11)
$CS_2 + h\nu \rightarrow CS + S$	$J_3^\dagger$	(23)
$CS_2 + OH \rightarrow COS + SH$	$k_4 = 1.9(-13)$	(11,15)
$CS_2 + O \rightarrow CS + SO$	$k_5 = 3.7(-11)exp(-700/T)$	(24)
$\rightarrow COS + S$		
$\rightarrow CO + S_2$	(branching ratios:	
	$f_a = .8, f_b = .1, f_c = .1)$	
$SO_2 + OH + M \rightarrow HSO_3 + M$	$k_7 = \dfrac{8.2(-13) \times (M)}{7.9(17) + (M)}$	(19)
$SO_2 \rightarrow$ precipitation & scavenging	$5.8(-7)$ $Z < 10$ km	(17)
$SO_2 + HO_2 \rightarrow OH + SO_3$	$k_8 < 2 \times 10^{-17}$	(21)

[a]The rate constants are in units of $cm^3 \ s^{-1}$ for two-body reaction, $cm^6 \ s^{-1}$ for three-body reaction and s^{-1} for photolysis and heterogeneous removal.

[†]The photolysis rates are calculated from measured cross-sections allowing for diurnal variation of solar intensity.

III. THE ATMOSPHERIC SULFUR BUDGET

Figure 1 shows schematically the roles of CS_2 and COS in atmospheric sulfur budget. Oxidation of CS_2 and COS could provide a tropospheric source for SO_2 as much as 6 MT (S) yr^{-1} (6,25). According to Friend, combustion of fossil fuel intro- duces approximately 65 MT (S) yr^{-1} into the atmosphere and probably accounts for the observed high concentration of sulfate and SO_2 over much of the industrial world (26,27). On the other hand, measurements at remote marine environments suggest that there is a persistent SO_2 background of about 0.09 ppb (26,8). Logan et al. recently argued that this background SO_2 might be attributed to in situ oxidation of COS and CS_2 (25).

Although much of the COS and CS_2 is removed in the tropo- sphere by reactions (1) and (2), a significant fraction (see Fig. 1) may be diffused into the stratosphere where photochemical oxidation of COS and CS_2 reactions (1 to 5) provides the dominant sulfur source for the Junge layer.

Figure 2 shows calculated profiles of CS_2 and COS. The cal- culated mixing ratio of CS_2 drops abruptly with altitude above $\sim$ 14 km, a feature reflecting the onset of the relatively fast reaction of CS_2 with O atoms (reaction 5). The calculated COS profile agrees fairly well with observations.

Figure 3 shows calculated altitudes of SO_2 along with data reported by Jaeschke et al. (28). The predicted SO_2 mixing ratio with CS_2 chemistry included appears to agree fairly well with observations. Clearly, more SO_2 data is needed to assess the role of CS_2 and COS in the stratospheric sulfur chemistry.

Figure 4 summarizes the various fluxes of sulfur compounds into the stratosphere. The total upward flux Φ_t of CS_2, COS, and SO_2 is about 2.1 x 10^7 molecules (S) cm^{-2} s^{-1} or 2.5 x 10^5 ton (S) yr^{-1}. This value is consistent with several independent

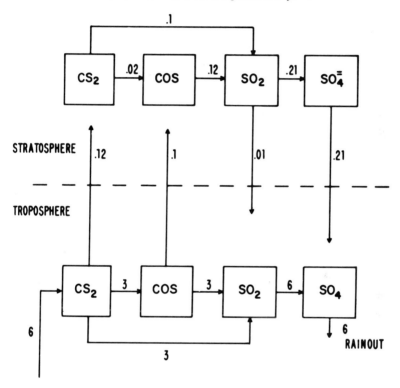

CS$_2$ and COS in atmospheric
sulfur budget

(transfer rate in Megaton (S) yr^{-1})

FIGURE 1.   The role of CS_2 and COS in atmospheric sulfur budget.   The transfer rates are in units of MT (S)/yr, of equivalent sulfur (1 MT = 10^{12} gm).

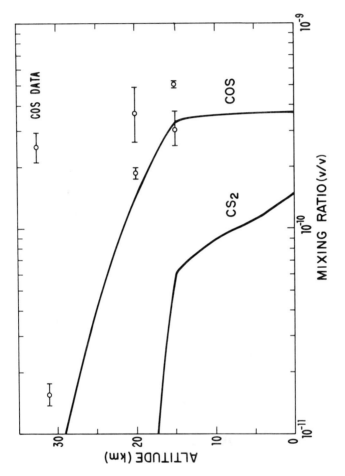

FIGURE 2. Calculated profiles of CS_2 and COS. The profiles are calculated using 1-D model described in Ref. 24. The boundary conditions at 0 km are 0.15 ppbv for CS_2 and zero flux for COS. The stratospheric COS data are plotted for comparison (10). Observed surface concentration of CS_2 and COS range from 0.07 to 0.37 ppbv and 0.3 to 0.5 ppbv (7-9).

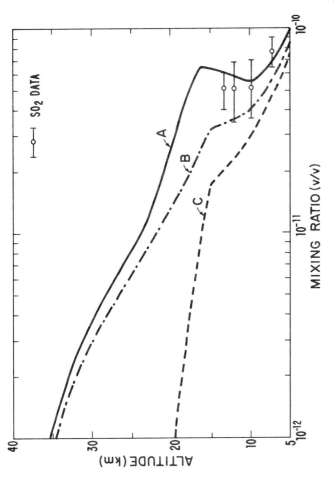

FIGURE 3. Calculated profiles of SO_2. The SO_2 data are from Ref. 23. Curve A is calculated using the full model described in Table I and the assumption that atmospheric COS is produced solely by CS_2 oxidation. Curve B is calculated with CS_2 chemistry omitted and prescribed surface concentration of 0.4 ppbv for COS. Curve C is calculated with both COS and CS_2 chemistry omitted. In all three cases a value of 0.3 ppbv is assumed for SO_2 at 1 km. This value is chosen such that the integrated tropospheric abundance of SO_2 is equal to 6×10^5 (S) ton, a value constant given in Refs. 21 and 22. The mixing ratio boundary condition at 1 km implicitly incorporates one effect of any production and removal processes occurring below 1 km (cf Ref. 15).

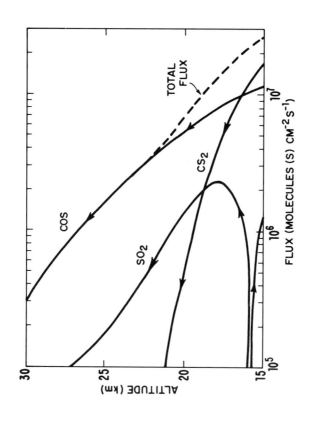

FIGURE 4. Calculated fluxes for CS_2, COS and SO_2. The flux of CS_2 has been multiplied by 2 to reflect that each molecule carries two S-atoms. The total flux Φ_t is set equal to $2 \times \Phi CS_2 + \Phi COS + \Phi SO_2$. The arrow indicates direction of flow.

estimates of $5 \times 10^6 - 2 \times 10^7$ molecules (S) $cm^{-2} s^{-1}$ necessary
to sustain the sulfate layer (30,31). Note that the major contribution to Φ_t comes mainly from CS_2 and COS with each gas
cominating at different altitude regions. The results in Fig. 4
apparently suggest that the persistence of the sulfate layer may
be attributed, in part, to the oxidation of CS_2 below $\sim$ 20 km
and, in part, to the oxidation of COS above $\sim$ 20 km. It therefore appears that the observed altitude ($\sim$ 20 km) at which
sulfate layer peaks may be modulated by the relative strengths
as well as by the altitude dependence of these two sources.

IV. CONCLUDING REMARKS

In conclusion, oxidation of CS_2 and COS might account for
the natural occurrence of sulfate and SO_2 in the remote troposphere and the stratosphere. Evidently, more data on COS, CS_2
and SO_2 are needed to substantiate this hypothesis. It is clear,
however, that a quantitative accounting of the natural background concentration of SO_2 and sulfate is important for the
assessment of the impact of sulfur emission associated with
anthropogenic activities.

REFERENCES

1. Junge, C. E., Chagnon, C. W., and Manson, J. E., J.
    Meteor. 18, 81 (1961).

2. Castleman, A. W., Jr., Munkelwitz, H. R., and Manowitz, B.,
    Tellus, 26, 222 (1974).

3. Junge, C. E., "Sulfur Budget of the Stratospheric Aerosol
    Layer." Review paper presented at the International
    Conference on Structure, Comparison and General Circulation
    of the Upper and Lower Atmospheres and Possible
    Anthropogenic Perturbations, Melbourne, Australia
    January 14-25, 1974.

4. Crutzen, P. J., *Geophys. Res. Lett. 13*, 73 (1976).

5. Sze, N. D., and Ko, M. K. W., *Nature, 278*, 731 (1979).

6. Sze, N. D., and Ko, M. K. W., *Nature, 280*, 308 (1979).

7. Hanst, P. L., Spiller, L. L., Watts, D. M., Spence, J. W.,
    and Miller, M. F., *J. Air Poll. Cont. Ass. 25*, 1220 (1975).

8. Maroulis, P. J., Torres, A. L., and Bandy, A. R., *Geophys.
    Res. Lett. 4*, 510 (1977).

9. Sandalls, F. J., and Penkett, S. A., *Atmos. Environ. 11*,
    197 (1979).

10. Inn, E. C. Y., Vedder, J. F., and Tyson, B. J., *Geophys.
    Res. Lett. 6*, 191 (1979).

11. Kurylo, M. J., *Chem. Phys. Lett. 58*, 238 (1978).

12. Atkinson, R., Perry, R. A., and Pitts, J. N., *Chem. Phys.
    Lett. 54*, 14 (1978).

13. Ko, M. K. W., and Sze, N. D., *in* "Environmental and
    Climatic Impact of Coal Utilization (J. J. Singh and A.
    Deepak, eds.) n.p. Academic Press, New York (1979).

14. Lovelock, J. E., *Nature, 248*, 625 (1974).

15. Kurylo, M. J., and Laufe, A. R., *J. Chem. Phys. 70*, 2032
    (1979).

16. Garland, J. A., *Atmos. Environ. 12*, 349 (1978).

17. Gravenhorst, G., Janssen-Schmidt, T., Ehhalt, D. H., and
    Röth, E. P., *Atmos. Environ. 12*, 691 (1978).

18. Calvert, J. G., Su, F., Bottenheim, J. W., and Strausz, O. P.,
    Atmos. Environ. 12, 197 (1978).

19. Moortgat, C. K., and Junge, C. E., *PAGEOPH, 115*, 759
    (1977).

20. Davis, D. D., *Can, J. Chem. 52,* 1405 (1974).

21. Burrows, J. P., Cliff, D. I., Harris, G. W., Thrush, B. A.,
and Wilkinson, J. P. T., World Meteorological Organization
Technical Note 511, pp. 25-28. Unipub, New York (1978).

22. Breckenridge, W. H., and Taube, H., *J. Chem. Phys. 52,*
1713 (1970).

23. Rabalais, J. W., McDonald, J. M., Schem, V., and McGlynn,
S. P., *Chem. Rev. 71,* 73 (1971).

24. Baulch, D. L., Drysdale, D. D., and Horne, D. G., "Evaluated
Kinetic Data for High Temperature Reactions," Vol. III.
Butterworths, London (1976).

25. Logan, J. A., McElroy, M. B., Wofsy, S. C., and Prather,
M. J., *Nature, 281,* 185 (1979).

26. Friend, J. P., *in* "Chemistry of the Lower Atmosphere"
(S. J. Rasool, ed.), pp. 177-201. Plenum Press, New York
(1973).

27. Mészáros, E., *Atmos. Environ. 12,* 699 (1978).

28. Jaeschke, W., Schmitt, R., and Georgii, H.-W., *Geophys.
Res. Lett. 3,* 517 (1976).

29. Sze, N. D., and Wu, M. F., *Atmos. Environ. 20,* 1117 (1976).

30. Cadle, R. D., *J. Geophys. Res. 80,* 1650 (1975).

31. Lazrus, A. L., and Gandrud, B. W., *J. Geophys. Res. 79,*
3424 (1974).

DISCUSSION

McElroy: There's a measurement which I guess is not yet out in
literature by Calvert.   I think it is of a remarkably fast
reaction rate for CH_3O_2 with SO_2.   Remarkably fast compared to
HO_2 and the brief look that we took at this it could be comparable
to OH in the troposphere as a sink if you accept his measurements
as given.   I'm not sure . . . let me ask the question.   Would
that have an effect on the stratosphere with your ideas about
the concentration of CH_3O_2?

Sze: I think it is going to have an effect.

McElroy: And the other remark was, I completely agree with the
concern you expressed about anthropogenic sources of COS and
CS_2 and potential impact on the stratosphere.   And just to under-
score that, if you look up chemical production of CS_2 for last
year or the last year, what 1974, was the last year I looked at
it, on a global scale, the books, which had not included the
Soviet Union and China, have a sulfur source of CS_2 manufactured
at 0.8 megatons.

Sze: That is good enough.   You make my point.

THE CS_2 AND COS BUDGET[1]

Malcolm K. W. Ko
Nien Dak Sze

Atmospheric and Environmental Research, Inc.
Cambridge, Massachusetts

Apart from oxides of sulfur, the atmosphere contains appreciable concentrations of CS_2 and COS. It is argued that CS_2 may be an important precursor for atmospheric COS in the light of new atmospheric and laboratory data. Uncertainties in the global budget of COS and CS_2 are highlighted and emphasis is placed on areas where there is need for additional data.

I. INTRODUCTION

The troposphere contains appreciable concentrations of carbonyl sulfide (COS) of about 0.5 ppb (1-3). Recent measurements by Inn et al. (4) indicated that COS is also present in the stratosphere with concentration of about 15 ppt at 31 km. According to present ideas (5-7), photolysis of COS,

$$COS + h\nu \rightarrow CO + S \tag{1}$$

and the reaction of COS with OH (8),

[1]*This work is partially supported by the Air Force Geophysics Laboratory under contract F19628-78-C-00215 and by the Manufacturing Chemists Association.*

323

$$COS + OH \rightarrow CO_2 + SH$$
$$\rightarrow CO + SOH \tag{2}$$

may provide an important source of sulfur for the formation of the stratospheric sulfate layer. More recent investigation suggests that oxidation of CS_2 may also contribute significantly to the stratospheric sulfur budget (7).

The origin of atmospheric COS is, however, uncertain. Crutzen estimated that reaction (1) may remove $\approx 5 \times 10^4$ ton (S) of COS per year and assumed that reaction (2) is relatively unimportant. He further cautioned that future industrial release of COS (especially in the area of fuel conversion technology) might perturb the atmospheric COS budget, with a consequent impact on the Junge layer.

Recent laboratory studies by Kurylo indicated that the rate constant k_2 of reaction (2) is moderately fast,[2] about 6.0×10^{-14} cm^3 s^{-1} (8). A simple 1-D calculation (6) suggests that reaction (2) may remove as much as 2×10^6 ton (S) of COS per year, a removal rate almost 10^2 times larger than that due to reaction (1). It is clear that a comparable source ($\approx 2 \times 10^6$ ton (S) yr^{-1}) is required to account for the observed concentration of COS.

[2]Atkinson et al. reported an upper limit of $7 \times 10^{-15} cm^3$ s^{-1} for k_2 (10). The apparent discrepancies between the two results are mainly attributed to the interpretation of the experimental data at 1 w and high pressures (8).

II. CS_2 CHEMISTRY

It appears that atmospheric oxidation of carbon disulfide (CS_2) may provide an important source for COS (6). Lovelock has detected the presence of CS_2 in seawater samples (9). More recent measurements showed that CS_2 has tropospheric concentrations ranging from 0.07 to 0.2 ppb (3).

The important atmospheric reactions involving the CS_2 molecule include the following:

$$CS_2 + h\nu \rightarrow CS + S \qquad \lambda < 2200 \overset{o}{A} \qquad\qquad (3a)$$

$$\rightarrow CS_2^* \qquad\qquad \lambda < 2200 \overset{o}{A} \qquad\qquad (3b)$$

$$CS_2 + OH \rightarrow \left(\begin{array}{c} S-C-S \\ | \\ OH \end{array} \right)^* \rightarrow COS + SH \qquad\qquad (4a)$$

$$\rightarrow CS + SOH \qquad\qquad (4b)$$

and,

$$CS_2 + O \rightarrow CS + SO \qquad\qquad (5a)$$

$$\rightarrow COS + S \qquad\qquad (5b)$$

$$\rightarrow CO + S_2 \qquad\qquad (5c)$$

Reactions (3a) and (5a), (5b), and (5c) appear to be relatively unimportant in the lower troposphere based on known kinetic data.[3] The rate constant for reaction (4) has been measured by Kurylo (8) who reported a value of $k_4 = 1.9 \times 10^{-13} cm^3 s^{-1}$. The reaction products for reaction (4) have not yet been identified.

[3]Photoexcitation of CS_2 (reaction (3b)) and subsequent reactions of CS_2^* could provide a sink for CS_2, as well as additional sources of COS (12).

Nevertheless, branch reaction (4a) is exothermic by 35 kcal mol^{-1}, while reaction (4b) is endothermic by 23 kcal mol^{-1}. Kurylo argued that the adduct

$$\begin{pmatrix} S-C-S \\ | \\ OH \end{pmatrix}^{*}$$

formed by reaction (4a) might be relatively long-lived and subsequent hydrogen migration could lead to COS formation (11). It is reasonable to assume that reaction (4a) is the dominant path for reaction (4).

III. THE CS_2 AND COS BUDGET

An approximate atmospheric balance in COS budget would require the quantity

$$\int_{0}^{\infty} P_{COS}(z) \, dz \approx 0 \tag{6}$$

where P_{COS} denotes the "net" local production rate ($cm^{-3} s^{-1}$) which may be given by

$$P_{COS}(z) \approx k_{4a}[OH][CS_2]$$
$$-k_2[OH][COS] - J_1(COS) \tag{7}$$

Equation (6) implies that

$$\int_{0}^{\infty} k_{4a}[OH][CS_2]dz \approx \int_{0}^{\infty} (k_2[OH] + J_1)[COS]dz \tag{8}$$

Recognizing that the contribution from reaction (1) is significantly smaller than that due to reaction (2), that is,

$$\int_{0}^{\infty} k_2[OH][COS]dz \gg \int_{0}^{\infty} J_1[COS]dz \tag{9}$$

Eq. (8) becomes

$$\int_0^\infty k_{4a}[OH][CS_2]dz \approx \int_0^\infty k_2[OH][COS]dz \tag{10}$$

From reaction (10), a ratio $[CS_2]_o/[COS]_o$ independent of OH concentration may be derived by noting

$$k_{4a}[OH]_o[COS]_o H_{4a} \approx k_2[OH]_o[COS]_o H_2 \tag{11}$$

where H_{4a} and H_2 are the appropriate scale heights for reactions (4a) and (2), respectively, and the subscript "o" denotes surface concentration. Since H_{4b} and H_2 are of similar order, the ratio $[CS_2]_o/[COS]_o$ is given by

$$R_A \equiv \frac{[CS_2]_o}{[COS]_o} \approx \frac{k_2}{k_{4a}} \equiv R_L \tag{12}$$

Equation (12) links an atmospheric quantity R_A with a laboratory quantity R_L. This relation must be approximately satisfied if CS_2 were indeed a precursor for atmospheric COS. By taking typical measured concentrations of 0.4 ppb and 0.1 ppb for COS and CS_2, respectively, the ratio R_A is about 0.25, a result in harmony with the quantity, $R_L = 0.3$ derived from laboratory measurement of k_2 and k_4 by Kurylo (8).

The calculated global flux of CS_2 is about 4×10^6 ton (S) yr^{-1}, and implies a CS_2 lifetime of about 0.2 year. The measured concentration of CS_2 in seawater samples reported by Lovelock (9) suggests that the ocean might provide a source of CS_2 of about 4×10^5 ton (S) yr^{-1}. In view of various uncertainties, the role of the ocean in the atmospheric CS_2 budget remains to be unraveled.

IV. CONCLUDING REMARKS

The reaction of OH with CS_2 may provide an important source for atmospheric COS of about 2×10^6 ton (S) yr^{-1}. It is estimated that present contribution of COS from human activities may approach 0.2×10^6 ton (S) yr^{-1}. Clearly, more atmospheric measurments of CS_2 and COS, as well as an accurate determination of rate constants for k_2 and k_4 are needed for model validation. It is also obvious that other sources of CS_2 and COS need to be iden-tified. Nevertheless, future increases in atmospheric release of CS_2 and COS could conceivably perturb the Junge layer with pos-sible adverse impact on long-term climate (13).

REFERENCES

1. Hanst, P. L., Spiller, L. L., Watts, D. M., Spence, J. W., and Miller, M. F., *J. Air Poll. Cont. Assoc. 25,* 1220 (1975).

2. Maroulis, P. J., Torres, A. L., and Bandy, A. R., *Geophys. Res. Lett. 4,* 510 (1977).

3. Sandalls, F. J., and Penkett, S. A., *Atmos. Environ. 11,* 197 (1977).

4. Inn, E. C. Y., Vedder, J. F., and Tyson, B. J., *Geophys. Res. Lett. 6,* 191 (1979).

5. Crutzen, P. J., *Geophys. Res. Lett. 4,* 510 (1976).

6. Sze, N. D., and Ko, M. K. W., *Nature, 278,* 731 (1979).

7. Sze, N. D., and Ko, M. K. W., *Nature, 280,* 308 (1979).

8. Kurylo, M. J., *Chem. Phys. Lett. 58,* 238 (1979).

9. Lovelock, J. E., *Nature, 248,* 625 (1974).

10. Atkinson, R., Perry, R. A., and Pitts, J. N., *Chem. Phys. Lett. 54,* 14 (1978).

11. Kurylo, M. J., and Kaufer, A. H., *J. Chem. Phys. 70,* 2032 (1979).

12. Wood, W. P., and Heicklen, J., *J. Phys. Chem. 75,* 854 (1971).

13. Cadle, R. D., and Grams, G. W., *Rev. Geophys. Space Sci. 13,* 475 (1975).

DISCUSSION

Miller: On what do you base your estimate of 100 plants? I am sure there are not anything approaching that number any more.

Ko: I think by now there are about 25 plants worldwide, with a few in the planning stage in the states. The number that I had in my head was the total coal consumption which was about 35 megatons of coal worldwide. And I thought that in the year 2000, perhaps 1000 megatons would not be too far off. So the number is a projected number.

McElroy: A comment about the conversation Jim Friend and I had this morning. It really is an intriguing possibility that perhaps the natural sulfur cycle taking away the anthropogenic component may be exceedingly small, as far as the atmospheric components are concerned. And it is therefore, as you pointed out, vitally important to get at the question of what are the sources of CS_2.

Ko: And to add to that there is one thing very important that I forgot to mention. It turns out that there is a natural source of CS_2 already identified by Lovelock. It turns out that the ocean contains CS_2 and by outgassing from the ocean using a simple piston model, the number that Lovelock gave in 1974 gives a flux of 0.1 megaton per year of CS_2 from the ocean.

McElroy: As a matter of fact, with your number for the concentration taking the low range, it is possible that there is no flux from the ocean. And that all that is happening is that the ocean is in equilibrium with the earth.

CARBONYL SULFIDE, STRATOSPHERIC AEROSOLS
AND TERRESTRIAL CLIMATE

R. P. Turco

R & D Associates
Marina Del Rey, California

R. C. Whitten, O. B. Toon, and J. B. Pollack

NASA Ames Research Center
Moffett Field, California

P. Hamill

Systems and Applied Sciences Corp.
Hampton, Virginia

The contribution of carbonyl sulfide (OCS) to the formation of stratospheric aerosol particles, and the effects of these particles on the Earth's radiation balance and average surface temperature have been studied. A recent measurement of the rate of reaction of OCS with OH places new limits on the likely atmospheric lifetime of OCS. Worldwide observations indicate a rather ubiquitous OCS distribution. Natural and anthropogenic sources of OCS are discussed and OCS measurements are compared with concentrations predicted theoretically. To calculate the effects of changes in OCS concentrations on ambient stratospheric particles, a detailed one-dimensional photochemical/microphysical model of sulfur-bearing gases and sulfuric acid aerosols is used. The aerosol model treats the physical processes of nucleation,

growth (and evaporation), coagulation, sedimentation, and
diffusion. Sulfurous gases can interact with aerosol particles
in the model. Atmospheric data and model simulations which
suggest that background concentrations of OCS may be largely
responsible for maintaining the ambient aerosol layer are considered. Also, calculations illustrating the alteration of
aerosol properties with 10 times the current amount of OCS in the
atmosphere are presented. Using a comprehensive radiation transport model (which accounts for solar energy insolation and
terrestrial albedo at all wavelengths, including the effects of
radiation absorption and scattering by gases and aerosols) shows
that a tenfold increase in OCS could decrease the average
global surface temperature by about 0.1 K. Possible future
anthropogenic perturbations of the OCS-aerosol-climate system
are discussed.

I. INTRODUCTION

Crutzen first suggested that carbonyl sulfide (OCS) might be
an important, if not the dominant, source of sulfur for the
stratosphere (1). As a result of sulfur influx, a distinct
layer of highly dispersed sulfate aerosols is formed in the
lower stratosphere. These particles reflect sunlight and affect
the overall global albedo in the visible and infrared wavelength
regions (2). If man, through his industrial and agricultural
activities, were to increase atmospheric OCS concentrations
significantly, climatic effects caused by changes in the stratospheric aerosol layer might occur.

Here, an evaluation of the possibility of future atmospheric
increases of carbonyl sulfide, and an estimation of their
potential effect on global climate, are made. To calculate
aerosol concentration changes following OCS accumulation, a
detailed model of gaseous sulfur compounds and stratospheric
sulfate aerosols is used (3,4). The climatic impact of aerosol
changes, in terms of the alteration of the Earth's average
surface temperature, is calculated using a comprehensive

radiation transport model (2). To the authors' knowledge, no other quantitative assessment of potential aerosol and climate effects attributable to OCS has yet been published.

II. SOURCES OF OCS

The sources of carbonyl sulfide in nature and industry are numerous, although the magnitudes of these sources are quite uncertain. For example, OCS has been detected in the vapors emanating from volcanoes and fumaroles (5, 6). Inasmuch as the contribution of volcanic sulfur to the total atmospheric sulfur budget is small (about 1% to 2% according to Friend, ref. 7), volcanoes should represent a negligible source of OCS. Carbonyl sulfide has also been measured over fresh cattle manure in concentrations as high as 6 ppmv (parts per million by volume) (8). Natural and agricultural fires may contribute to the OCS background. It has been estimated, for example, that fires generate more than 7000 tg (tg $\equiv$ teragrams or 10^{12} grams) of CO_2 each year, although controlled burning experiments indicate a negligible fraction of SO_2 emission from burning vegetation (OCS was not measured in these tests, but it would presumably have a much smaller concentration than SO_2) (9).

Natural biological decay may be a more important source of OCS. For example, biospheric (ocean and land) production of H_2S (including a small contribution from urban sewage treatment) is of the order of 50 to 100 tg/yr (7). Under anaerobic conditions, the release of small amounts of other reduced sulfur compounds such as OCS and CS_2 seems likely. In fact, carbon disulfide and dimethyl sulfide have been found in small amounts in air and in lake and ocean waters (10 to 14). Unfortunately, the direct contribution of natural decay to the OCS budget is difficult to estimate.

Biologically (and industrially) produced carbon disulfide may represent a large indirect source of OCS, however. Recently, Kurylo has studied the reaction

$$CS_2 + OH \rightarrow OCS + SH \tag{1}$$

and measured a rate coefficient of 1.9×10^{-13} cm^3/sec at 296 K (15). Because daytime OH concentrations are about 1 to 2 x $10^7/cm^3$ in the boundary layer (the first 1-1/2 km of the atmosphere) and 1 to 2 x $10^6/cm^3$ in the clean troposphere at 6 km, the CS_2 lifetime against conversion into OCS is of the order of a week (taking into account 24-hour averaging of OH concentrations); this value is comparable to the lifetimes estimated for other CS_2 oxidation processes, which yield OCS at a lower efficiency of about one-third (16, 17).

The CS_2 surface concentrations measured by Sandalls and Penkett at Harwell, England are in the range of 0.07 to 0.37 ppbv (parts per billion by volume), with an apparent large variability (10). Interestingly, these amounts are consistent with the quantity of CS_2 found dissolved in the Atlantic Ocean by Lovelock, if equilibrium between the atmosphere and oceans is assumed (11). Adopting a uniform CS_2 concentration of 0.2 ppbv in the boundary layer, a CS_2 lifetime of 2 weeks, and unit conversion of CS_2 into OCS, it is estimated that as much as 10 tg/yr of OCS could be generated from CS_2. This result implies a source of atmospheric CS_2 of about 12 tg/yr, which seems large in view of the fact that Lovelock finds the oceanic "source" of CS_2 to be an insignificant fraction of the total natural sulfur budget (11). Until more definitive measurements of the CS_2 distribution are available, and the magnitude of its natural sources ascertained, an estimate of 1 to 5 tg/yr of OCS from the decomposition of natural CS_2 is favored.

In the United States in 1974, about 0.35 tg of CS_2 were
produced by the five largest chemical manufacturers (17, 18).
Most of the CS_2 is used as a solvent in other industrial
processes (18) and the amount which eventually volatilizes
could be 30% or more (17). Accordingly, the contribution of
industrial CS_2 to the OCS budget may be of the order of 0.1 to
0.3 tg/yr.

Despite the fact that commercial production of OCS is negligible, anthropogenic sources of OCS, besides the ones mentioned
above, are common. Two examples of minor sources of OCS are
cigarette smoke and vapors from cooking grain mashes (19, 20).
In paper manufacture, wood pulp is usually treated with sulfate
or sulfite compounds; Kraft mills are notorious for their odor
produced by organic sulfur emissions (21). Worldwide, as much
as 200 tg of paper are manufactured each year, but the amount of
OCS released as a byproduct is unknown.

Potentially more serious atmospheric emissions of OCS can
occur during fuel refining and combustion processes. In terms
of total sulfur released, these processes are currently estimated
to emit about 100 to 150 tg/yr of SO_2 as a result of oxidation of
the small sulfur component of most organic fuels (7, 9). From
published data, average sulfur residues (by mass) of roughly 1%
in coal and 0.2% in petroleum are estimated; among individual
fuel lots, the residue can range from 0.1% to 5% sulfur (22, 23).
Although most of the sulfur is oxidized into SO_2, G. E. Moore,
of Massachusetts Institute of Technology in 1976 (unpublished),
has determined that coal combustion alone--accounting for the
emissions of desulfurization plants--could generate up to 5 tg/yr
of OCS. This estimate is in accord with one made by assuming a
total rate of sulfur emission due to combustion of 50 tg S per
year, of which roughly 6% is emitted as OCS (17). The 6% figure

is taken as an average upper-limit value for existing sulfur
recovery systems. Of course, worldwide, only a small fraction
of combustion emissions are processed for sulfur removal.
Untreated stack gases generally have very low OCS concentrations
(17).

Peyton *et al.* have made a detailed study of sulfur removal
and recovery techniques, and the gaseous emissions of coal and
oil processing plants (17). They point out that tail gas treatment systems that are based on sulfur vapor reduction have the
capacity to generate large amounts of OCS. During coal gasification and liquification, highly reducing conditions hold, and OCS
may be produced in fractional quantities as large as 10^{-3} of the
total gas volume. After scrubbing and treatment of the tail
gases, however, emissions of OCS from these plants should
(typically) amount to less than 1% of the sulfur in the coal
processed (17).

Certain catalysts used to reduce nitrogen oxides in stack
gases can also convert a substantial fraction of the accompanying
SO_2 into OCS. One study indicates 6% conversion to OCS by non-
selective reducing catalysts (17). This situation does not apply
to the catalytic converters used in automobiles, however; these
converters oxidize carbon monoxide and hydrocarbons to CO_2 and
H_2O. Occasionally, in fuel-enriched streams, large amounts of
H_2S and OCS may be generated, imparting their characteristic
odor to the exhaust (24). Although automotive gasolines
typically have very low sulfur contents--values of 0.03% to
0.06% S by weight are common--all of the sulfur is released to
the atmosphere (17). Thus, a small contribution to the global
OCS budget can be expected from gasoline consumption.

Currently, some oil-producing nations (particularly those in
the Middle East) incinerate vented natural gas at the well head.
This probably represents a substantial source of atmospheric SO_2,
but the contribution to the OCS budget is undetermined.

According to the previous discussion, OCS may comprise 1% to 10% of the sulfur released to the atmosphere as a result of organic fuel combustion. Based on an expanding rate of coal gasification and increasing utilization of reducing catalysts, the 10% figure is probably more conservative. Note, however, that in many industries 90% to 95% of the sulfur in fuels is recovered before venting to the atmosphere. Nevertheless, the residual sulfur emission could have a large OCS component. In untreated stack gases, on the other hand, the OCS concentration is quite small. Hence, even though OCS may comprise as much as 10% of certain industrial sulfur emissions, current overall OCS release probably represents 1% or less of the total sulfur residue of fuels.

Thus it may be concluded that the total global source of atmospheric OCS is probably in the range of 1 to 15 tg/yr. A large part of this, perhaps 50% or more, may result from anthropogenic activities, most notably activities related to fuel processing and consumption. Fig. 1 illustrates the likely atmospheric budget of OCS (as well as that of sulfur other than OCS). As will be shown shortly, the estimated emission rate of OCS is consistent with evaluated lifetimes and measured concentrations of OCS. (Also see Fig. 1.)

Man's contribution to the global OCS budget over the next half-century can be projected. Rotty has estimated that worldwide fossil fuel consumption rates may rise to as much as 23,000 tg of carbon per year by 2025 (25). In addition, because of the inevitable depletion of petroleum reserves and an increasing dependence on coal, the average sulfur content of fuel could be expected to rise significantly. Rapidly increasing, largely uncontrolled, emissions from the developing nations of the world are also forecast. Accordingly, a tenfold increase in sulfur emission rates in the next half-century may be expected.

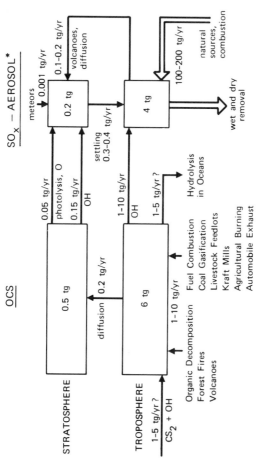

FIGURE 1. *Atmospheric budgets of carbonyl sulfide and other sulfur compounds. Masses are given in teragrams (tg).*

Of course, this situation could become intolerable, and severe measures might have to be taken to remove the sulfur from fuels and tail gases. If these measures were unwisely based on SO_2 reduction schemes which lead to substantial OCS leakage, large enhancements in OCS emissions to the atmosphere might result. Therefore, a tenfold increase in future OCS concentrations is assumed as the basis for the perturbation calculations. Whether such increases actually materialize or not depends largely on man's environmental strategy for the use of his remaining carbonaceous fuel reserves.

III.   ATMOSPHERIC SINKS AND CONCENTRATIONS OF OCS

Crutzen has pointed out that OCS photolysis above 20 km is a sink for OCS and a source of sulfur for stratospheric aerosols (1). This process, as studied by Turco et al., leads to an overall OCS atmospheric lifetime of about 50 years, however (3). The reactions of OCS with ground state and excited oxygen atoms lead to even longer lifetimes.

Hydrolysis and washout in rainfall appears to be an unimportant sink for OCS. Although hydrolysis of OCS is rapid in basic solutions, it is quite slow (weeks to months) in neutral and acidic solutions (17, 26). The pH of rainwater is typically 5 to 6, while occasional episodes of more acidic rainfall (with a pH of 4 to 5) are known to occur. By assuming that the total water vapor content of the atmosphere is precipitated daily, and utilizing an OCS solubility of about 1.4 g/ℓ H_2O (STP), it is estimated that less than 0.01 tg of OCS per year would by hydrolyzed in rainwater or washed from the atmosphere. Likewise, OCS hydrolysis in sulfuric acid aerosols should be negligible.

Seawater has a pH greater than 7, and a high salt content,
which could lead to efficient OCS hydrolytic decomposition
(F. S. Rowland, private comm.). The actual rate of hydrolysis
is unknown, however, and its contribution to the OCS cycle is
difficult to determine. However, Rowland (private comm.) has
estimated that oceanic hydrolysis could lead to an OCS lifetime
as short as a year.

The dominant atmospheric sink for OCS may be its reaction
with OH. Kurylo has measured the rate coefficient for the
process,

$$OCS + OH \rightarrow CO_2 + SH \tag{2}$$

and obtained a value of 5.7×10^{-14} cm^3/sec at room temperature
(15). (In the atmosphere, the SH radical formed in reaction
(2) is quickly oxidized to SO_2.) Newly measured OH concentrations (averaged over 24 hours) imply roughly a one-half
year atmospheric lifetime for OCS due to reaction (2)
(16). However, because reaction (2) is likely to be slower at
reduced atmospheric temperatures, and because there are
indications that tropospheric OH concentrations may be somewhat
smaller than directly measured values, a 1 year OCS lifetime
has been adopted as a basis for arguments about its atmospheric
budget (27-29). This new estimate of the OCS lifetime is
probably accurate to within a factor of 3 to 4, considering
that two independent loss processes (reaction with OH and
hydrolysis in the seas) are estimated to have lifetimes of this
magnitude.

Carbonyl sulfide has been measured extensively in the
troposphere (30-34). OCS has also been measured at several
heights in the lower stratosphere (30). Figure 2 compares
measurements with a typical model prediction for OCS; the agreement is generally satisfactory (3). OCS is observed to have a
ubiquitous tropospheric concentration of about 0.5 ppbv with

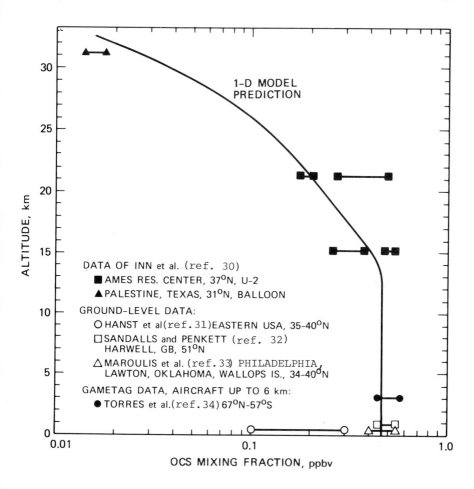

FIGURE 2.   Measurements of OCS.   Also shown is a one-dimensional model prediction for comparison.

no apparent contrast between the boundary layer and the "clean" upper troposphere (34).   OCS also has a small ($\approx$ 5%) inter-hemispheric contrast (34).   In addition, OCS variability (from one measurement to another) is found to be less than 10% (32-34). All of these observations are consistent with an OCS lifetime of the order of 1 year.

Referring to Fig. 1, it can be seen that the 6 tg of OCS measured in the atmosphere is compatible with an OCS source of 6 tg/yr and a tropospheric lifetime of 1 year, as has been estimated. Accordingly, a fairly detailed picture of the global OCS balance appears to be available, although confirming measurements are needed.

The budgets of stratospheric OCS and oxidized sulfur are also shown in Fig. 1. Turco *et al.* and Toon *at al.* discuss the stratospheric sulfur balance more thoroughly, in the context of a model sensitivity study (3, 4). They conclude that the ambient stratospheric aerosol layer is maintained in a large part by OCS diffusing upward from the troposphere. The OCS profile measurement of Inn *et al.* confirms the existence of a substantial source of sulfur for aerosols due to OCS decomposition between 20 km and 30 km (30).

Except during intense volcanic activity, when large amounts of SO_2 may be lofted into the stratosphere, SO_2 diffusing upward from the troposphere to the stratosphere is trapped and quickly oxidized near the tropopause level, where aerosol residence times are short. Thus, SO_2 contributes mainly to aerosol formation just above the tropopause. Other sulfur-bearing gases such as H_2S, CS_2, and $(CH_3)_2S$ are expected to have small concentrations in the upper tropo-sphere and should not affect stratospheric aerosols.

Figure 3 illustrates the influence of different strato-spheric sulfur sources on the mixing ratio of large particles (3, 4). OCS is seen to have the dominant effect on large particle abundances. Consequently, a tenfold increase in OCS concentrations could have a large impact on stratospheric aerosols.

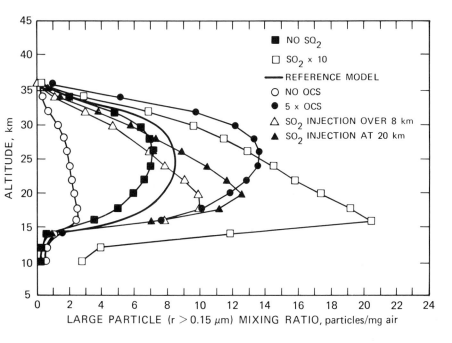

FIGURE 3. Effects of various gaseous sulfur sources on the
stratospheric large particle mixing ratio (i.e., the mixing
ratio of particles with radii greater than 0.15 μm).

IV. THE AEROSOL MODEL: OCS PERTURBATIONS

The model which was utilized to make aerosol perturbation
calculations is described at length in references 3 and 4. Both
sulfur photochemistry and aerosol microphysics are treated, and
the gaseous and condensed phases are interactive. The model has
been subjected to numerous validation and sensitivity tests and

has been shown to predict most of the observed properties of
natural aerosols. The model simulates the atmosphere between
the ground and 58 km with a 2-km grid spacing. Thirty-five
particle sizes are considered; the particle radii increase
geometrically from 0.001 to 2.56 μm so that particle volume
doubles between sizes. The aerosol composition (i.e., the
sulfuric acid weight percentage of the droplets, and the fractional volume occupied by nonacid "core" materials) is also
predicted. Accordingly, the model yields high resolution
particle size-altitude-composition distributions for radiation
transfer calculations.

Figure 4 shows predicted model distributions for the major
sulfur gases in the stratosphere and compares them with

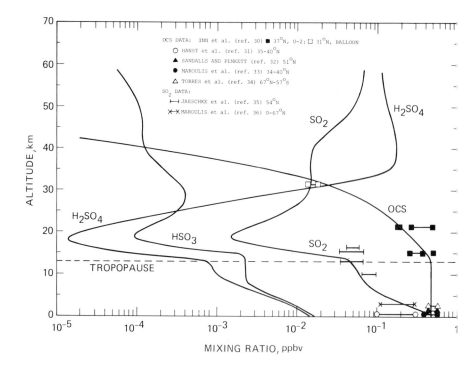

FIGURE 4. Typical model profiles for the dominant
atmospheric sulfur gases. Data are compared.

measurements. It is noteworthy that the interaction of sulfur gases with aerosols greatly modifies the gas distributions, and this modification should be accounted for in a model. The calculations reported here were made before Kurylo (15) had accurately determined the rate coefficient for reaction (2). Based on earlier measurements of a much smaller rate coefficient, this reaction has been ignored (37). The inclusion of reaction (2) (with an assumed 2 kcal/mole activation energy) causes the following changes. First, the OCS profile in Fig. 2 moves into somewhat better agreement with Inn's observation near 30 km because of the enhanced OCS loss rate through the stratosphere (30). Second, there is an additional source of sulfur below 25 km from OCS decomposition by OH, which further emphasizes the role of OCS relative to SO_2 in the lower stratosphere. Thus, it is believed that the present calculations may slightly underestimate the importance of OCS perturbations.

Changes in the large aerosol particle mixing ratio due to a tenfold increase in OCS or SO_2 concentrations are illustrated in Fig. 5. The large particles are the most optically active ones. Clearly, even though OCS is increased by a factor of 10, the number of large particles is increased by only a factor of about 2.

The corresponding effect on the particle size distribution at 20 km is shown in Fig. 6. The average particle size has obviously increased with sulfur added to the stratosphere. One consequence of the size increase is to reduce the average particle residence time in the layer. Thus, increasing the particle radius rapidly increases the sulfate mass loss rate from the stratosphere because the particle settling velocity increases as r and the mass carried by each particle as r^3. This explains why very few additional large particles are needed to affect the overall mass balance. Hence, for a

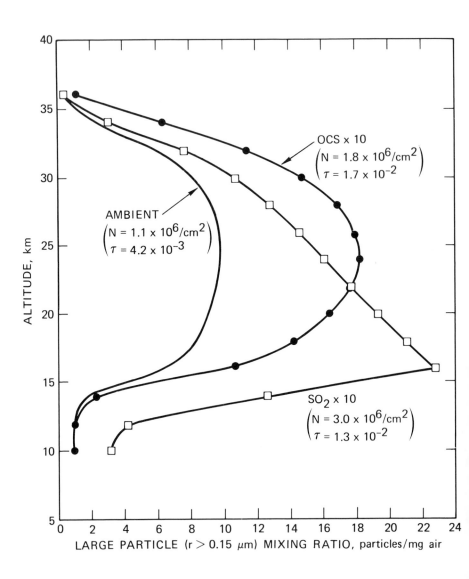

FIGURE 5. Predicted effect of a tenfold increase in OCS or SO₂ on the large particle mixing ratio. The total column abundances of large stratospheric particles (N) and the optical depth of the layer at 550 nm (τ) are indicated in parenthesis.

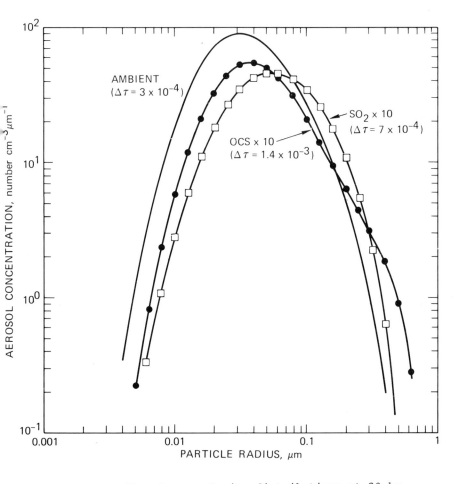

FIGURE 6. Predicted aerosol size distributions at 20 km for the same cases as in Fig. 5. The optical depth of the aerosols in a layer 1 km thick centered at 20 km (Δτ) is given in parenthesis.

tenfold increase in the sulfur source strength, only a (roughly)
70% increase in the average particle radius is expected. This
is found to hold, for example, in the case of a tenfold
increase in SO_2. For a tenfold increase in OCS, however, the
size distribution tends to become bimodal in character near
20 km. The larger particles are formed at, and settle from,
higher altitudes where OCS is decomposed, and the smaller
particles originate near the tropopause.

The ambient aerosol layer has an optical depth of about
0.004 at 550 nm (38). For a tenfold increase in OCS, the
estimated optical depth is about 0.014. This value can be
compared with the optical depths observed following volcanoes
Fuego (0.02 to 0.03) and Agung (0.02 to 0.04 in the northern
hemisphere, and 0.2 to 0.3 in the southern hemisphere). Thus, a
tenfold increase in OCS produces an effect comparable with that
of a large volcano. The OCS effect, however, would be semi-
permanent, lasting as long as OCS emissions persisted, but would
decay rapidly (within 1 to 2 years) should emissions cease.

V.   CLIMATE EFFECTS

To calculate the climate impact of aerosol layer perturbations, predicted altitude-dependent aerosol size spectra were
incorporated in the "doubling" routine described by Pollack
et al. (2). The doubling calculations are highly accurate
multiple scattering computations that explicitly account for
solar energy absorption by CO_2, O_3, O_2, and H_2O, and absorption
and scattering by aerosols in the stratosphere and troposphere.
For stratospheric aerosols the optical constants of a 75%
sulfuric acid solution are used; for tropospheric aerosols
the model of Toon and Pollack is used (38). Once the solar
energy deposition rate profile is determined, an infrared

calculation is performed to achieve radiative-convective equilibrium accounting for infrared emission and absorption by H_2O, CO_2, O_3, and aerosols. The technique differs from that of Pollack et al. in that the routine is iterated until a stable radiative-convective temperature profile is obtained (2).

Applying the radiative transfer model to the case when OCS is increased by a factor of 10, a possible decrease in the average global surface temperature of about 0.1 K is found. Such a temperature change is at the threshold of climatic significance. However, OCS, like CO_2, can create a "greenhouse" effect in which far infrared radiation emitted from the earth is trapped in the lower atmosphere. The OCS greenhouse effect will cause some surface warming and will partially offset the surface cooling due to aerosol enhancement. In reality, neither the warming nor the cooling trends would be uniformly distributed over the globe, and the overall climate impact might still be severe. Careful calculations of the greenhouse effect need to be carried out.

With regard to climate impacts, if large OCS increases were to materialize from fuel consumption, there would be an enormous attendant increase in the emissions of other combustion products such as CO_2, CO, and SO_2; the related environmental and climatic problems might overshadow those caused by carbonyl sulfide.

VI. CONCLUDING REMARKS

By a study of the atmospheric OCS budget, the 6 tg of OCS observed in the atmosphere appears to be consistent with a global OCS source of about 6 tg/yr and an average OCS lifetime of about 1 year, both of the latter quantities being derived from independent data. Because the current global source of OCS is small in comparison with the total amount of sulfur released by man to the environment, and because modern reliance

on fossil fuels is likely to grow in the future, it is possible
that substantial increases in OCS emissions could occur by the
turn of the next century. Based on model calculations, it was
found that a tenfold increase in OCS concentrations might
double the large aerosol particle concentration of the stratosphere and lead to a reduction in the average global surface
temperature of 0.1 K, which is climatically significant.

Several factors, not considered in detail in this work,
could alter this result. Projected increases in atmospheric
OCS abundances are quite uncertain because the amounts of OCS
generated during fuel processing and combustion are still not
precisely known and could be easily controlled; sulfur recovery
technology has, in fact, been making rapid advancements.
Equally important, the competitive natural sources and sinks
of OCS, which limit the effect of anthropogenic emissions,
are still in considerable doubt. An OCS greenhouse effect might
also cause some compensating warming as OCS levels rise. Finally
more serious environmental implications of expanding rates of
fuel combustion may involve the emissions of species such as
CO and CO_2. Additional research is obviously needed before an
accurate assessment of the total environmental impact of man's
growing energy consumption can be made.

REFERENCES

1. Crutzen, P. J., *Geophys. Res. Lett. 3*, 73 (1976).

2. Pollack, J. B., Toon, O. B., Summers, A., Van Camp, W.,
    and Baldwin, B., *J. Appl. Meteor. 15*, 247 (1976).

3. Turco, R. P., Hamill, P., Toon, O. B., Whitten, R. C.,
    and Kiang, C. S., *J. Atmos. Sci. 36*, 699 (1979).

4. Toon, O. B., Turco, R. P., Hamill, P., Kiang, C. S., and
    Whitten, R. C., *J. Atmos. Sci. 36*, 718 (1979).

5. Stoiber, R. E., Leggett, D. C., Jenkins, R. F., Murrmann, R. P., and Rose, W. I., *Bull. Geol. Soc. Am. 82*, 2299 (1971).

6. Momot, J., *Bull. Mens. Soc. Linnéene Lyon, 33*, 326 (1964).

7. Friend, J. P., *in* "Chemistry of the Lower Atmosphere" (S. I. Rasool, ed.), p. 177, Plenum Press, New York (1973).

8. Elliot, L. F., and Travis, T. A., *Soil Sci. Soc. Amer. Proc. 37*, 700 (1973).

9. Bach, W., *Rev. Geophys. Space Phys. 14*, 429 (1976).

10. Sandalls, F. J., and Penkett, S. A., *Atm. Env. 11*, 197 (1977).

11. Lovelock, J. E., *Nature, 248*, 625 (1974).

12. Lovelock, J. E., Maggs, R. J., and Rasmussen, R. A., *Nature, 237*, 452 (1972).

13. Maroulis, P. J., and Bandy, A. R., *Science, 196*, 647 (1977).

14. Zinder, S. H., and Brock, T. D., *Nature, 273*, 226 (1978).

15. Kurylo, M. J., *Chem. Phys. Lett. 58*, 238 (1978).

16. Philen, D., Heaps, W., and Davis, D. D., *Trans. Am. Geophys. Union, 59*, 1079 (1978).

17. Peyton, T. O., Steele, R. V., and Mabey, W. R., Carbon Disulfide, Carbonyl Sulfide: Literature Review and Environmental Assessment, Stanford Research Institute Report 68-01-2940 (1976).

18. Austin, G. T., *Chem. Engineering, 54*, 125 (1974).

19. Guerin, M. R., *Analytical Lett. 4*, 751 (1971).

20. Ronkainen, P., *J. Inst. Brew. 79*, 200 (1973).

21. Sivela, S., and Sundman, V., *Archs. Microbiol. 103*, 303 (1975).

22. Watkins, J. S., Bottino, M. L., and Morisawa, M., "Our Geological Environment," p. 257, W. B. Saunders and Co., Philadelphia, Pennsylvania (1975).

23. Smith, H. M., Qualitative and Quantitative Aspects of Crude Oil Composition, pp. 7-21, Bull. 642, U.S. Bureau of Mines, Washington, D.C. (1968).

24. Cadle, S. H., and Mulawa, P. A., Sulfide Emissions from Catalyst-Equipped Cars, Technical Paper 780200, Soc. Automotive Eng., Inc. (1978).

25. Rotty, R. M., The Atmospheric CO_2 Consequences of Heavy Dependence on Coal, Institute for Energy Analysis Research Memorandum 77-27, Oak Ridge, Tennessee, Dec. 1977.

26. Ferm, R. J., *Chem. Rev. 57*, 621 (1957).

27. Lovelock, J. E., *Nature, 267*, 32 (1977).

28. Singh, H. B., *Geophys. Res. Lett. 4*, 101 (1977).

29. Campbell, M. J., Sheppard, J. C., and Au, B., *Geophys. Res. Lett. 6*, 175 (1979).

30. Inn, E. C. Y., Vedder, J. F., Tyson, B. J., and O'Hara, D., *Geophys. Res. Lett. 6*, 191 (1979).

31. Hanst, P. L., Speller, L. L., Watts, D. M., Spence, J. W., and Miller, M. F., *J. Air Poll. Control Assoc. 25*, 1220 (1975).

32. Sandalls, F. J., and Penkett, S. A., *Atmos. Env. 11*, 197 (1977).

33. Maroulis, P. J., Torres, A. L., and Bandy, A. R., *Geophys. Res. Lett. 4*, 510 (1977).

34. Torres, A. L., Maroulis, P. J., Goldberg, A. B., and Bandy, A. R., *Trans. Am. Geophys. Union, 59*, 1082 (1978).

35. Jaeschke, W., Schmitt, R., and Georgii, H. W., *Geophys. Res. Lett. 3*, 517 (1976).

36. Maroulis, P. J., Torres, A. L., Goldberg, A. B., and Bandy, A. R., *Trans. Am. Geophys. Union, 59*, 1081 (1978).

37. Atkinson, R., Perry, R. A., and Pitts, J. N., *Chem. Phys. Lett. 54*, 14 (1978).

38. Toon, O. B., and Pollack, J. B., *J. Appl. Meteor. 15*, 225 (1976).

DISCUSSION

McElroy: I take it, although I didn't follow the details of your chemistry, that your calculations essentially confirm the results of Sze and Ko on the COS-CS$_2$ connection?

Turco: With regard to COS, our calculations seem to support Paul Crutzen's idea that carbonyl sulfide is an important contributor to the stratospheric sulfate aerosol layer. Questions still remain, of course, about the global cycle of carbonyl sulfide, including its sources and its tropospheric lifetime. Nevertheless, Crutzen's fundamental premise, that carbonyl sulfide is an important precursor of background stratospheric aerosols, appears to be sound. With regard to the importance of CS$_2$ as a precursor of COS, it is quite difficult to reach a firm conclusion because there is only one highly uncertain measurement of carbon disulfide, and the coupling chemistry is still controversial. As we point out in our paper, CS$_2$ could be a large, if not the dominant, source of atmospheric COS. Another point which is discussed in the paper is that the reaction of COS with OH is a source of oxidized sulfur in the lower stratosphere which competes with other sources such as SO$_2$ diffusing upward from the troposphere, and direct CS$_2$ decomposition in the lower stratosphere.

McElroy: The other was a question. On one of your charts, you had a list of the various sources of CS$_2$ on which you gave a summary remark. Would you like to break those source strengths down into organic decomposition? Would you break them down somewhat?

Turco: The source breakdown is written up in the paper in some detail. Take, for example, the possible source due to organic decomposition. A few researchers have estimated that 50 tera-grams (tg) per year of H$_2$S can be produced by natural decom-position under anaerobic conditions. Thus, there is a strong possibility of similar sources of COS, CS$_2$ and other reduced sulfur compounds. Of course, dimethyl-sulfide has been measured in lake and ocean waters, as has carbon disulfide. It has been pointed out that the amount of carbon disulfide measured in the Atlantic ocean by Lovelock (1974) appears to be roughly in equilibrium with the amount measured in the atmosphere by Sandalls and Penkett (1977). Therefore, it is difficult to say whether the oceans are a net source or sink of carbon disulfide. In relation to most of the possible sources of carbonyl sulfide and carbon disulfide, the data is not available with which to make accurate quantitative estimates.

McElroy: But the various numbers that you had particularly
concerning your reference to cattle--was that a guess or was
that actual data?

Turco: We haven't put firm numbers on individual sources, but
rather have given a range of values which we feel brackets the
total source. For example, I pointed out that several parts per
million of COS have been detected over cattle manure. However,
the total quantity released on a global scale would be extremely
difficult to calculate. Our estimate of a total COS source of
1-10 tg/yr is partly based on the fact that COS has a lifetime
of about 1 year while the amount in the atmosphere is about 6 tg;
therefore, we know roughly what the total source must be. On
the other hand, we can make some rough estimates of individual
sources. For example, the overall combustion source of sulfur
is nearly 50 tg S per year. It is not unreasonable to assume
that 1% to 5% of the sulfur is emitted as carbonyl sulfide.
There have been detailed studies of OCS emissions which support
this viewpoint. Peyton *et al.* reviewed the industrial sources
of carbon disulfide and carbonyl sulfide and concluded that for
certain industrial processes which utilize reducing catalysts
as much as 6% to 10% of the sulfur in the fuel could be emitted
as carbonyl sulfide.[1] In other words, there is likely to be a
substantial combustion-related source of carbonyl sulfide.

Ko: I would like to make a remark concerning the amount of
sulfur that could possibly come out in a coal gasification
process. Just based on a thermodynamic equilibrium calculation,
if you gasify coal with no cleanup whatsoever, you get about
3% of sulfur coming out as COS. And if you clean it up, you
get a lot less. So I think the estimate of 1% to 5% coming out
of COS is really way too high.

Turco: As I mentioned, Peyton *et al.* have found cases involving
reducing catalysts where you often get quite a bit more.
Actually, I stated that 1% to 5% of the emitted sulfur might be
COS. During coal gasification, most of the sulfur is in a
highly reduced state and COS can comprise 10% or more of the
total sulfur in the coal processed.

Ko: But I think the 1% to 5% number quoted by you corresponds to
the case where there is no clean up of the tail gas. Since
most of it is cleaned up, I think the number should be a lot
less.

---

[1] *Peyton, T. O., Steele, R. V., and Mabey, W. R., "Carbon
Disulfide Carbonyl Sulfide: Literature Review and Environmental
Assessment," Stanford Research Institute Report 68-01-2940 (1976).*

Turco: Yes. In fact, in our paper, we conclude that, conservatively, perhaps 1% of the total sulfur in coal and oil fuels could be emitted as COS. The actual amount released depends upon the gasification and clean up processes employed, as you point out.

Sze: I have a couple of questions. First, is the reaction COS + OH included in your model? In your presentation?

Turco: It is included in the present model, but it was not included in these calculations.

Sze: Second, what did you use for the SO_2 boundary condition?

Turco: SO_2 was set to about 0.4 parts per billion by volume at the ground, and was given a tropospheric rainout lifetime of several days. In other words, it was treated in the typical way that tropospheric species such as HNO_3 and H_2O_2 are treated in one-dimensional models.

Sze: What is the scientific basis of choosing 0.4 ppb?

Turco: The predicted SO_2 profile that I showed agrees fairly well with the Jaeschke *et al.* (1976) measurements in the upper troposphere.

Sze: That is exactly my point. That you are actually adjusting your boundary condition. Your solution is driven by a boundary condition. You pick 0.4 ppb in order to agree with the SO_2 data. However, the SO_2 burden according to Jim Friend and according to Mérzáros suggests an SO_2 burden of about 0.2 megatons. If you transform it to a mixing ratio, it is about 0.1 ppb or maybe 0.2 ppb. So you are using a mixing ratio about four times as large. You can always choose the boundary condition to agree with anything. Okay, your 0.4 ppb boundary condition is not consistent with the existing SO_2 data on a global basis.

Turco: As I said, SO_2 at the surface is 0.4 parts per billion. In addition, the SO_2 concentration near 2 km is close to the measurements made during GAMETAG as well as agreeing with the Jaeschke measurements in the upper troposphere. I should also point out that the SO_2 concentration at the ground has little relevance to our stratospheric calculations, as long as the amounts in the upper troposphere are correct, because tropospheric SO_2 does not contribute significantly to the stratospheric aerosols in the calculations that we made.

Sze: The other thing is, as you also notice, your calculated SO_2 mixing ratio dropped much faster than the observed data. So you do not explain the shape of your SO_2 data.

Turco: We were not trying to explain the shape of the SO_2 profile in the troposphere; that would be a difficult problem at this point, as the GAMETAG data indicates large SO_2 variability, both locally in time and with geographical location. Thus, it does not appear to be a problem that can be solved easily with a one-dimensional model. If you are referring to the steep decrease of SO_2 in the lower stratosphere, this depends on several factors: the vertical transport rate near the tropopause level, the SO_2 lifetime in the lower stratosphere, and the presence of local SO_2 sources such as CS_2 decomposition. The problem is obviously very complicated and requires further study. Interestingly, Jaeschke *et al.* concluded that their SO_2 data showing nearly uniform SO_2 mixing in the lower stratosphere were influenced by a tropopause folding event.

LONG RANGE TRANSMISSION OF COAL COMBUSTION
EMISSIONS AND THEIR OPTICAL EFFECTS[1]

R. B. Husar
D. E. Patterson

Department of Mechanical Engineering
Washington University
St. Louis, Missouri

Coal consumption in the eastern United States, especially by
electric utilities, has undergone strong regional and seasonal
increases in recent decades. The associated SO_x emissions
partially convert to aerosol sulphate with residence time of 3-
to 5-days and an impact scale of 1,000 km. Ambient SO_4 concentrations display geographic pattern and recent trends which
coincide with regions of high coal combustion. The optical
effect of coal-burning plumes is increased haziness.
Large-scale hazy air masses formed from many such plumes contain
elevated concentrations of sulfate and ozone which may impact
upon normally clean regions of the eastern United States. The
regional trends of haziness and turbidity are consistent with
corresponding coal consumption trends, with evidence of at least
two times higher sulfate conversion during summer than winter.

[1]This research has been supported by the Federal Interagency

Energy/Environment Research and Development Program through

U.S. Environmental Protection Agency Grant No. R803896. The

work was conducted as part of projects MISTT (Midweat Interstate

Sulfur Transformation and Transport) and STATE (Sulfur

Transformation and Transport in the Environment) with the close

cooperation of Dr. William E. Wilson, Jr., Project Officer.

I. INTRODUCTION

Coal consumption in the eastern U.S., especially by electric
utilities, has undergone strong regional and seasonal increases
in recent decades. The associated SO_x emissions partially convert to aerosol sulfate, which has a residence time of 3- to
5-days and an impact scale of 1,000 km. Ambient SO_4 concentrations display a geographic pattern which coincide with regions
of high coal combustion.

The optical effect of coal-burning plumes is increased
haziness. Large-scale hazy air masses formed from many such
plumes contain elevated concentrations of sulfate which may
impact upon normally clean regions of the eastern U.S. The
regional trends of haziness and turbidity are consistent with
the corresponding coal consumption trends.

II. DISCUSSION

Man-made air pollutants emitted over the eastern U.S. are
primarily due to combustion of the fossil fuels: coal, oil
products and gas. The trend in coal combustion indicates a
shift from heaviest demand in the winter to a more pronounced
summer peak (Fig. 1 taken from Ref. 1).

More instructive is the examination of the state-by-state
spatial trend of yearly coal consumption data available since
1957 (Fig. 2).

In the Ohio River region (Indiana, Michigan, Ohio,
Pennsylvania) the coal consumption was high in the late 1950's
and the early 1960's and has further increased since 1960.
New England coal consumption has declined since about 1965,
and reflects the fuel shift from coal to oil. This was also
the case for the northeast megalopolis states of New York,
Maryland, and Virginia. The strongest increase in coal

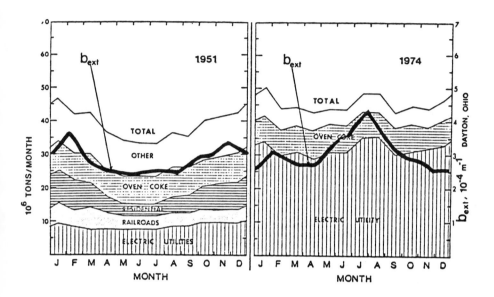

FIGURE 1.   In the 1950's, the seasonal U.S. coal consumption
peaked in the winter primarily due to the increased residential
and railroad use.   By 1974, the seasonal pattern of coal
usage was determined by the winter and summer peak of utility
coal usage.   The shift away from a winter peak toward a summer
peak of coal consumption is consistent with the shift in
haziness from a winter peak to a summer peak at Dayton, Ohio, for
1948-1952 and 1970-1974.   (Data from U.S. Bureau of Mines,
Mineral Yearbooks 1933-1974.)

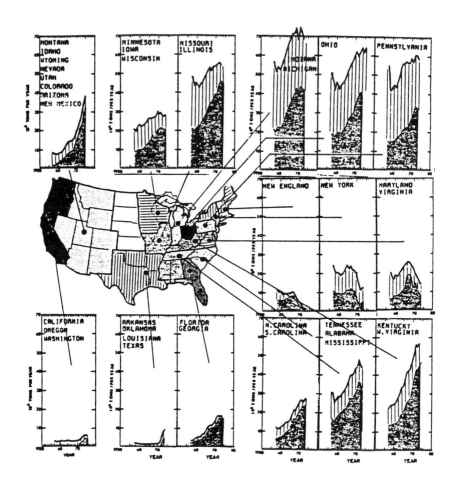

FIGURE 2. U.S. coal consumption.

consumption was recorded for the states of Kentucky, West
Virginia, North and South Carolina, and Tennessee, defined as
the Smoky Mountain region. In that region, essentially all
of the growth was due to electric utility coal (dark shading,
Fig. 2). In the Midwest region, Missouri and Illinois, there
has been a moderate increase, again due to the rising demand
for utility coal. In the West (Montana, Idaho, Wyoming,
Nevada, Utah, Colorado, Arizona, New Mexico), coal demand has
grown rapidly and is now becoming appreciable. The Pacific
coast states have negligible coal use.

With current particulate control equipment, the optical
effects of coal combustion emissions beyond the first few tens
of kilometers are determined by fine (submicron size)
particulates. Of the fine particles, sulfate appears to be the
most important species (see, for example, Refs. 2 to 4). The
coarse primary particulates normally settle out from the lower
atmosphere within minutes or hours; their major removal
mechanism is wet deposition by precipitation. The residence
time of secondary particles is the sum of the time required to
create them and the subsequent time for removal from the
atmosphere. The flow diagram of sulfur transmission through
the atmosphere (Fig. 3) illustrates, for example, that most
of the SO_2 emission is removed or transformed to sulfate within
the first day of atmospheric residence, but the atmospheric
burden of the secondary sulfate may remain relatively high for
several days of resident time.

An example of the optical properties of power plant plumes
is illustrated in Fig. 4. The optical properties of the
Labadle plume in St. Louis were sampled at 0600, 0930, and
1430. The figures represent the measured optical depth,
τ (integral of plume excess light scattering coefficient
measured by the nephelometer), during horizontal traverses
across the plume. In the morning, lack of dispersion yelds an

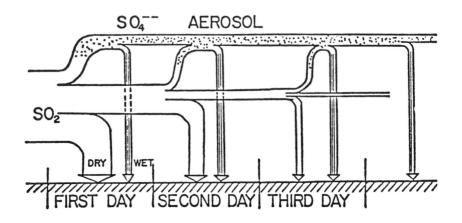

FIGURE 3.   *Flow diagram of sulfur transmission through the*
atmosphere.   Over half of the SO_2 *is removed or transformed to*
sulfate within the first day of its atmospheric residence.   Most
of the sulfate may remain for several days of residence in the
atmosphere.

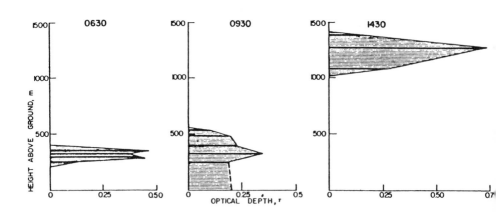

FIGURE 4.   *Horizontal integrals of plume excess b*$_{scat}$
(optical thickness τ*) at 20 km obtained 14 July 1976.   At 0630,*
0930, and 1430 CDT.   The heavy horizontal lines show the height
of traverses: the contours of the vertical profile of τ *are*
obtained by interpolation and extrapolation of the data.

optical depth of about $\tau = 0.4$, but the vertical plume extent
was only about 200 meters. By 0930 the plume had been diluted
somewhat, such that its optical depth was only about 0.25.
Also, the gradient at the plume top was not so sharp as in the
early morning plume. Hence, its contrast to the sky was less
sharp. By 1430 the horizontal plume excess optical depth had
increased to $\tau = 0.75$ and the plume had reestablished itself as
a coherent elevated layer. The overall increase of light
scattering aerosol concentration at 1430 is attributed to
secondary aerosol formation at the 20-km distance from the
stack.

   It is more intriguing to examine the long-range effects of
the Labadie power plant plume. Figure 5 illustrates the plume
geometry and the measured sulfur dioxide concentrations
attributed to the Labadie plume for two long range sampling
days July 9 and July 18, 1976. On both days the plumes were
tracked to about 300 km from the source. The optical properties
of the aging power plant plume are shown in Fig. 6. The bottom
figure for each date indicates the cross-plume optical depth at
different distances away from the source, sampled at the
indicated altitudes. On both dates the crosswind plume optical
depth at around 50 km distance from the source, sampled during
midday, was about $\tau = 0.5$.

   As the plume aging and light scattering aerosol formation
continued over the day, the horizontal optical depth had
increased to about $\tau = 2$ on July 9 and to $\tau = 3$ on July 18,
1976. The fivefold increase in the horizontal depth at 150 km
to 200 km distance from the source is attributed to the formation
of secondary sulfates in the plume. The emphasis here is that
most of the light scattering aerosol has been formed 100 or
more kilometers from the source.

   The International Symposium on Sulfur in the Atmosphere
concluded that the typical atmospheric residence time of sulfate

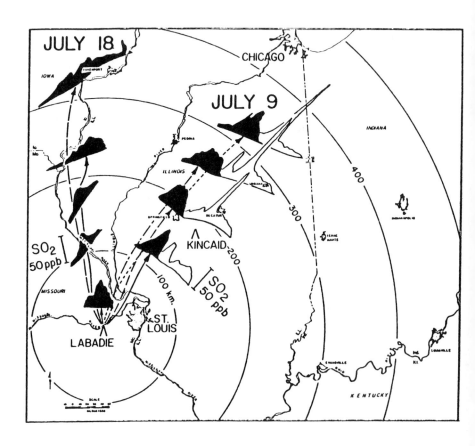

FIGURE 5. *Horizontal profiles of* SO_2 *during selected constant-altitude aircraft traverses on July 9 and July 18, 1976. July 9 traverses are at about 450 m AGL, and July 18 traverses are at about 750 m AGL. The Labadie plume sections are shaded. Also shown are backward trajectories for the Labadie plume (5).*

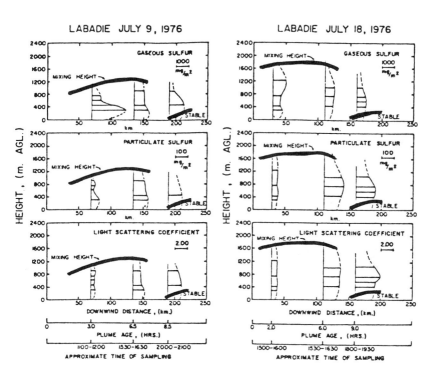

FIGURE 6. Vertical profiles of the crosswind integrals of excess plume concentrations of gaseous sulfur S_0, particulate sulfur S_p, and light scattering coefficient b_{scat} for the transport of the Labadie plume of July 9 and 18, 1976. Broken portions of the vertical profiles represent extrapolations (5).

aerosol is on the order of 3- to 5-days, depending on precipitation patterns (6). Mean winds of 500 km per day are not uncommon in the eastern U.S., so that a reasonable scale of sulfate impact from an individual SO_2 source is on the order of 1000 km to 2000 km (7). The implication is that ambient light scattering sulfate aerosol can result from the additive effects of tens or even hundreds of SO_2 plumes of different atmospheric ages. Long range transport of fine aerosol can cause

degradation of visual air quality on a multi-state regional
scale. Altshuller noted the anomaly of decreasing urban SO_2
concentrations with increasing rural sulfate trends as evidence
for long range transport as the cause of regional sulfate
levels (8).

Contours of yearly average sulfate concentration reveal
that the maximum concentrations occur in the region of high
SO_2 emission density and of highest coal consumption. In the
Ohio River Valley, for instance, the yearly average sulfate
concentration exceeds 15 $\mu g/m^3$; it is between 10- and 15-$\mu g/m^3$
over a large part of the eastern U.S. (Fig. 7). The seasonal
pattern displays a greater geographical extent of the high
sulfate concentration in the summer than in the winter months
(Fig. 8).

FIGURE 7. Yearly average sulfate concentrations.

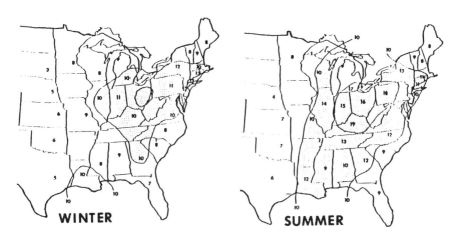

FIGURE 8. Seasonal sulfate concentration patterns.

Since the emission of precursor gases remains relatively
constant from day to day, the most important factor governing
sulfate concentration is the regional scale dilution of the
emissions. Therefore, stagnation of a regional scale air mass
can cause high pollutant concentrations and extreme haziness.
Such episodes are key contributors to the personal discomfort
and the annual average sulfate and haziness. An example of
one such episode over a 2-week period in June–July, 1975, is
presented in Fig. 9. Inspection of the sequence of contour
maps reveals that multistate regions are covered by a haze
layer in which noon visibility is less than 10 km (b_{ext} > 4 x
$10^{-4}m^{-1}$, outer contours).

The air mass of June 25, 1975, defined by the shaded region
of visibility 10 km, was of maritime origin in the Gulf of
Mexico and had traversed LA, IL, and IN. Between June 25 and
27, relative stagnation prevailed in the Great Lakes region,
and lead to increasing haziness in this high emission density
area. The tropical storm "Amy" moving up off the northeast

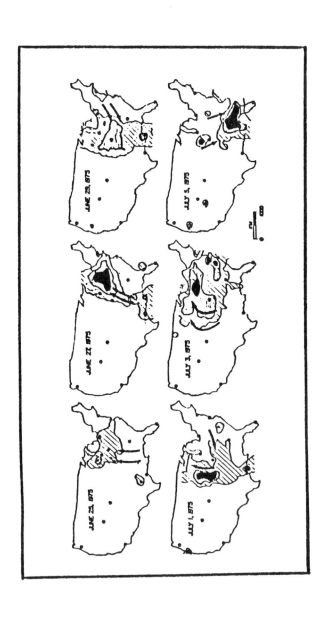

FIGURE 9. Sequential contour maps of noon visibility for June 25–July 5, 1975 illustrates the evolution and transport of a large-scale hazy air mass. Contours correspond to visual range 6.5 km to 10 km (light shade), 5 km to 6.5 km (medium shade), and 5 km (black).

coast blocked the eastern motion of the hazy air mass as a
front approached the Great Lakes from the northwest. Between
June 28 and June 30 an easterly flow developed which caused the
hazy air mass to drift slowly westward passing over St Louis,
Missouri, on June 28 and June 29 and continuing across
Missouri and Kansas by June 30. From June 30 to July 2 the
haze moved in a clockwise pattern up and around to the southern
Great Lakes. By July 3, the advancing Canadian front formed
the northern border of a massive hazy air mass occupying most
of the midwestern and northeastern U.S. The hazy air mass
again passed over St. Louis at this time. During July 3 to
July 5 the cold front advanced rapidly to the south, pushing
the heavily polluted air mass across the southeast and out to
sea.

The apparent motion of the haze was confirmed by geostation-
ary satellite photographs (9). The region of haziness on
July 30 may be clearly seen where the turbid air mass is visible
over the states of Arkansas, Missouri, Kansas, Iowa, and
Minnesota.

Two passages of the hazy air mass over St. Louis, Missouri,
resulted in sharp increases of b_{scat} over the entire metropolitan region (Fig. 10a). Sulfate concentration also increased
during the haze episode, from about 9- to 33- g/m^3. The
spatial coherency of the haziness is seen in the correspondence
of the extinction coefficient at St. Louis, Missouri, and
Springfield, Illinois (Fig. 10b), again confirming that the
observed haziness was primarily due to inflow of the polluted
"background" material of the hazy air mass rather than to local
contributions.

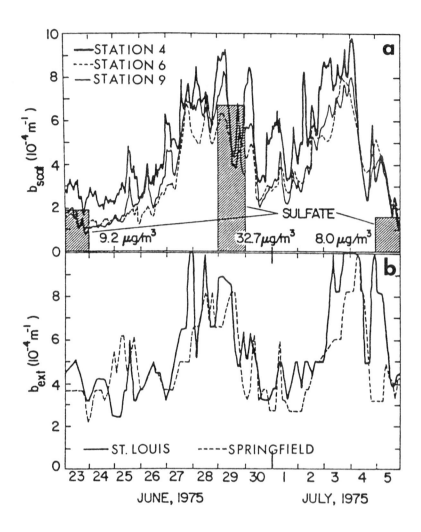

FIGURE 10. Local monitoring data in the St. Louis, Missouri, area during the June-July 1975 haziness episode. (a) Light scattering coefficient (b_{scat}) recorded at three widely spaced locations in the St. Louis metropolitan area and daily average sulfate concentrations. (b) Extinction coefficients (b_{scat}) obtained from visibility observations at St. Louis, Missouri, and Springfield, Illinois, 150 km apart.

Sufficient sulfate data was available from the National
Aerometric Data Bank for comparison with visibility on 2-days
during the episode period. Figure 11 indicates the substantial
correspondence of the regions of highest sulfate and lowest
visibility.

Husar, Patterson, Paley, and Gillani reported that in June-
August 1975 there were at least six episodes similar to the one
cited (10). The work of other investigators confirms that
episodes of regional scale hazy air masses are not rare in the
U.S. Yet, at present, only the qualitative features of such
episodes are understood--the observed effect on visibility, the
composition in terms of secondary sulfate and ozone, and the
apparent motion of the haze.

Important questions remain to be answered about regional
scale episodes of haziness, including the following:

1. Quantifying the effects of superimposing multiple SO_2
plumes and urban reactive plumes.

2. Quantifying the actual residence time of fine particulates in the atmosphere during such episodes; it may be, for
example, that lack of precipitation leads to extremely long
sulfate lifetime.

Evidence of the increasing importance of long-range transport of pollutants in causing local visibility reduction may be
extracted from historical observations taken by National Weather
Service sites over the past three decades (Figs. 12a and 12b).

The spatial-temporal trend in visibility over the eastern
U.S. (as average extinction coefficient) is illustrated in
Fig. 13. The shift from a winter peak toward a summer peak in
coal combustion is reflected in the worsening of summer
visibilities. The Smoky Mountain states, which have undergone
the strongest rise in coal use, have also exhibited the strongest
increase in average haziness.

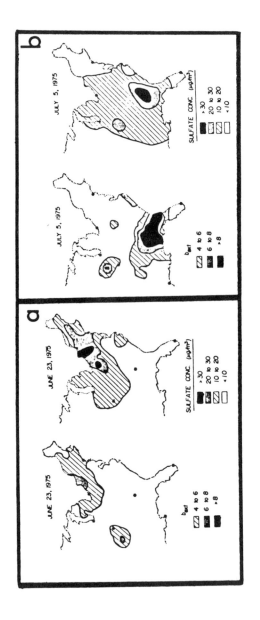

FIGURE 11. Comparison of noon extinction coefficient and daily mean sulfate concentration on June 23 and July 5, 1975. The regions of highest sulfate concentrations coincide with areas of lowest visibility.

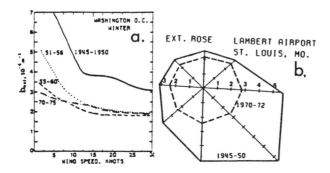

FIGURE 12. Wind speed influences the dispersion by increasing the ventilation. If the haziness is due to local sources,
low wind speeds are associated with higher extinction levels.
Wind direction dependence of air pollutants has been used
extensively in the past to construct "pollution roses" as
directional pointers to local sources. (a) In the winters of
1945-1950 at Washington, D. C., high extinctions were associated
with low wind speeds. By the winters of 1970-1975 the level
haziness declined and the wind speed dependence has essentially
vanished. The obvious interpretation is that the winter haziness in 1970-1975 is due to remote rather than local sources.
(b) A haziness rose for Lambert Airport in St. Louis, Missouri,
shows that in 1945-1950 the highest extinction levels were
observed when the wind was from the city direction; by 1970-
1972 the haziness is essentially independent of wind direction.
Here again, the conclusion is drawn that the haziness is
currently due to distant rather than local sources.

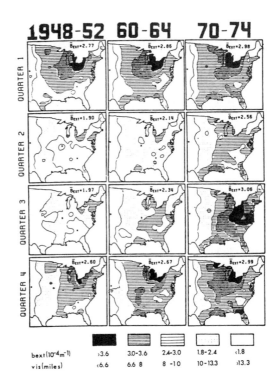

FIGURE 13.   The spatial distribution of a 5-year average
extinction coefficients show the drastic increases of quarter
3 extinction coefficients in the Carolinas, Ohio River Valley,
and Tennessee-Kentucky area.   In the summers of 1948-1952, a
1000 km size multistate region centered around Atlanta, Georgia,
had visibility greater than 15 miles, which was declined to
less than 8 miles by the 1970's.   The spatial trend of winter
(quarter 1) visibility shows improvements in the northeast
megalopolis region and some worsening in the sunbelt region.
Both spring and fall quarters exhibit moderate but detectable
increase over the entire eastern U.S.

The comparison of the eastern U.S. summer coal consumption
and summer average extinction over the entire eastern U.S. is
shown in Fig. 14. Although statistically a high correlation
could be extracted from the coal consumption-haze trend data,
a cause-effect relationship cannot be established from the trend
analysis alone.

Two independent measures of the relation of sulfate to
degradation of the optical environment may be obtained by
examination of the recent behavior of turbidity and solar
radiation. The seasonal patterns of sulfate and turbidity
(Fig. 15) are consistent, given the increased mixing depths
during summer months. Evidently, the sulfate burden within the
mixed layer is 2- to 3-times higher in summer than in winter.
The trend in turbidity over the eastern U.S. indicates a
pronounced summer increase, as was noted for extinction
coefficient (Fig. 16).

Angell and Korshover also reported substantial decreases in
the total hours of solar radiation over the eastern U.S.
(Fig. 17) which could be due to haze "delaying sunrise" and
"advancing sunset" (from the point of view of an on-off solar
radiation detector) (11). Most of these observed anomalies
could be due to an increase of daytime cloudiness, which may or
may not be related to human activities. The robust nature of
the planetary atmosphere, with its numerous apparent and hidden
feedback mechanisms, also provides a warning. Hence, prudence
dictates that with our current experience and capability, no
conclusion is claimed as to whether the observed climatological
changes over the east central U.S. are primarily due to an
increase in anthropogenic haziness.

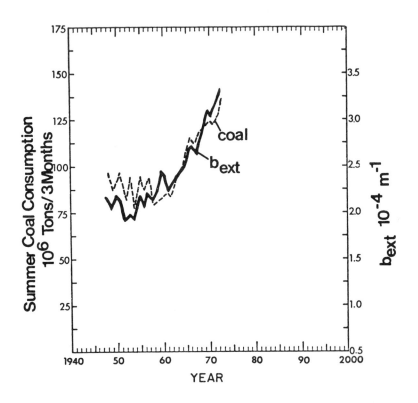

FIGURE 14.   Summer trends of U.S. coal consumption (dashed
line) and eastern U.S. average extinction coefficient, or
haziness (solid line).

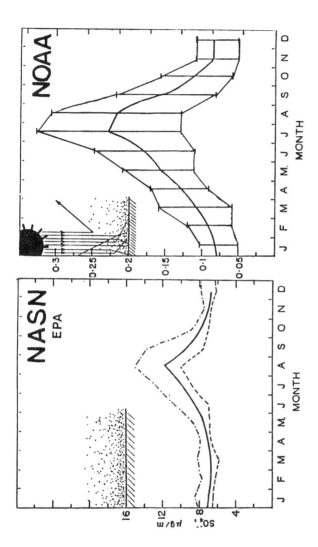

FIGURE 15. (a) Monthly average sulfate concentration ($\mu g/m^3$) 1970-1974 for 18 eastern nonurban sites (———). The seasonal pattern at six stations in the industrialized northeast (— —) is more pronounced than the seasonal pattern of 12 peripheral stations in the midwest and southeast (– – –). (b) The seasonal pattern of turbidity for 26 eastern U.S. sites.

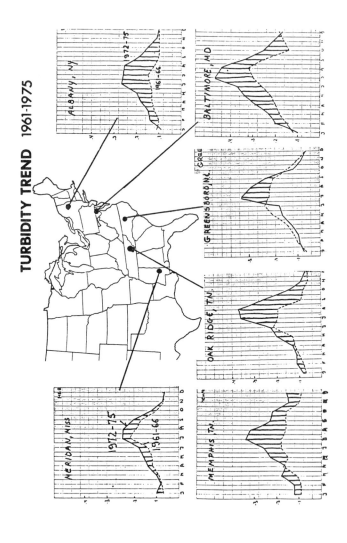

FIGURE 16. The seasonal turbidity trend for eastern U.S. stations is illustrated by comparing the patterns for the period 1961-1966 to 1972-1975. Current summer turbidities exceed the 1961-1966 values by as much as 50%.

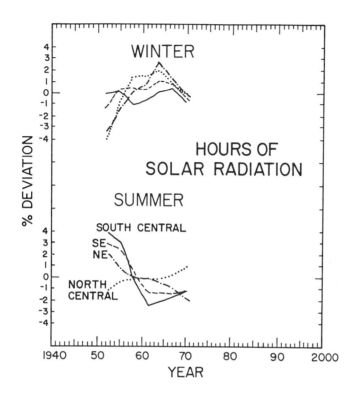

FIGURE 17. The analysis of Angell and Korshover of the hours of solar radiation shows an increase of winter radiation since the 1950's and a decrease of summer solar radiation over the eastern U.S. There may be several causes for this trend, including an increase of cloudiness; some of the change may be due to haze (11).

III. CONCLUDING REMARKS

In summary, the optical effect of coal-fired power plant
plume has been measured at 200 km downwind of the stack, where
the cross plume optical depth was $\tau > 2$. An episode of extreme
haze covering multistate regions of the eastern U.S. appears to
have been largely due to secondary sulfate aerosol. Finally,
the spatial and seasonal trends of coal combustion, sulfate,
light extinction coefficient, turbidity, and solar radiation
all exhibit reasonably consistent trends.

The purpose of this paper was to demonstrate that there were
regional changes in the pattern of coal demand. For some regions
these trends were consistent with the trend of regional extinction coefficient and other parameters. It is also intended to
point out that for proper trend analysis, the spatial scale of
emission pattern has to be matched to the atmospheric transport
distance of visibility-reducing aerosols (on the order of 1000 km
to 2000 km). Finally, it must be emphasized that coal combustion
is just one of the fuels that result in the formation of visibility reducing aerosols. Until the appropriate study of the
pattern for the consumption of gasoline and other oil products
is incorporated, such a cause and effect analysis is incomplete
and prohibits any conclusions being made.

REFERENCES

1. U.S. Bureau of Mines, "Minerals Yearbooks, 1933-1974."
    U.S. Department of Interior, Washington, D.C. (1975).

2. Dzubay, T. G., Chemical Element Balance Method Applied to
    Dichotomous Sampler Data, *in* "Proc. Symp. on Aerosols:
    Anthropogenic and Natural Sources and Transport." New York
    Acad. of Sci., New York (1979).

3. Lewis, C. W., and Macias, E. S., Composition of Size-Fractioned Aerosol in Charleston, West Virginia, *in* *Atmos. Environment* (1979).

4. Trijonis, J., and Yaun, K. "Visibility in the Northeast: Long Term Visibility Trends and Visibility/Pollutant Relationships." EPA-600/3-78-075, Washington, D.C (1978).

5. Gillani, N. V., *Atmos. Environment, 12,* 569-588 (1978).

6. Husar, R. B., Lodge, J. P., Jr., and Moore, D. J. (eds.), "Sulfur in the Atmosphere." Pergamon Press, Oxford, England (1978).

7. Blumenthal, D. L., Ogren, J. A., and Anderson, J. A., *Atmos. Environment, 12,* 613-620 (1978).

8. Altshuller, A. P., *J. Air Pollut. Control Assoc. 26,* 318-324 (1976).

9. Lyons, W. A., and Husar, R. B., *Monthly Weath. Rev. 103,* 1623-1626 (1976).

10. Husar, R. B., Patterson, D. E., Paley, C. C., and Gillani, N. V., Ozone in Hazy Air Masses. Paper presented at the International Conference on Photochemical Oxidant and Its Control, Raleigh, North Carolina, September 12-17 (1976).

11. Angell, J. K., and Korshouer, J. *J. Appl. Meteor. 14,* 1174-1181 (1975).

DISCUSSION

Green: What was your source of the historic data going back to 1948? How do you get these numbers?

Patterson: The National Weather Service, primarily at airport sites, has been collecting visibility observations along with wind speed, temperature, and a number of meteorological variables. The extinction coefficient we take simply by the Koschmeider relationship and work with that. There are a number of questions about how valid that is. But it is the only index available for haziness that goes back on that kind of a time scale.

Green: Did you ever correlate with the turbidity network of McCormick?

Patterson: Well, we haven't really done it site by site. Some other people have been doing that. We have obtained turbidity information that was collected by Argonne. The turbidity information is pretty sparse. One thing we are finding also is that when you are looking at things on the scale of the eastern US geographic average, you come up with excellent correlations with sulfate. We were doing this with the August 1977 SURE data and our visibility data. If you are looking at the geographic pattern, at least on the size of our regions (the Smokey Mountain region, the Ohio River Valley) the correlation is very good. If you look at individual sites, sometimes it is quite poor. We don't really understand what is going on in that case.

Hamill: You mentioned some studies on individual plumes, and I was reminded about the work that Komp and Auer[1] did on the St. Louis urban plume, in which they saw that the visibility was worse at a point several hours downwind of St. Louis; however, the number of particles peaked much earlier. In other words, the peak in particle number was not the same as the peak in poor visibility. I was wondering if you have any comment on that? Have you done any studies relating particle number density with visibility?

Patterson: Well, I don't have any such data with me, but that's essentially true. The condensation nucleii count which we had on the aircraft clearly indicates that the particle number peaks tend to show up more in the fresh plumes. The b_{scat} and the aerosol charge, which is another aerosol measurement we had, tend to build up later on. The number esentially disappears when you get out at 300 km, downwind for instance. There is really a very low number concentration.

[1]*Komp, M. J. and Auer, A. H., J. Appl. Meteorol. 17, 1352 (1978).*

Hamill: Do you have any notion on the size range when the visibility gets bad, when you have great degradation of visibility?

Patterson: Well, no. I have some ideas; I have no data.

Singh: Have you thought of what effects the increased use of Eastern coal may have on the visibility? Has the high sulfur content of the Eastern coal shown any impact on the visibility pattern recently?

Patterson: At the moment, we are trying to build up our past analysis. We are at about 1975 right now in terms of looking at things in detail, so actually all of the work that we are talking about is still in progress. Haven't seen the effect yet. We did see that, looking at the eastern US index, there does seem to be a drop right around 1974. The average haziness levels went down slightly, which corresponded to a very slight decrease in the coal use at the same time. But we haven't seen more recent data yet.

Kornegay: If you look at the EPA trends data, they show a rather significant drop in SO_2 emissions throughout the early 70's. Your analysis of a slight decrease in turbidity around 1974 seems to correlate well with that. Do you anticipate, or will your work continue on through the '77 and '78 data that are available? Because if indeed there is a correlation, we should begin to show a real increase in visibility or decrease in atmospheric turbidity in your next three years of analysis.

Patterson: This work is continuing on a number of different levels. That's actually about all I can say right now.

Kornegay: One other thing that is interesting is that the Smokey Mountains' got their name because they do have and always have had a haziness problem. It is interesting that you show them to be the cleanest area in the country or in the East in the early 50's, because they have always been a very hazy area and that is how they got their name.

Patterson: Right, it was a little surprising, actually, to find that they did have an average extinction coefficient which was quite low in the summer months back then. But that's what the data says.

Beadle: Have you considered any analysis of the trends based on mobile source parameters, such as gasoline usage, automobile density, tourism?

Patterson: Yes, that's one of the reasons why I was trying to make a strong point about similar patterns in coal combustion and extinction coefficient not really meaning much at the moment.

From the plume data, it is obvious that mixing, for instance, the power plant plumes with urban plumes can lead to a very great increase in the conversion rate. So to try and establish the cause for these visibility trends, we have to look at least at the corresponding trends for hydrocarbons and their spatial arrangement, and presumably the whole answer is much more complex than that. But certainly the hydrocarbon trend, the mobile source and the urban area, we have to look at the contributions from them.

Pack: How do you handle the known variability--not only case by case but year to year in the macrometeorology? As you are well aware, some periods even for 5 to 10 years, will have a predominance of, say, the Bermuda high dominance, much more moist air, southwesterly trajectories, so on and so forth. How are you trying to take this into account because some of your trends might very well be explained away just by the meteorology alone and not by the emissions.

Patterson: Well, certainly the meteorology, the number of these extremely hazy situations, is going to have a lot of effect on an individual year or an individual season with only about 90 days in it. The spatial patterns that I showed were 5-year averages. That was chosen as an attempt to smear out the meteorology. Also the individual trend lines that were shown, although I didn't mention it, were 3-year moving averages. But that's really about all that we have been able to do at the moment. That's one of the reasons that I said we have to go back on a case by case basis to try to understand what's really going on. You can't just look at the trends and say there's an answer there.

Pack: Korshover now has published data through 1975 on the spatial temporal frequency of stagnation episodes. It would be extremely interesting to compare that time series to your frequency of low visibility episodes.

Patterson: Right. We were just at a meeting down in Durham, NC, the EPA PEPE Seminar (Persistent Elevated Pollution Episodes), and there are a lot of people working in this area right now. Dr. Gerald Watson from, North Carolina(NCSU at Raleigh), is doing precisely that, and so far he was indicating that the agreement isn't that good. But nevertheless, you can observe whether the trend in a given season comes from a relatively low level that's consistent or whether it comes from the episodes. That tells you essentially the same thing. You can look at the daily visibility isopleths and focus on the episodes and make some assumptions on that basis, as well as looking at more classical stagnation definitions. By the way, these episodes are not necessarily associated with low wind speeds. There can be reasonably high local wind speeds while they are building up as long as there is an anticyclonic situation in which the air mass is recirculating. Well, there's a major EPA project on several levels right now that is focusing on these same issues.

Deepak: You didn't mention about the terpenes in the Smokey Mountains. Is there any relationship?

Patterson: We really have no idea. But one of the surprising parts was simply the trend in the Smokey Mountains. If the terpenes are doing it, they may have been contributing a large part of what happened back in the earlier times. But unless you can establish that there has been a gigantic increase in terpene emissions in the past three decades, then presumably they don't have much effect.

CARBON DIOXIDE AND CLIMATE: COMPARISON
OF ONE-DIMENSIONAL AND THREE-
DIMENSIONAL MODELS

Ruth A. Reck

Physics Department
General Motors Research Laboratories
Warren, Michigan

*Calculations indicate that there may be a potential
climatic problem due to the release of CO_2 from burning fossil
fuels, but it is believed there is not sufficient scientific
evidence to predict the impact because the available models are
not well enough understood and do not take into account many
feedback mechanisms. Therefore, the atmospheric temperature
changes with latitude have been calculated with the simpler
Manabe-Wetherald radiative-convective model and compared with
the results of their general circulation model. In addition,
the changes in several physical quantities, such as cloud
abundance and surface albedo, required to compensate the
temperature rise related to increased carbon dioxide concentrations have been determined.*

I. INTRODUCTION

The present paper compares calculated results of the steady-
state temperature profile for a doubling of CO_2 by using the
Manabe-Wetherald one-dimensional radiative-convective atmospheric
model (RCA) including suspended particles with the previous
results of Manabe and Wetherald using their general circulation
model of the atmosphere. In 1975 they published their calculations on the effects of doubling the atmospheric CO_2

concentration on the climate of a general circulation model
(GCM) and undoubtedly this continues to be the most definitive
study on this subject (1,2). Their model includes not only
radiative and convective effects that are present in the one-
dimensional RCA model as detailed in Fig. 1, but also the transport of energy, momentum, and water vapor as well as a very
detailed hydrologic cycle including changes in rain, snow, and
ice. Because of the complexities of the three-dimensional
model, it is difficult to understand the detailed relationships
which lead to the final results.

The Manabe and Wetherald radiative-convective atmospheric
model has been used by the author for a number of years and
the radiative effects of suspended Mie scattering particles have
been added (3). Because the RCA model is much simpler in detail
in comparison with the GCM, the RCA model can help to sort out
the various mechanisms which may be represented by the models.
Hence, the purpose of this paper is to compare the temperature
changes predicted by the two models. A comparison of the GCM
and RCA models as used for this work is shown in Table I.

II. DESCRIPTION OF THE RCA MODEL

The RCA model has been used for several years (4). It
was developed over many years by workers at the Geophysical
Fluid Dynamics Laboratory at Princeton University and one version
was brought to its present form by Manabe and Wetherald (5).
It was modified by us to include the role of airborne particles.
In this RCA model, solar radiation is considered to be absorbed
by ozone, carbon dioxide, and water, and absorbed and backscattered by water clouds, airborne particles, and the earth's
surface (see Fig. 1). The ozone, carbon dioxide, water, water
clouds, airborne particles, and the earth's surface also
radiate and absorb infrared radiation. A temperature is

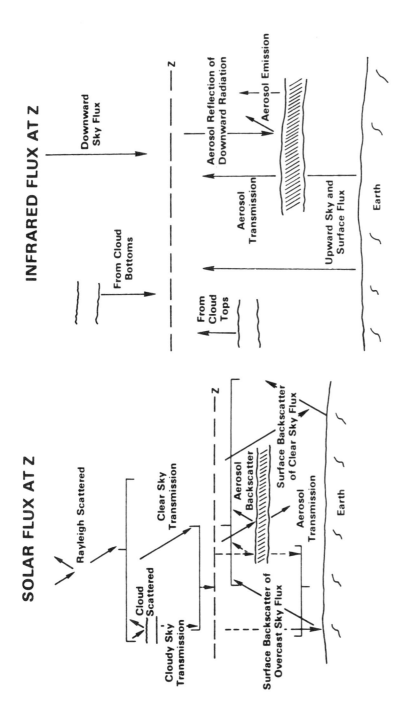

FIGURE 1. Typical up and down components of the solar and infrared fluxes.

TABLE I. *Comparison of the GCM and RCA Models for Computing*
Temperature as a Function of Latitude and Altitude.

Quantity Considered	GCM	RCA
General Circulation	Yes	No
Input Data	Yearly average in latitudinal bands	Latitudinal bands for April
Convection	Yes	Yes
Fixed relative humidity	No	Yes
Detailed hydrology-- water balance	Yes	No
Ice feedback (change ω_s)	Yes	No
Rain--snow	Yes	No
Variable clouds	No	No
Ocean heat transport	No	No
Advective heat transport	Yes	No
Number of CO_2 infrared bands considered	1	4
Airborne particles	No	Yes

initially assumed for each of nine vertically aligned points
in the atmosphere. If the temperature decrease with altitude
is greater than the critical adiabatic lapse rate, energy is
removed from the lower altitudes and placed higher up. A
forward time integration of the solar and infrared flux
imbalance is performed until a radiative-convective steady-state
temperature is asymtotically approached at each of the nine
points.

The water vapor content is allowed to vary with temperature
during the time integration so that the relative humidity
remains fixed. The particle layer was assumed to have an
extinction coefficient of 0.1 km^{-1} because this is in the
range of the average global value. It is 650 meters thick and
located between 0.1 km and 0.75 km.

III. DESCRIPTION OF THE MANABE-WETHERALD GENERAL CIRCULATION
      MODEL.

The three-dimensional GCM solves the primitive equations
requiring energy conservation on a Mercator projection.
Details of the calculation are given in the original paper (1).
The radiative portion is similar to that of the previously
described RCA model. The distribution of land and ocean is
idealized (Fig. 2), but the model does not contain a separate
ocean computation. It assumes instead an infinite source of
soil moisture. A detailed hydrologic cycle is introduced with
snow-albedo feedback. A diagram indicating the major components
included in the model is shown in Fig. 3. Airborne particles
were not included in the three-dimensional case and neither
model included interactive clouds.

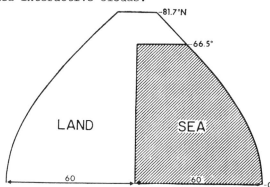

FIGURE 2. Diagram showing idealized land and sea distribution in the Manabe-Wetherald GCM (1).

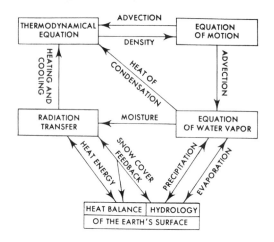

FIGURE 3. Major components in the Manabe-Wetherald GCM (1).

IV. PRESENT RCA CALCULATIONS

The latitudinal data for April (furnished by R. T. Wetherald) was used to calculate the present standard April temperature profiles. Next, within the RCA model, the abundance of atmospheric CO_2 was doubled (from 300 ppmv to 600 ppmv) and a new set of temperature profiles calculated. All RCA calculations assumed that the relative humidity is fixed, the standard background of particle loading that has been used in the previous work, and average cloudiness for April (6). The particle layer is assumed to exist everywhere in a layer between 991.1 mb pressure and 925.9 mb and has a solar energy optical thickness of 0.065 (equivalent to dust particle density of about 150 $\mu g/m^3$ with a mean visible index of refraction of n = 1.5 - 0.1i and a r^{-4} particle size distribution with the radius r between 0.1 μm and 10 μm; the ratio of particle extinction in the infrared to that in the visible is 0.108).

V.   RESULTS

Figure 4 shows at the top the global-average temperatures
calculated by Manabe and Wetherald with their GCM and at the
bottom temperatures for April calculated by the author with
the RCA (1). Figure 5 (top) shows the calculated temperature
difference for doubling CO_2 published by Manabe and Wetherald
for their GCM and at the bottom shows the comparable result
for the RCA model calculated by the author (7). Finally, in
Fig. 6 is shown the calculated mean temperature at the lowest
level (991 mb) obtained from both the GCM and RCA models for
the present atmosphere and for doubling of CO_2. Also shown are
the observed mean annual temperatures as a function of
latitude.

VI.   DISCUSSION OF RESULTS

Comparison of the two sets of curves in Fig. 4, shows
relatively similar results for both models for latitudes between
5° to 35°N latitude. North of 35° the RCA model results show
lower temperatures than those of the GCM. This difference is
principally due to the neglect of energy transport in the RCA
which is included in the GCM. A comparison of the two sets of
curves in Fig. 5 also shows similar results for the two models
for latitudes between 5° and 35°N latitude. The heating in the
RCA model, however, is somewhat less. Moving poleward the GCM
predicts massive heating at the surface whereas the RCA model
predicts a cooling at the most northern latitudes. By comparing various energy flux terms in the RCA model, it has been
possible to understand the differences in the results of the
two models.

Let us first discuss the RCA model for doubling of CO_2.
Carbon dioxide will absorb both solar radiation and infrared
radiation (at 1.6, 2.0, 2.7, 4.3, 13.7, 15 μm) from the earth.

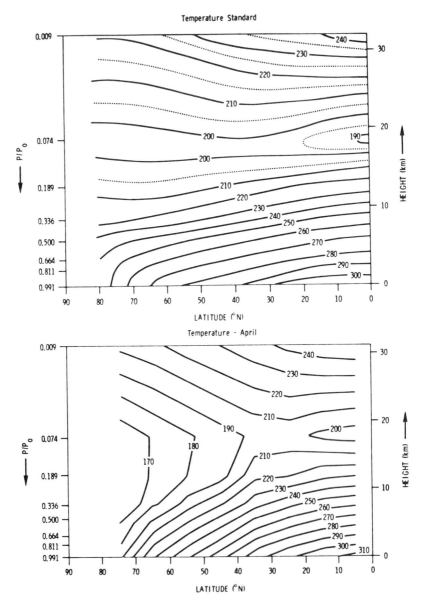

FIGURE 4. Calculations of the atmospheric temperatures
using average global parameters in latitudinal bands using a
global circulation model and a radiative-convective atmosphere
model. Top: Lines of constant temperature calculated by Manabe
and Wetherald using their GCM (1). Bottom: Lines of constant
temperature calculated with the radiative-convective (RCA)
model with particles, assuming constant relative humidity (7).

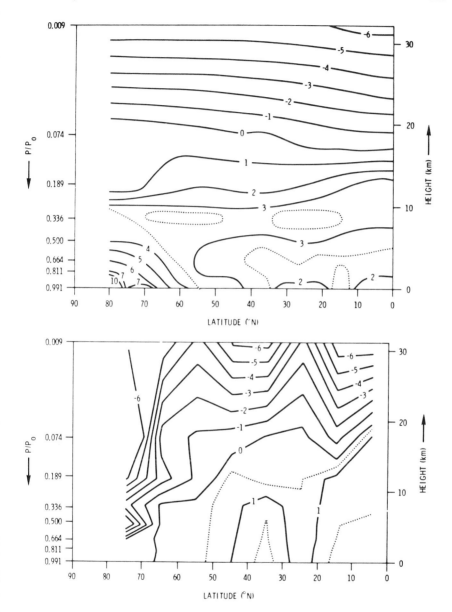

FIGURE 5. Lines of constant temperature change for 2 x CO_2.
Top: Lines of constant temperature change from 2 x CO_2, calculated by Manabe and Wetherald using their GCM (1). Bottom: Lines
of constant temperature change for 2 x CO_2, calculated with the
radiative-convective (RCA) model with particles, assuming constant relative humidity (7).

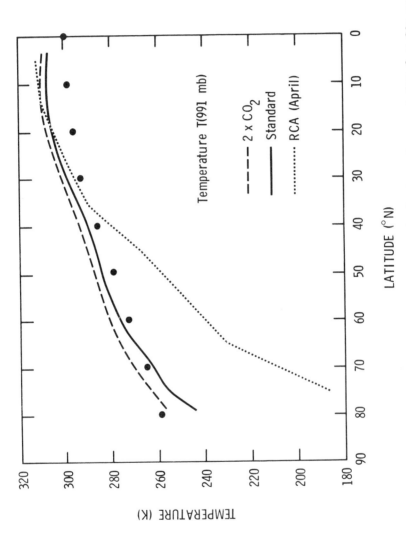

FIGURE 6. Temperatures at 991 mb for the standard atmosphere and for 2 x CO₂ calculated by Manabe and Wetherald using their GCM and for the standard April conditions using the radiative-convective model. Dots refer to observed annual values of temperature (1, 7).

However, as CO_2 never reaches a temperature above 320 K, it will emit mainly in the infrared (at 13.7 and 15 μm). In the stratosphere, far above the surface of the earth, CO_2 emits more infrared energy than it absorbs. (See Fig. 7.) Therefore, when CO_2 is doubled, it will cool the stratosphere both over the polar atmosphere and over the equatorial atmosphere. Near the surface of the earth in the polar region there is little radiant energy from the surface of the earth, and because of the large zenith angle, the solar radiation is small so that the CO_2 emits more infrared radiation than it absorbs. Hence, when CO_2 is doubled, it also cools the polar troposphere. However, in the equatorial troposphere there is more solar energy because the zenith angle is small, the surface of the earth is hotter and radiates much more infrared, and the water vapor is more plentiful and radiates infrared to the 13.7 and 15 μm CO_2 bands. Thus, here CO_2 absorbs more energy than it can emit in the infrared without increasing in temperature. Therefore, doubling CO_2 causes an increase in temperature. (See Fig. 7.)

The large heating which is obtained from the CO_2 increases at the polar latitudes by the GCM is the result of the general circulation shift producing a very stable atmosphere with little convection and a very low tropopause height [1]. This condition restricts the energy to a very small volume near the surface and leads to massive heating. Decrease in surface albedo with melted ice also contributes to the large temperature increase.

What is obviously missing in the GCM? Principally, it is (1) the seasonal variation in parameters, (2) the effects of particles, (3) variability of clouds, (4) oceanic heat transport, and (5) the detailed topography of the northern and southern hemispheres. To illustrate the relative importance of changes in global parameters, the RCA model has been used to

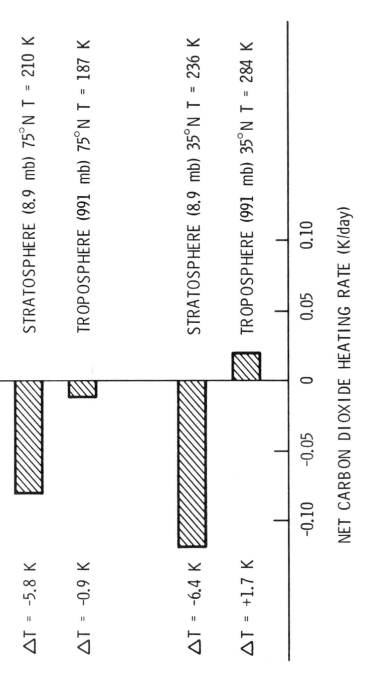

FIGURE 7. Net carbon dioxide heating rate due to change in the radiation balance when CO_2 is doubled (7).

calculate the change in five parameters (low cloud abundance, surface albedo, mean particle extinction, high cloud abundance, and surface relative humidity) necessary to compensate for a 50% and 100% CO_2 increase at 35°N latitude. (See Table II.) The extreme sensitivity of the RCA model to surface albedo and low clouds points out the importance of monitoring all parameter changes simultaneously.

VII. LIMITING FACTORS IN PREDICTION OF FUTURE TEMPERATURE INCREASES FROM CO_2

A. Feedback Links

Although the Manabe-Wetherald GCM includes the feedback between albedo and ice-line as well as a very detailed hydrologic cycle, there are other well-known feedback links which could alter the results but which have not yet been able to be included in their calculations. Some of these feedback mechanisms are (1) suspended particle-surface albedo effects; (2) water vapor amplification; (3) cloud variability; (4) CO_2 distribution between ocean, atmosphere and biosphere;

TABLE II. Changes in Physical Parameters Necessary to Compensate CO_2 Increase at 35°N Latitude.

Parameter	% CO_2 increase	
	50	100
Abundance of low clouds	+1½% (abs)	3% (abs)
Abundance of high clouds	-2½% (abs)	-5% (abs)
Surface albedo	+0.88% (abs)	1.8% (abs)
Suspended particles	+150% (rel)	300% (rel)
Relative humidity	-9% (abs)	18% (abs)

(5) albedo-circulation changes; (6) continental ice-albedo-
sea level changes; and (7) direct albedo effects.

The radiative effects of suspended particles has been
studied for some time with the RCA one-dimensional model but
insufficient experimental information on the spatial and size
distribution and properties is available to adequately determine
their climatic role (4). With the RCA model it has been shown
that particle thermal effects can be either positive or
negative depending on the amount of solar radiation backscattered to space as opposed to that backscattered to the
earth's surface. (See Fig. 8.) Since higher values of surface
albedo backscatter more of the incoming solar radiation, the
net effect is for more radiation to be trapped between the
particles and the earth's surface and a shift in the direction
of heating. Hence, an increase in atmospheric temperature
that changes the surface albedo or that redistributes the
suspended particles through changes in circulation would cause
a change in the thermal contribution of the particles in
various regions with greatest cooling effect near the equator.

Any CO_2-produced change in temperature also produces a
change in the thermal contribution from water vapor. The
intricate detail of the hydrologic cycle was developed to
include much of this feedback in the three-dimensional model.

Additional water vapor effects not included, however, are
(1) independent changes in the water vapor profiles appropriate
for clear and cloudy skies, and (2) effects of the water vapor
dimer. Experimentally it is known that the water vapor
profiles found in clear and cloudy skies are different. The
profiles assumed in RCA models are based on a constant distribution of relative humidity and does not include these differences
(5). They may be shown to yield a precipitable water amount
for global conditions that is about 30% lower than the measured
global value of about 2.6 cm (8). As noted by Hummel and Kuhn

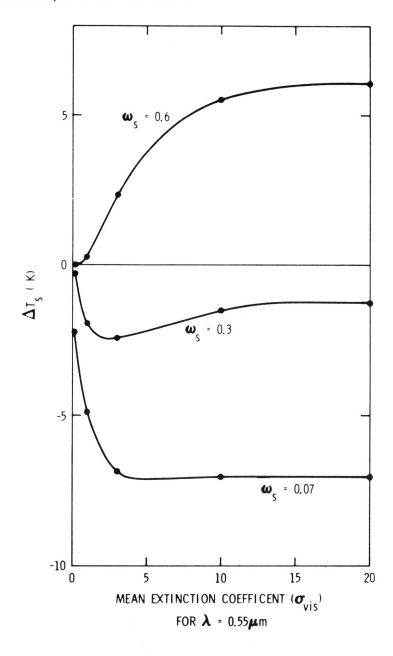

FIGURE 8.   Total anthropogenic release of CO_2 to the atmosphere due to biospheric and fossil fuel sources as calculated
by Stuiver (10).

neglecting these differences can lead to large underestimations of the radiative importance of water vapor and sometimes to even change the sign of the temperature effect of a parameter variation (9). In RCA model predictions from CO_2 doubling, Hummel and Kuhn, and Hummel and Reck find that the fixed cloud and constant relative humidity model of Manabe and Wetherald yields a surface temperature increase somewhat less than the predictive model used by Hummel and Kuhn (9,11).

The water vapor dimer can also alter the results for CO_2 doubling. (Note: The water vapor dimer is two water vapor molecules linked together by a weak hydrogen bond and in tropical regions its concentration is believed to be high enough to influence the radiative energy budget (12)). Hummel and Kuhn, and Hummel and Reck have demonstrated that the dimer can add as much as a degree or two to the surface temperature differences and can control both the sign and magnitude of surface temperature sensitivities (9,11).

Cloud contributions to the thermal structure could also change with increased temperatures (as demonstrated by Hummel and Kuhn (9)). Higher temperatures lead to increased evaporation which puts more water vapor into the atmosphere. This in turn could lead to either more or thicker clouds being produced or to increased precipitation (11). Since the GCM result allows for changes in precipitation but not in cloud cover, the net result of the temperature increase was a precipitation increase. Changes in temperature can also lead to changes in the altitude of existing clouds, to changes in the altitude where clouds are formed, and also possibly to changes in the optical properties of the clouds (e.g., if ice clouds are formed as opposed to liquid drops or to different fluid drops) (13,9,14). Cess has found cloud effects to be small while further work by Schneider et al. shows a strong regional dependence (15,16).

The distribution of CO_2 between the various reservoirs can be altered by changes in temperature. An increase in the biospheric growth rate (for example, from an extended growing season) would increase the rate of CO_2 uptake whereas an increase in oceanic temperature would lead to less physically absorbed oceanic CO_2. Changes in the amount of dissolved CO_2 (as carbonate ion) are also possible.

Albedo changes could occur from increases (or decreases) in the biosphere, changes in the soil moisture content (e.g., the dust bowl) or to changes in ice or snow covers. This, in turn, would alter the quantity of solar radiation absorbed and would produce secondary effects on the circulation and large scale eddies. For example, Bryson argues a 1°C rise at 60°N and none at 20°N gives a decrease in the north-south gradient of 0.24°C per 1000 km which (using a criterion for the latitude at which wave number one of the zonal flow becomes unstable) predicts a half degree change in the latitude of anticyclones and a change in West African rainfall of 270 mm/yr, an amazingly large difference in rainfall (17).

B. Additional Anthropogenic Perturbations

Additional anthropogenic perturbations may occur such as (1) direct greenhouse effects; (2) albedo effects (industrial trace gases, water vapor effects); (3) increases in particle extinction, scattering and absorption; and (4) direct release of waste heat.

Additional trace gases (besides CO_2) of anthropogenic origin, when released to the atmosphere, may absorb radiation in the infrared atmospheric window and lead to additional greenhouse contributions (see Table III). Whereas for current abundances most of these are very small, their significance may become greatly enhanced as their release rates are increased (18-20).

TABLE III. *Calculated Temperature Changes Using the RCA Model*

	Present value	Units	ΔT_s	Relative rank
ω_s	14.01	%	- 0.14000	1.0000
R	7.00	%	- 0.07800	0.5600
h^*	74.00	%	0.06500	0.4600
CO_2	330.00	ppmv	0.02000	0.1400
σ_v	0.10	km^{-1}	- 0.00630	0.0570
N_2O	637.00	ppbv	0.00440	0.0310
O_3	0.37	atm-cm	- 0.00320	0.0230
CH_4	1.60	ppmv	0.00200	0.0140
h_s	3.50	ppmv	0.00120	0.0086
$CHCl_3$	2×10^{-4}	ppmv	0.00100	0.0071
NH_3	6×10^{-3}	ppmv	0.00090	0.0060
HNO_3	$10^{-3} - 10^{-2}$	ppmv	0.00060	0.0040
$CFCl_3$	2.3×10^{-4}	ppmv	0.00035	0.0025
SO_2	2.00	ppmv	0.00020	0.0014
CF_2Cl_2	1.3×10^{-4}	ppmv	0.00019	0.0014
C_2H_4	2×10^{-4}	ppmv	0.00010	0.0007
CH_3Cl	5×10^{-4}	ppmv	0.00010	0.0007
CCl_4	1×10^{-4}	ppmv	0.00010	0.0007

ΔT_s = *surface temperature change (K) for 1% increase in the present average value.*

Extensive irrigation and the formation of man-made lakes can
also alter the water vapor effects on a local and/or regional
scale.

Land development can alter the surface albedo and local
climate as is well documented in the "heat island" effects of
cities. These, in turn, would lead locally to differing
effects for increasing CO_2 as well. Williams *et al.* have
studied the direct heating from energy parks using a general
circulation model and show locally much enhanced heating over
the effects of CO_2. Bryson has for many years studied the
variations in Mie scattering particles from natural sources
such as volcanoes and from direct man-made sources such as
fossil fuel usage, soil disturbances, increased forest fires,
and slash and burn farming (17). There is no doubt that during
their residence time in the atmosphere, these particles will
tend to alter the radiation balance and probably cool the
atmosphere. Generally, their lifetimes are small compared with
CO_2 and their contribution is not likely to be monotonically
increasing as rapidly, if at all.

C. Errors in Model Input Data

Errors in the input data can also change the calculated
results of the atmospheric models. To compare the surface
temperature response T_s of the RCA model to the different
model parameters, the $\partial T_s / \partial Q$ ranking for a 1% change in the
present value of each parameter, with $\partial T_s / \partial S$ taken as unity has
been calculated. The ranking is shown in Table IV.
Also listed are the present values of the parameters, a likely
probably error in their measurement, and the uncertainty in
T_s due to the probably error. These results indicate that the
calculated surface temperature using this RCA model is an
order of magnitude more sensitive to changes in the surface
albedo and the surface relative humidity than to changes in

TABLE IV. Calculated Uncertainty in Model Input Parameters Producing a 0.1 K Uncertainty in Surface Temperature

Parameter	Relative Sensitivity Ranking	Parameter Changes to Cause 0.1 K Changes	Units of Quantity	Measured Parameter Value	Probable Error in Measured Value	Probable Uncertainty in T_s due to Probable Error
Solar constant	1.00000	0.002	cal/cm^2-min	1.9500	0.010	0.500
Zenith angle	0.98000	0.040	degrees	59.2000	0.040	0.100
Daylight time/day	0.90000	0.700	min	720.0000	0.700	0.100
Dry heat capacity	0.68000	0.001	joules/K·g	1.0048	0.002	0.200
Surface albedo	0.11000	0.100	%	14.0000	0.500	0.500
Surface relative humidity	0.11000	1.100	%	74.0000	5.000	0.500
Critical lapse rate	0.10000	0.020	K/km	6.5000	0.030	0.100
Surface atmospheric pressure	0.07700	11.000	mb	1000.0000	4.000	0.030
Rayleigh scattering	0.06900	0.090	%	7.0000	0.1(?)	0.100
Carbon dioxide	0.01800	23.000	ppmm	496.0000	2.000	0.010
Particles extinction coefficient	0.00550	0.016	km^{-1}	0.1000	0.002	0.010

TABLE IV. Continued

Parameter	Relative Sensitivity Ranking	Parameter Changes to Cause 0.1 K Changes	Units of Quantity	Measured Parameter Value	Probable Error in Measured Value	Probable Uncertainty in T_s due to Probable Error
Ozone concentration	0.00290	31.000	%	100.0000	1.500	0.005
Dry limit pressure	0.00160	11.000	mb	20.0000	10.0(?)	0.100
Dry limit concentration	0.00093	3.000	ppmm	3.5000	1.000	0.030

the atmospheric constituents CO_2, particles, and O_3. Changes in the cosine of the zenith angle, fraction of daylight hours, specific heat, the critical lapse rate, the atmospheric pressure and Rayleigh scattering are high on the scale but there is not as much concern over the possibility of large uncertainties in these quantities. The large heat capacity sensitivity is probably important only at stratospheric levels where species may spend a significant portion of their lifetime in an excited energy state.

These results suggest that the greatest effort to improve input data and lessen the calculated uncertainty should be applied in representations of the surface albedo, incoming solar radiation, and relative humidity. Recent parameterizations of lapse rate as a function of altitude and latitude have become available (22). Although the results are tentative, it appears they may substantially increase the model sensitivity to CO_2 doubling.[1]

D. Prediction of Natural Climate Change

Since the atmosphere behaves essentially as a heat engine, any natural changes in the controlling external variables will alter climate. The multiplicity of combinations which can lead to climate change are vast and scores of theories of natural climate change have been developed. Basically, they must all include one or more changes in the following: (1) flux of solar radiation reaching the atmosphere, (2) transmittance of the atmosphere, (3) planetary albedo of the earth atmosphere system or surface albedo of the earth, and (4) abundance of trace constituents such as gases and particles. Volcanic activity is believed by some scientists to have a sporadic effect on climate through this latter effect.

---

[1] Hummel, private communication.

Effects of CO_2 increases must be considered over and above
any natural perturbation of external variables. Because of the
large seasonal changes and large noise level of climatic
variables, a CO_2-related change is likely to be very difficult
to detect at an early stage. Although natural perturbations
always have been cyclic, the present CO_2 perturbation is likely
to be monotonic on the scale of man's lifetime.

VIII. PREDICTIONS OF FUTURE CO_2 INCREASES AND THEIR SIGNIFICANCE

Stuiver has calculated the relative contributions of
atmospheric CO_2 from different sources and has found between
1850 and 1950, two thirds was derived from biospheric sources
(Fig. 8) and one third from fossil fuel combustion (10). He
names the deforestation of the Great Lakes area and logging
of the Pacific Northwest as major examples of biospheric sources.
However, over recent decades he believes fossil fuel combustion
to be the dominant source. Unfortunately, many assumptions
about the stability of the ocean reservoir and the net biospheric
flux must be made to develop any CO_2 mass balance model and
these are highly uncertain at best. Hence any scenario to
predict future effects of fossil fuel combustion must be viewed
with much caution. It is likely, however, that since 1950 the
amount of anthropogenically produced atmospheric CO_2 has
increased at a rate of 4.3%/yr. If this rate of increase is
correct and continues, the curve plotted in Fig. 9 would
result. Figure 10 shows the corresponding surface increase
(calculated by using the RCA model) caused by this CO_2 increase,
provided none of the other global parameters change simultaneously (assuming constant relative humidity and fixed clouds).
On this basis the calculated temperature change since 1950
due to CO_2 increase is 0.16°C, which is not large enough to

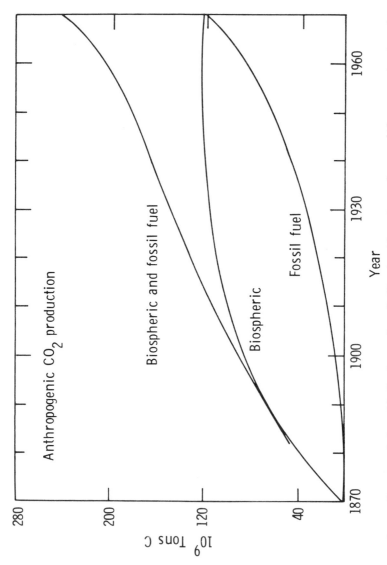

FIGURE 9. Release of fossil-fuel produced CO_2 to the atmosphere with a constant rate of increase of 4.3%/yr (7).

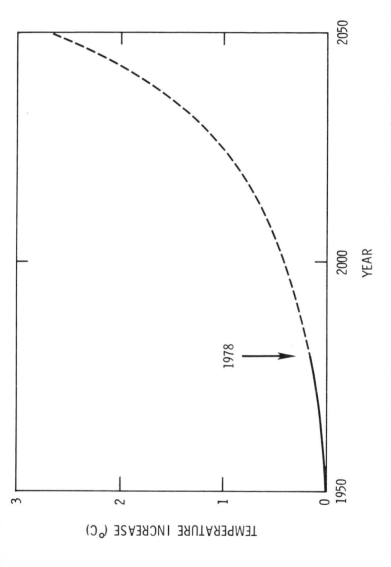

FIGURE 10. Calculated temperature increase at near earth's surface (99/mb) due to 56% of fossil-fuel-released CO_2 remaining in the atmosphere, using the RCA model with particles and relative humidity (7).

separate it from presently observed climate fluctuations. However, if the calculations are assumed to be correct and man continues to produce CO_2 at the rate shown in the figure, before the year 2050 the base-line temperature change may be recognizable.

ACKNOWLEDGMENT

The author wishes to express her sincere gratitude to Richard T. Wetherald for his continued patience and help in the use and interpretation of the Manabe-Wetherald radiative-convective model. She also wishes to thank him for furnishing the input data.

REFERENCES

1. Manabe, S., and Wetherald, R. T., *J. Atmos. Sci. 32*, 4 (1975).

2. Smagorinsky, J., *in* "Studies in Geophysics: Energy and Climate," pp. 133-139. National Academy of Sciences, Washington, D.C. (1977).

3. Reck, R. A., *in* "Report of JOC Study Conference on Climate Models: Performance Intercomparison and Sensitivity Studies" (W. L. Gates, ed.), p. 947. Global Atmospheric Research Program, World Meteorological Association, Geneva, Switzerland (1979).

4. Reck, R. A., *Atmos. Env. 8*, 823 (1974).

5. Manabe, S., and Wetherald, R. T., *J. Atmos. Sci. 24*, 241 (1967).

6. Reck, R. A., *Atmos. Env. 10*, 611 (1976).

7. Reck, R. A., Carbon. Dioxide and Climate: Comparison of One-Dimensional and Three-Dimensional Models. *Environ. International, 2*, n.p. (1979).

8.   Starr, V. P., Peixoto, J. P., and McKean, R. G., *Pure and Applied Geophys. 73,* 85 (1969).

9.   Hummel, J. R., and Kuhn, W. R., An Atmospheric Radiative-Convective Model with Interactive Water Vapor Transport and Cloud Development. *Tellus* (1979).

10. Stuiver, M., *Science, 199,* 253 (1978).

11. Hummel, J. R., and Reck, R. A., Carbon Dioxide and Climate: The Effects of Cloud Specifications in Radiative-Convective Models. *Geophys. Res. Letters* (1979).

12. Cox, S. K., *Quart. J. Roy. Met. Soc. 99,* 669-679 (1973).

13. Reck, R. A., Comparison of Fixed Cloud-Top Temperature and Fixed Cloud-Top Altitude Approximations in the Manabe-Wetherald Radiative-Convective Atmospheric Model. *Tellus* (1979).

14. Reck, R. A., "Thermal Effects of Cloud Parameter Variations in the Manabe-Wetherald Radiative-Convective Atmospheric Model." GMR-2820, General Motors Research Laboratories, Warren, Michigan (1978).

15. Cess, R. D., *J. Quant. Spectrosc. Radiat. Transfer, 11,* 1699 (1972).

16. Schneider, S. H., Washington, W. M., and Chervin, R. M., *J. Atmos. Sci. 35,* 2207 (1978).

17. Bryson, R. A., *Science, 184,* 753 (1974).

18. Wang, W. C., Yung, Y. L., Lacis, A. A., Mo, T., and Hansen, J. E., *Science, 194,* 685 (1976).

19. Reck, R. A., *in* "CO_2, Climate and Society" (J. Williams, ed.). Pergamon Press, New York (1978).

20. Reck, R. A., and Fry, D. L., *Atmos. Env. 12,* 2501 (1978).

21. Williams, J., Kromer, G., and Gilchrist, A., "Further Studies of the Impact of Waste Heat Release on Simulated Global Climate: Part 2." International Institute for Applied Systems Analysis RM-77-34, Baden, Austria (June 1977).

22. Rennick, M. A., *J. Atmos. Sci. 34,* 854 (1977).

DISCUSSION

Potter: I wanted to know if you thought about the temperature feedback in the stratosphere by increasing CO_2 on ozone production

Reck: Well, it's certainly there. I mean, as the stratosphere cools, the ozone production decreases.

Stokes: I was intrigued by the possible inclusion of other green house gases, but I noted in the GCM model you included only one carbon dioxide band. What affect do you feel it would have in the GCM model if additional bands were included?

Reck: I have looked at that rather carefully with a one-dimensional model and it does not seem to be too sensitive. My own personal preference is to use four bands, computer time permitting.

Stokes: I understand. It seems that that would give you some idea as to what the effect of the other bands might be.

Reck: Yes, it's fairly small.

Shaw: CO_2 is fairly well mixed in all of these models, but particulates are not. How do you account for that?

Reck: We didn't. I don't think their spatial distribution is well enough known in a temporal sense. Also their optical properties are also not well enough known. Likewise in reaching a steady state, you would need the aerosol parameterization on an eight-hour basis for a year as a functional altitude, and that's certainly not known.

Shaw: Manabe has said that his error band is about as big as the temperature change, and that was plus or minus a few degrees. What would you put as an error band for the aerosols?

Reck: I have not looked into that problem. But I think it would be large--probably comparable in magnitude or larger.

Shaw: You are talking about a half of a degree?

Reck: A half a degree and about plus or minus half a degree.

Singh: You had the temperature range of the order 1-1/2 to 3 degrees by the year 2050. Would you care to say any more about its implications on future climate?

Reck: I shall defer that question to Dr. Mitchell. He will probably answer that question in detail. I think there are things to be said on that topic, but the purpose of my talk was to look at the model results and to look at how you might get a different result if you took a different model.

APPROACHES TO THE STUDY OF CLIMATE SENSITIVITY

G. R. North

NASA-Goddard Space Flight Center
Greenbelt, Maryland

A survey is given of the various techniques used in estimating the sensitivity of the earth's large scale climate to perturbations induced by agents usually defined as external to the system. Examples of such agents are changes in the light intensity reaching the earth, concentrations of trace gases in the atmosphere by aerosols, etc. The fundamental sensitivity is defined as the change in global average temperature for a 1 percent change in solar constant. That the fundamental sensitivity is proportional to the response to changes in any other parameter is shown. Several members of the climate model hierarchy are reviewed, such as the Budyko-Sellers models, other statistical dynamical models, the radiative convective models, and finally, the general circulation models. Particular emphasis is placed upon the limitation of each approach, and an attempt is made to assess the state-of-the-art of estimating climate sensitivity. A discussion of climate noise and a new approach to sensitivity studies based upon study of the system's own natural fluctuations is included.

I. INTRODUCTION

The ordinary variables which are associated with weather, such as instantaneous temperature or rainfall, fluctuate on many different time and space scales. It is generally thought that if all external boundary conditions, such as the sun, the earth's orbital parameters, the atmosphere's chemical composition, and others are held fixed, the seasonally adjusted statistics of the

417

weather variables would reach a steady state. For example,
average January temperatures for one decade would be about the
same as those for another, long time averages being necessary
because of sampling errors. Such a time average statistic might
be used to describe "climate". This method is improved if
spatial as well as temporal averages are taken, since the standard error due to sampling is further reduced.

Models of atmospheric motion suggest that nonlinearities
which tend to couple different space scales of motion are
responsible for the persistent fluctuations in the weather
variables. The atmospheric fluid is in a turbulent state, a
weather variable at one point becoming statistically uncorrelated
with itself after only a few days (1). These few days may be
thought of as the "memory" of the mechanical system, and it is
a rough measure of how far into the future weather might be
predicted with a simple statistical model (using numerical
solutions of the equations of motion for the atmosphere may
theoretically extend the predictability to a few weeks) (2).
Another characteristic time in the system is the thermal radiative relaxation time which may be of the order of 50 days. This
is roughly the time for a column of air in contact with the
surface to adjust itself thermally to a equilibrium state.
This characteristic time is responsible for the phase lag of
about 1 month between the seasonal solar heating cycle and the
corresponding thermal response cycle. This longer characteristic
time suggests that some form of "climate forecast" might be
possible up to a season into the future. Other longer time
constants involve the slower modes of the system such as the
oceanic mixed layer (few years), deep ocean (10^3 years), ice
packs and glaciers (10^4 years).

In defining climate as a time average, it is important to
know these lag correlation times (memories) in estimating the
sampling error incurred in averaging over a finite interval,

since the total number of independent samples is roughly equal to the averaging interval divided by the lag correlation time (1, 3). These sampling errors lead to fluctuations in monthly averages and these fluctuations might be termed the natural variability or noise in such averages. Any climate change estimates (signal) will have to be comparable with (or larger than) this noise level to be detectable.

The purpose of this paper is to briefly survey a special type of climate forecasting--the question of how the climate responds to a given change in one of the external boundary conditions. Such a change is usually referred to as a "forcing" with the change in climatic variables as the response. This type of climate change is easier in principle to study because it does not require the detailed history of the change, only the difference between initial and final states. Presumably, it is necessary to wait longer than all relevant characteristic times in the system since these times also probably measure the time necessary for the system to adjust completely to the forcing. It is, of course, an assumption that the answer is unique (path independent) at least for small changes.

There are three basic approaches to estimating climate response to external forcing:

(1) Monte Carlo Method

(2) Statistical Dynamics

(3) Leith's Method (Fluctuation-Dissipation Theorem)

Before describing and appraising each of these approaches, consider a very simple example of a sensitivity calculation. First a simple (toy) climate model is constructed. Then an estimate of the change in global temperature can be made for a 1 percent change in solar luminosity (the "fundamental" sensitivity, χ_o).

Let the rate of heat absorbed by the earth-atmosphere system be equal to the rate at which infrared radiation is emitted to space (steady-state condition). The condition may be written

$$\pi R^2 S_o (1 - \bar{\alpha}) = \sigma T_k^4 4\pi R^2$$

where R is the earth's radius, $\bar{\alpha}$ is the average reflectivity
(albedo), of the earth, S_o is the solar constant (perpendicular
energy flux reaching earth), α is Stefan's constant, T_k is the
Kelvin temperature of the earth. Thus,

$$\frac{S_o}{4}(1 - \bar{\alpha}) = \sigma T_k^4 \simeq A' + B'T_c$$

where $T_c = T_k - 273°$ is the Celsius temperature, A' and B' are
the coefficients in a Taylor series of σT_k^4. The "fundamental"
sensitivity is defined as

$$\chi_o \equiv \frac{S_o}{100} \frac{dT_c}{dS_o}$$

It is straightforward to show that for this model $\chi_o = 0.6°$ C.
The planet described, however, has no atmosphere. If the
"greenhouse" effect (absorption and re-emission of infrared by
atmospheric constituents such as clouds, O_3, H_2O, and CO_2) is
taken into account, A' and B' may be estimated from either
detailed radiative transfer calculations or from correlations of
observed temperatures with satellite measurements of the
infrared. In that case χ_o is larger

$$\chi_o \text{ (greenhouse)} \simeq 1.1° \text{ C}$$

(see refs. 4 and 5). Semiempirical addition of the greenhouse
effect makes the climate more sensitive. When such a mechanism
increases the sensitivity, it is said to be a "positive feedback
mechanism." Another well-known positive feedback is the ice-
feedback mechanism. In this case if the solar constant is
increased, the highly reflective ice caps tend to shrink, and
thus the planet absorbs more radiation and the temperature
increases still further. The ice-feedback mechanism may increase

the sensitivity to about 1.8° C; however, the estimate of this effect (as well as the greenhouse effect) are still extremely tentative. Clearly, these feedbacks operate to amplify the response due to any forcing, and the solar constant change has been chosen merely as a standard for comparison. It can be shown that for many models the response to any forcing is proportional to the fundamental sensitivity (refs. 6 and 7; also private communication from Salmun). For example, in the present model if ΔF is added to the energy balance equation, the corresponding response is

$$\Delta T = \Delta F/B = K\chi_o \Delta F$$

where K is a constant.

The greatest challenge confronting those wishing to compute climate sensitivity is the proper inclusion of all relevant feedback mechanisms. Allusion has been made to two more important and well-studied ones, but there must surely be others, perhaps even more important, that have not even been discovered yet. To mention a few other known ones: biological changes in albedo due to changes in climate (8); changes in atmospheric chemistry due to changes in solubility of gases in the oceans or other reservoirs brought about by climate change; changes in oceanic circulation due to climate change. Perhaps the most perplexing and important but least understood is cloud feedback.

No model estimates of climate sensitivity can be taken as final until progress is made on the problem of cloud feedback. The problem has two distinct parts: (1) how do cloud parameters change when climate changes? (2) do the changes lead to a positive or negative feedback? The answer may even be region dependent. Presently, the earth is about 50 percent covered with clouds, which account for most of the planetary albedo. Increasing the cloud cover increases the albedo (cooling) but simultaneously increases the greenhouse effect (warming). The two effects almost cancel each other--whichever is larger

determines whether the feedback is positive or negative. Of
course, it is not known whether the cloud cover increases or
decreases in response to a thermal forcing. Several years may
pass before noticeable progress can be made on this problem which
should be attacked from all possible directions--observational
and theoretical.

II.   APPROACHES

A. *Monte Carlo Method*

The approach here is to set up Newton's second law for the
relevant geophysical fluids and integrate them forward in time
as an initial value problem. One may integrate an ensemble of
randomly chosen initial conditions and average the results of
these runs to estimate the change in climate or equivalently
(ergodicity) average a single run over a long time. In principle
this method should work since classical mechanisms and thermodynamics are well understood and large computers are available
for the computations. In addition, considerable expertise has
developed over the last 25 years since these are essentially
the equations integrated for the conventional weather forecast.
There are, however, some differences between the present problem
and a weather forecast. It is known that the radiative properties of the system become much more important after long
integrations; for example, in weather forecasts it is reasonable
to neglect diabatic heating altogether and concentrate on the
mechanics of advective processes; whereas, in climate, adiabatic
processes dominate. Therefore, more accurate and detailed radiative packages have to be included in the programs. Similarly,
such previously unstudied components as clouds and oceanic
coupling must be important in the new application. All these

new problems lead in the direction of more complicated programs
which must run more slowly than the old programs used in short-
range forecasting.

The large general circulation models (GCM's) of the atmosphere can have as many as a quarter of a million degrees of
freedom and run on state-of-the-art computers at a rate only 10
or 20 times faster than nature itself. Leaving out the slow
components such as the oceans means that doing a sensitivity
experiment should run several (say five) enfolding times
(5 x 50 = 250 days) for the radiative thermal relaxation to
allow the transient effects to die out. This effect was demonstrated by the simplified GCM runs by Manabe and Wetherald (9)
and Wetherald and Manabe (10) where hundreds of days were
required before a satisfactory steady state was reached. In
computing sensitivities, it is necessary that all transients have
decayed to very small magnitudes compared with the small climate
change being estimated. If several members are run in the
ensemble or if long time averages (several radiative relaxation-
auto-correlation times) are made, climate modeling experimentation will be very expensive even with models which have no ocean
dynamics included.

For this reason, several groups (NCAR, Oregon State, GISS)
are experimenting with initial value models which have a
simplified dynamics but treat the radiation very carefully.
Usually, the simplification is through a coarser resolution in
the numerical integration scheme. It remains to be seen whether
these models generate the same climate and climate sensitivity as
their larger counterparts.

Recently, it has become apparent that mean annual boundary
conditions imposed upon the models may lead to different sensitivities than those computed from seasonally driven models. The
seasonal driving introduces a whole new class of difficulties
involving the ocean mixed layer. The heat capacity of "thermal
inertia" of this layer (about 100 m deep) adds considerably to

the thermal relaxation time of the system. Over the ocean a
column of atmosphere-mixed layer has a relaxation time of about
5 years rather than the 50 days for the atmosphere alone. This
value suggests that one may have to run for 25 model years to
get the transient effects out of the system.

Even with the difficulties listed here, the GCM is probably
the ultimate tool in understanding climate. Although the large
models are expensive and their output almost as hard to interpret
as climate data itself, they have the advantage of control-
lability of constraints, a feature which allows experimentation.
Furthermore, the GCM can be used as a standard in the development
of more simplified models or approaches.

B. *Statistical Dynamics*

Since climate is the time average over the shorter term
fluctuations of weather variables, it is interesting to formulate
equations for the ensemble or time average quantities themselves.
Models of this type also frequently average over large spatial
extents. The example used to illustrate the fundamental sensitivity concept is perhaps the simplest statistical dynamical
model (SDM) one might imagine. The entire range of SDM's has
recently been reviewed by Saltzman (11).

Perhaps the next more comprehensive examples after the zero-
dimensional models are the Budyko-Sellers models (12, 13, 14)
which have one horizontal dimension, latitude. These models have
the zonally averaged mean annual surface temperature field as
their solution. All quantities entering the heat balance equation are assumed to be derivable from this temperature field.
The greenhouse effect is included by an empirical formula relating the flux of terrestrial radiation to surface temperature.
The ice-albedo effect is included by having an icecap whose
equatorward edge is determined by a critical isotherm (13); this

latter makes the models nonlinear. The horizontal divergence of
heat is typically taken to be a linear form either algebraic (13)
or diffusive (14, 15, 16, 17).

These models reproduce the zonally averaged temperature
remarkably well by merely adjusting one parameter such as the
thermal diffusion coefficient. Early estimates of the sensitivity (14, 13, 17) indicated that the ice-albedo feedback mechanism increased the sensitivity to about 4.00° C and only about
a 2 percent lowering of the solar constant was necessary to cause
a complete ice cover of the earth. In recent years, however,
new data from satellites has been reduced (18) so that new
estimates of albedo and infrared radiation are available. These
new results have been used to recompute the empirical coefficients in the model parameterizations (5, 19). The new results
generally suggest (20) that the sensitivity including these
effects (ice and greenhouse) only should be about 1.70° C. Even
though these results are in reasonable agreement with those of
the simplified GCM experiments by Wetherald and Manabe (10),
there is always the possibility that the agreement is fortuitous
(21).

Obviously, more work needs to be done with these models in
the areas of parameterization theory (e.g., see ref. 22), and in
questions concerning the estimation of empirical coefficients.
Further developments with the models include a seasonal extension
(23, 24, 25). Although these works differ somewhat in detail,
they generally agree that with the feedbacks assumed, there is
little difference in sensitivities computed from the mean annual
versus the seasonal versions of the same model. Theoretical
advances with the Budyko-Sellers models (number of solutions,
stability theory, etc.) have been recently reviewed by North
(26).

Another class of simplified low order models are the radiative convective models (RCM's), which were recently reviewed by
Ramanathan and Coakley (27). These models attempt to treat the

radiative transfer accurately and therefore require good vertical
resolution. The horizontal dependences or advective processes
are generally ignored or parameterized. Since an atmosphere in
radiative equilibrium is usually unstable near the ground, some
form of convection must be included in the vertical to obtain
realistic profiles in the lower troposphere. These models are
especially useful in estimating the effects of changes in atmospheric composition in either the stratosphere or troposphere;
they can be readily merged with detailed chemical models.

Other more detailed (two- and three-dimensional) SDM's have
been largely due to Saltzman and his coworkers. The reader is
referred to Saltzman's (11) comprehensive review for more
information and a complete list of references. One difficulty
is common to all the SDM's--the closure crisis. In the
statistical formulation of any nonlinear problem, this crisis
arises. As an illustration consider the simple nonlinear equation

$$\frac{dx}{dt} = Ax^2$$

where x represents the dependent variable with time denoted by t.
An ensemble average can be taken

$$\frac{d}{dt} <x> = A<x^2>$$

which contains $<x^2>$. An equation for $<x^2>$ can be readily obtained

$$1/2 \frac{d}{dt} <x^2> = A <x^3>$$

which unfortunately contains $<x^3>$. The process of terminating
this infinite sequence in a sensible way is called closure. Most
modern theories of turbulence are simply methods of dealing with
this problem. Whether a given procedure works or does not
depends upon the physics. For example, in systems with widely

separated time scales, one can often effect a closure by a
stochastic method--i.e., replace rapidly fluctuating variables
occurring in slow variable equations by random variables with
suitable statistics (28). This procedure may work, for example,
if the slow subsystem is a thermal variable (averaged over a
large volume) and the velocity field fluctuations are the fast
variables. In some such "Brownian motion" cases, the transport
(in ensemble average) may be modeled by diffusion. This effect
may explain why the Budyko-Sellers models give reasonable estimates of the temperature field (17).

C. Leith's Method (FDT Method)

Leith (2) has suggested that the fluctuation-dissipation
theorem (FDT) from statistical mechanics (29) be used for
estimating climate sensitivity. Basically, the theorem states
that for certain systems the sensitivity to externally induced
perturbations can be estimated by a systematic study of the
system's own natural fluctuations. In particular, if the system
relaxes irreversibly back to the norm from one of its own
fluctuations in the same way it relaxes after being "kicked" to
the same state by a force, it is possible to compute the sensitivity from the natural fluctuation data.

As an example of the technique, consider the zero-dimensional
model studied earlier but modified to include time dependence

$$C\frac{dT}{dt} + A + BT = Qa(T) + f(t)$$

where $f(t)$ is a random internal forcing variable with "white
noise" characteristics:

$$<f(t)f(t + \tau)> = F_o\delta(\tau)$$

where $\delta(\tau)$ is the Dirac function. If the fluctuations are small,
the equation may be linearized about its mean value to obtain

$$\frac{dT'}{dt} + \frac{T'}{\tau_o} = \frac{f(t)}{C}$$

where τ_o is the characteristic relaxation time (inverse of the stability eigenvalue)

$$\tau_o^{-1} = (B - \varrho(\frac{da}{dT})_o)C^{-1}$$

It is straightforward to show that the lag correlation function is given by

$$<T'(t)T'(t + \tau)> = \frac{F_o e^{-\tau/\tau_o}}{C^2} = \frac{F_o \rho(\tau)}{C^2}$$

Now consider the same basic system without the random internal forcing but instead an external forcing $\delta(t - t_o)$. The linear response $g(t, t_o)$ to this forcing satisfies

$$\frac{d}{dt} g(t,t_o) + \frac{g(t,t_o)}{\tau_o} = \frac{1}{C}\delta(t - t_o)$$

the solution of which is

$$g(t,t_o) = \frac{1}{C} e^{-|t - t_o|/t_o}$$

or relating to the lag correlation function

$$g(\tau) = \rho(\tau)/C$$

with $|t - t_o| = \tau$. Since $g(\tau)$ is the Green's function the response to an arbitrary (small) function $h(t)$ may be found by integration

$$\delta T(t) = \int_{-\infty}^t dt'g(t - t')h(t')$$

The illustration presented here is, of course, highly simplified compared with the real climate. In particular, the system is linear. However, the theorem can be proven for some nonlinear statistical systems; therefore, there is the hope that it might work at least approximately for climate. The best way of finding out is by testing the theorem on initial value climate models. Such a program is now being conducted by T. Bell and the author at Goddard Space Flight Center.

The FDT method faces many obstacles in its eventual implementation with real data. First it must be tested with a variety of model climates. If the tests are positive, the question arises as to how long are the lags τ necessary in estimating $\rho(\tau)$ so that a reasonable estimate of $g(\tau)$ can be obtained. Surely the answer depends upon the time scale of the phenomenon of interest. Finally, the theorem only applies to infinitesimal perturbations and how good this approximation may be is not known, especially at large τ. In short, this approach is a long shot, but it may be the best because of its (hopeful) model independence. Presumably, for example, all feedbacks are included.

III.   CONCLUDING REMARKS

In conclusion, there are a variety of approaches to the estimation of climate sensitivity. It is generally felt that the knowledge of climate is still too meager to take any current estimates seriously. The largest gaps still are in understanding cloud and ocean interactions with the atmospheric system and fear of omitting unknown feedback mechanisms.

REFERENCES

1.   Leith, C. E., *J. Appl. Meteor. 12,* 1066 (1973).

2.   Leith, C. E., *J. Atmos. Sci. 32,* 2022 (1975).

3.   Madden, R. A., *Mon. Wea. Rev. 104,* 942 (1976).

4.   Budyko, M. I., *Tellus, 29,* 193 (1977).

5.   Cess, R. D., *J. Atmos. Sci. 33,* 1831 (1976).

6. North, G. R., *in* "Proceedings of the First International Conference on Mathematical Modeling," Vol. IV, p. 2291. University of Missouri, Rolla, Missouri (1977).

7. Salmun, H. M.S. Thesis, University of Missouri, Physics Department, St. Louis, Missouri (1979).

8. Cess, R. D., *J. Atmos. Sci. 35*, 1765 (1978).

9. Manabe, S., and Wetherald, R. T., *J. Atmos. Sci. 32*, 3 (1975).

10. Wetherald, R. T., and Manabe, S., *J. Atmos. Sci. 32*, 2044 (1975).

11. Saltzman, B., *Adv. in Geophys. 20*, 183 (1978).

12. Budyko, M. I., *Meteorol. Gidrol. 2*, 3 (1968).

13. Budyko, M. I., *Tellus, 21*, 611 (1969).

14. Sellers, W. D., *J. Appl. Meteor. 8*, 392 (1969).

15. Held, I., and Suarez, M., *Tellus, 36*, 613 (1974).

16. North, G. R., *J. Atmos. Sci. 32*, 1301 (1975).

17. North, G. R., *J. Atmos. Sci. 32*, 2033 (1975).

18. Ellis, J. S., and Vonder Haar, T. H., Zonal Average Earth Radiation Budget Measurements from Satellites for Climate Studies, *Atmos. Sci. Pap. 240*, Colorado State University, Fort Collins, Colorado (1976).

19. Lian, M. S., and Cess, R. D., *J. Atmos. Sci. 34*, 1058 (1977).

20. Coakley, J. A., *J. Atmos. Sci. 36*, 260 (1979).

21. Coakley, J. A., and Wielicki, B., *J. Atmos. Sci. 36*, n.p. (1979).

22. Stone, P. H., *J. Atmos. Sci. 30*, 521 (1973).

23. Sellers, W. D., *J. Appl. Meteor. 12*, 241 (1973).

24. Thompson, S. L., and Schneider, S. H., *J. Geophys. Res. 84*, 2401 (1979).

25. North, G. R., and Coakley, J. A., *J. Atmos. Sci. 36*, 1189 (1979).

26. North, G. R., Recent Work on Heat-Balance Models, *in* "Report
     of the JOC Study Conference on Climate Models" (L. W. Gates,
     ed.), n.p. WMO (GARP Series), Geneva, Switzerland (1979).

27. Ramanathan, V., and Coakley, J. A., *Rev. Geophys. and
     Sp. Phys. 16,* 465 (1978).

28. Hasselmann, K., *Tellus, 28,* 473 (1976).

29. Callen, H. B., and Greene, R. F., *Phys. Rev. 86,* 702 (1952).

DISCUSSION

Potter: I was just wondering if you were talking about the different mechanisms like clouds and oceans. Is it better to have fixed clouds or some parameterization of clouds?

North: Traditionally, people hold the clouds fixed. The problem is that the competing effects almost cancel, as you know and since we don't know what else to do, we are holding them fixed. In general circulation models, there are now attempts to put in changing clouds. One problem is that we won't know even if we have done it correctly. There isn't enough observational data, and that's an additional weakness. I'm looking into this question prsently at NASA-Goddard Space Flight Center, trying to unravel time series of various properties of clouds from satellite data so that when the GCM people model them they will have something to compare with. You know they presently have only monthly mean climatology to compare with and that is not sufficient. You must examine the time series for cloud fraction, cloud height, etc. This problem is so difficult that it will be years before we make noticable progress in my opinion. By the way, that is one nice thing about Leith's method--presumably all feedbacks are included in the analysis of real data. Of course, even if it works that also is years off.

Singh: Can you take care of increased evaporation by just having increased relative humidity and leave the cloud cover constant?

North: Well, of course not, but it's the only game in town at this time.

Edwards: You spoke of time averages and spatial averages, and if we are trying to make some actual measurements in order to detect trends to see if they would fit your model, there is a sampling problem in order to get the right averaging. Could you comment on that?

North: Well, there are several different kinds. One has been discussed lately by Madden and also Leith, and that is the sampling problem which comes about from finite length time averages. One might ask about estimating the January mean temperature from a single month's record. It might be expected that since there are 31 numbers to average we would have a good estimate of the mean. However, the autocorrelation time is 3 or 4 days and Leith has shown that in a single month there are only about 5 independent samples. Therefore our estimate of the mean January temperature can be rather poor due to this sampling error. A real change in climate (signal) has to be comparable to or greater than the sampling fluctuations (noise) to be detectable - or if predicted useful. Of course, a similar concept arises with spatial averages.

ZONAL MODEL CALCULATION OF THE CLIMATIC
EFFECT OF INCREASED CO_2[1]

G. L. Potter

University of California
Lawrence Livermore Laboratory
Livermore, California

The increase in atmospheric carbon dioxide expected in the next few decades may result in a general increase in global surface temperatures. The purpose of this presentation is to demonstrate how a two-dimensional zonal atmospheric model can be used to test the possible atmospheric response to various carbon dioxide concentrations. The response of one-dimensional models and a three-dimensional general circulation model with a stylized surface is well documented but, to date, no experiments have been attempted with a comprehensive two-dimensional model. In the model, increased CO_2 warmed the troposphere and reduced the intensity of the polar surface inversion which, in turn, magnified the temperature response with latitude. This response appears to be much the same as that caused by ice-albedo feedback reported by researchers at GFDL. Precipitation, evaporation, and the total atmospheric water content increased with increasing CO_2 concentrations.

---

[1]This work was performed under the auspices of the U.S. Department of Energy by Lawrence Livermore Laboratory under contract no. W-7405-Eng-48.

433

I. INTRODUCTION

The response to increased atmospheric CO_2 of a three-dimensional general circulation model with simplified topography is well documented, but to date no comprehensive experiments have been attempted with a comprehensive two-dimensional model (1). This paper presents only the preliminary results that will serve to provide a basis for future improvement of the model and further understanding of the feedback mechanisms that respond to changing atmospheric carbon dioxide concentrations.

II. MODEL DESCRIPTION

The two-dimensional zonal atmospheric model developed at the Lawrence Livermore Laboratory uses the primitive form of the conservation equations and presently computes prognostic variables at nine vertical levels and at $10°$ intervals of latitude (2-4). The surface at each latitude is divided pro-portionally into land (of various types and elevations) and ocean (open and partial ice cover). The fractional representa-tion allows a somewhat realistic treatment of the surface energy balances, although without the spatial coherence of a continen-tal structure. The surface treatment allows fluxes of sensible heat and moisture to be calculated from vertical gradients of temperature and water vapor, respectively, the surface wind velocity and the appropriate bulk transfer coefficients.

In most studies using this model, clouds are calculated at four levels as a function of relative humidity and calibrated with data from London, England (5). Convection is a function of cloud overlap, vertical lapse rate, and moisture content. The mechanism relates the lapse rate adjustment (convective intensity) to the departure from the local moist adiabatic lapse rate.

The land surface is divided into as many as 10 layers of
variable depth to facilitate matching the thermal inertia to
diurnal and seasonal forcing. The number of snow and ice layers
depends on the total accumulated depth. The ocean is treated as
an isothermal layer of prescribed depth corresponding approximately to that of the top of the thermocline. The ocean
temperature depends on the surface energy balance and a prescribed meridional ocean heat flux. The model also calculates
sea ice depth and lateral extent depending on the energy
balance of the sea ice and ocean at that latitude.

III. MODEL PERFORMANCE

Although the model version with annual average solar flux
is used in this experiment to reduce computation time, the
overall model sensitivity has been validated against the observed
seasonal cycle, against other models, and against observations
to understand better the effect of utilizing an annual average
solar radiation. The annual average version also differs from
the seasonal version in having a reduced ocean depth to permit
more rapid convergence to a near-equilibrium state.

Figures 1 to 4 demonstrate some aspects of the model's
performance for the month of January using the seasonal version
of the model (Potter gives more complete performance data (6)).
Comparisons in these figures are based on data from Refs. 7 to
17. Figure 1 shows the zonally averaged cloud cover as
compared with observations and various other models.[1] The

---

[1] In all comparisons, general circulation models (GCM) are
labeled with **, radiative convective models with *, the
Lawrence Livermore Laboratory (LLL) zonal model with ZAM2 and
observations by a reference alone.

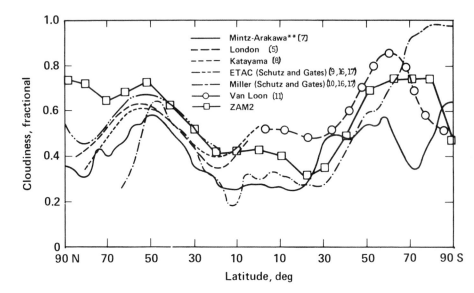

FIGURE 1.   Zonally averaged total cloud cover, from Kahle
and Haurwitz (7).

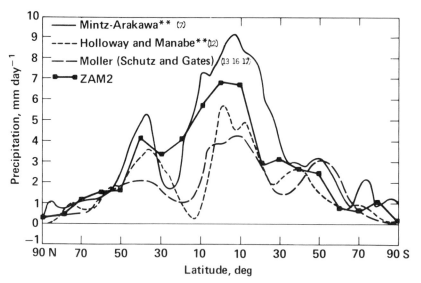

FIGURE 2.   Zonally averaged precipitation, from Kahle and
Haurwitz (7).

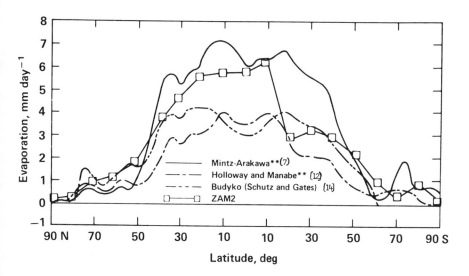

FIGURE 3.   Zonally averaged evaporation, from Kahle and Haurwitz (7).

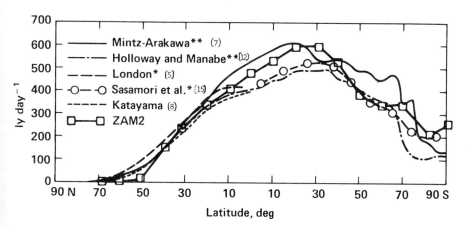

FIGURE 4.   Zonally averaged solar radiation absorbed by the earth's surface, from Kahle and Haurwitz (7).

divergence from the observations in the high northern latitudes
is believed to be due to difficulties in simulating the relative
humidity fields. The 600 mb cloud amount in polar latitudes
also tends to be too great. The maximum cloud amount in polar
regions should occur at the 850 mb level. Figure 2 shows how
the zonal precipitation compares with results from two general
circulation models and with observations. The reason for the
exaggerated precipitation (Fig. 2) and evaporation (Fig. 3) in
the tropical latitudes may be a result of a too active Hadley
cell.

In terms of radiation, Fig. 4 shows the absorbed solar
radiation at the earth's surface. The results indicate that
the surface albedo is properly treated and that the general
influence of cloud cover in terms of short-wave radiation is
reasonable compared with other model results and with observations.

IV. THE CO_2 EXPERIMENTS

To decrease the computer time requirements and to increase
the number of model experiments undertaken for this presentation,
the annual average version of the model is used in this preliminary study. The computational time required for the model
is approximately 20 seconds per model day (on a CDC 7600 computer). Starting from initial conditions similar to observed
zonally averaged temperature and moisture fields, the model was
integrated for 200 days. The carbon dioxide perturbations (2
times present CO_2 and 4 times present CO_2) and the control run
were then integrated for an additional 200 days. Near-
equilibrium was assumed to be reached when the net radiation
at the top of the atmosphere approached zero.

A. *Temperature*

Meridional sections of the difference in temperature between the perturbed atmosphere and the control case are shown in Figs. 5 and 6 for 2 x CO_2 and 4 x CO_2, respectively. Increased CO_2 requires a decrease in temperature in the stratosphere before longwave radiation comes into balance with solar absorption. In the troposphere, increased CO_2 leads to warming due to reduced loss of heat to space.

Figures 7 and 8 show the surface temperature differences for the two cases (Figs. 5 and 6 extend down to only the lowest prognostic level which, in most cases, is 1000 mb). The surface temperatures display the exaggerated polar response to increased

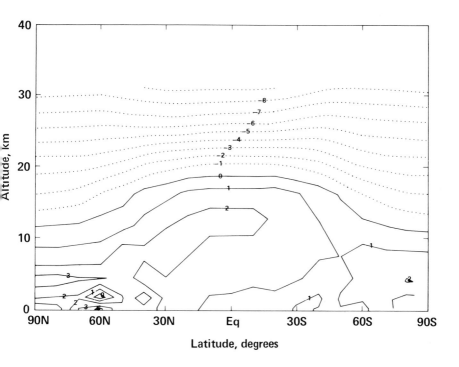

FIGURE 5.   *Temperature difference (2 x CO_2--control)*
contour interval = 1.00 K.

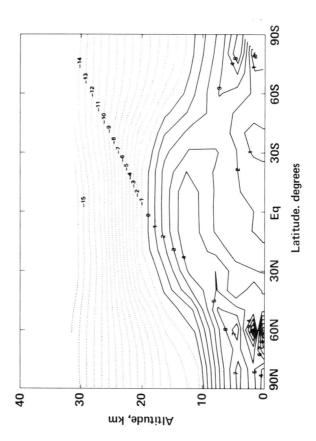

FIGURE 6. Temperature difference (4 x CO$_2$ --control) contour interval = 1.00 K.

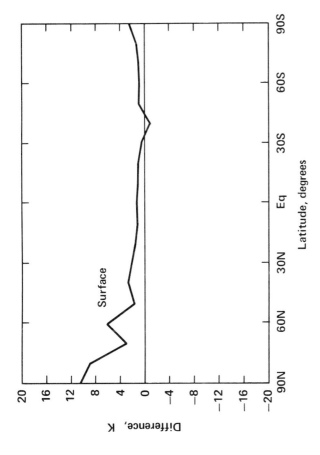

FIGURE 7. *Temperature difference (2 x CO₂--control).*

441

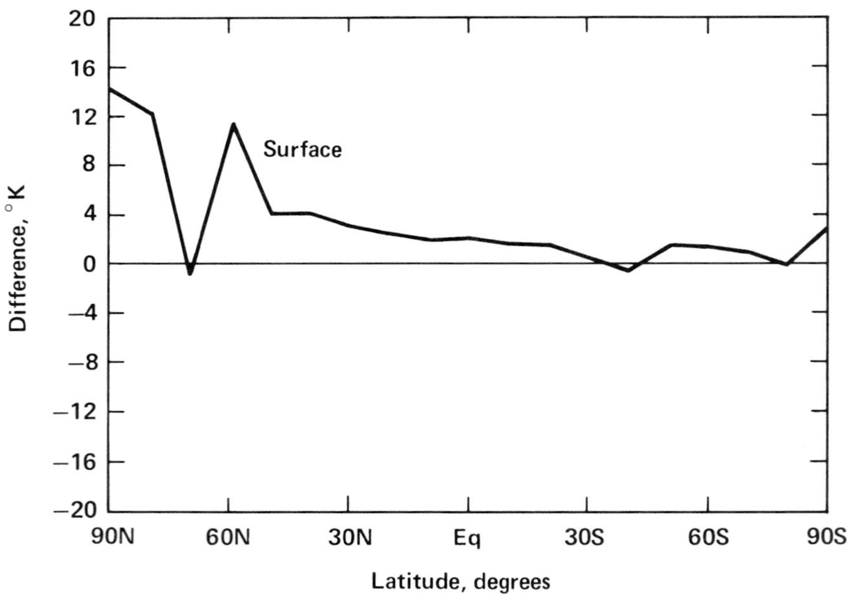

FIGURE 8. Temperature difference (4 x CO_2 --control).

atmospheric CO_2 but for apparently different reasons than those reported by Manabe and Wetherald (1). This response results in part from insufficient poleward eddy transport of heat and water vapor in the control simulation. This condition in turn produces too steep a poleward temperature gradient and extremely cold polar temperatures. With increased CO_2 the polar warming is confined to the shallow atmospheric layer near the surface where only a slight increase in radiant energy results in an amplified warming because of the small heat capacity of the subinversion layer. The amplified polar temperature response results almost entirely from the weakening of the surface inversion whereas the ice-albedo feedback mechanism was essentially ineffective in amplifying temperatures in these model runs. Another feature of the model that may limit sensitivity to changing CO_2 concentrations are the calculated changes in snow and sea-ice depths (and extent). The snow, which continuously accumulates in the annual average version, has the effect of insulating the sea ice, and thus limits the sensitivity in high latitudes.

From a subsequent series of sensitivity studies it was found that with fixed (as opposed to computed) clouds the sea-ice margins receded with enhanced atmospheric CO_2. The cloud fractions were based on climatological values and diverged from the model computed cloud fraction primarily in high latitudes and at the cloud levels. The global climate response of using fixed clouds is beyond the scope of this study, but it suffices to say that the dynamic sea-ice parameterization used in the model is quite sensitive and deserves a much more detailed study.

B. *Hydrology*

Figure 9 shows the total precipitation for the control and
the perturbation simulations. As CO_2 increased, so did global
precipitation. The globally averaged precipitation for the
control case was 0.269 cm day^{-1}; for 2 x CO_2, 0.285 cm day^{-1}
(6% increase). The increased tropospheric downward longwave
radiation from CO_2 increased the surface energy available for
evaporation, and thus increased the precipitable water. This
condition, in turn, further increased the downward longwave
radiation from the atmosphere.

C. *Albedo*

As mentioned earlier, the snow and ice cover did not change
appreciably in the CO_2 experiments. Yet because of the moisture
dependence of the surface albedo (wet soil is darker than dry
soil) the Northern Hemisphere surface albedo (a_n) decreased as
the CO_2 content increased. For the control a_n = 22.4%;
2 x CO_2, a_n = 22.2%; and 4 x CO_2, a_n = 19.1%. The 4 x CO_2 case
displayed some decrease in snow extent at 60°N and resulted in
a further drop in surface albedo. The Southern Hemisphere
displayed no significant changes in surface albedo, which in a
large part was due to the complete lack of sea ice in the high
southern latitudes. However, in the fixed cloud experiments,
sea ice did develop in the Southern Hemisphere. Given the lack
of ice cover (in the free cloud cases) in the Southern Hemisphere,
the warming from CO_2 and water vapor produced no ice-albedo
feedback.

To more fully evaluate the relative importance of surface
albedo changes in the global response to increased CO_2, an
analysis of the radiative fluxes was performed. The method
involved computing the change in fluxes at the top of the

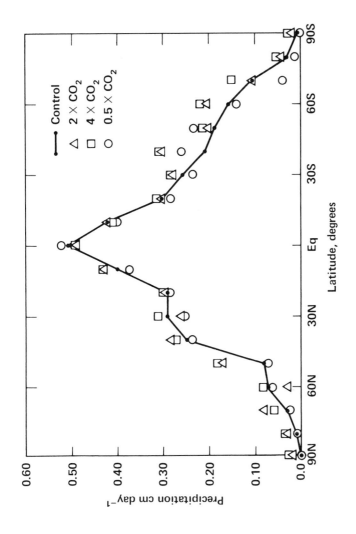

FIGURE 9. Total precipitation, cm day^{-1}.

atmosphere initiated by the individual contribution of each of
the relevant factors. The net radiation change at the top of
the atmosphere resulting from doubling CO_2 can be expressed by

$$\delta R = \delta S - \delta F$$

where δS and δF are the changes in the net solar and longwave
fluxes at the top of the atmosphere. The δS is the sum of the
changes due to such individual factors as water vapor, clouds,
and surface albedo whereas δF reflects the combined effects of
changes in CO_2, temperature, water vapor, and cloud amount. The
factors that provide the largest effects are CO_2, temperature,
and water vapor. The smallest contribution in the zonal model
experiments comes from changing cloud amount and surface
albedo.

D. *Heat Balance*

   The area mean balance for various concentrations of CO_2 is
shown in Fig. 10.[2] The net longwave radiation loss at the surface decreased with higher CO_2 concentrations. Including solar
radiation, for the 2 x CO_2 case, the net radiation increased
4.2%; for the 4 x CO_2 case, it increased 6.3%. This effect can
be attributed to both the increased solar absorption (lower
surface albedo) and the reduced loss of longwave radiation
(because of increased atmospheric back-radiation from CO_2 and
water vapor). Because of the warmer surface temperatures, the
Bowen ratio (sensible heat/latent heat) decreased since evaporation is more effective in removing heat from the surface than
turbulent heat exchange.

   Planetary albedo (Fig. 11) compares quite favorably with
observations compiled by Rashke *et al* (18). Recent data from

---

[2]*Also shown in Fig. 10 is the energy balance for the 1/2
times present CO_2.*

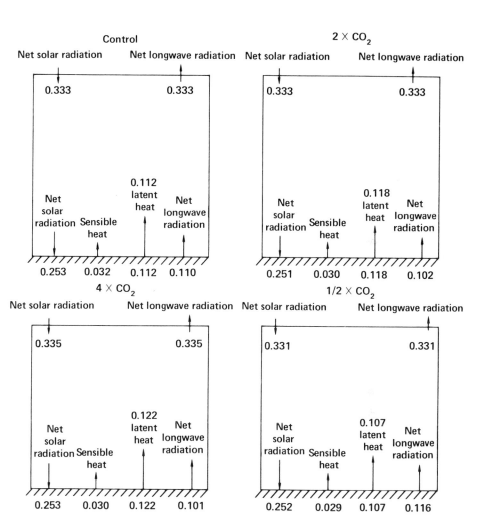

FIGURE 10.   Area mean heat balance (ly min^{-1}).

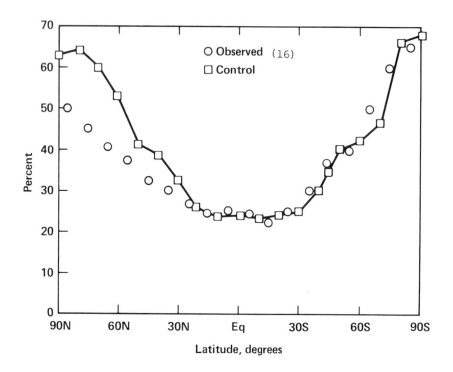

FIGURE 11.   Planetary albedo.

Ellis (personal communication) suggest that the values from
Rashke *et al.* are quite low in the high northern latitudes and
that the zonal model results are quite similar to observations
(18). Figures 12 and 13 show the change in planetary albedo
resulting from increased CO_2 (perturbed minus control).
Increasing the CO_2 concentrations reduced the planetary albedo
at almost all latitudes. Figure 13 (difference in planetary
albedo between the 4 x CO_2 case and the control simulation)
shows the influence of the reduction in snow cover on one land
surface type fraction at 60°N. The areas of strongest response
were at 30°S and 40°S where an increase in cloud cover caused
the planetary albedo to also increase.

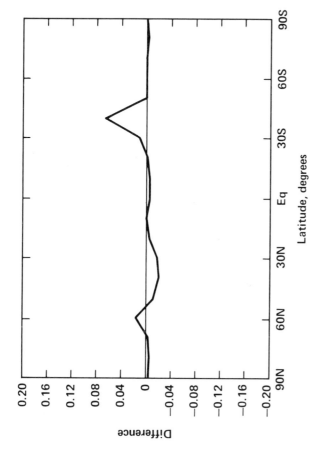

FIGURE 12. Difference in planetary albedo (2 x CO_2 -- control).

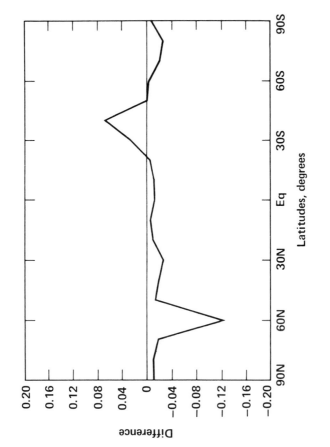

FIGURE 13. Difference in planetary albedo (4 x CO_2 -- control).

Only those Southern Hemisphere changes in planetary albedo
noted appear to have exhibited a significant effect on surface
temperatures. This result indicates that the model is possibly
too sensitive to change in cloud cover since the decrease in short-
wave radiation from increased clouds should be nearly balanced
by the increased downward longwave radiation from the clouds (19).

V. CONCLUDING REMARKS

This study was a preliminary numberical experiment to test
the effect of changing atmospheric CO_2 concentrations on the
climate simulated by a zonal atmospheric model including
numerous feedback mechanisms. The main results of this analysis
may be summarized as follows:

1. Increased CO_2 concentrations warm the troposphere and
reduce the intensity of the polar surface inversions. This
effect caused the magnification of the temperature response
with latitude. This response was similar to the Manabe and
Wetherald computation, although different mechanisms appear
to be largely responsible (1). In the case of the zonal model,
polar temperatures were very cold, and the warming associated
with the increased CO_2 and water vapor only partially compen-
sated for the intense sub-freezing temperatures. Consequently,
no snow or ice melted even though the surface temperatures
increased.

2. The magnitude of the stratospheric temperature difference
increased with altitude as CO_2 concentrations were increased.

3. Precipitation, evaporation, and the total atmospheric
water content increased with increasing CO_2 concentrations.
The Bowen ratio (H/LE) decreased with increasing CO_2.

ACKNOWLEDGMENTS

I would like to thank Dr. H. W. Ellsaesser and Dr. M. C.
MacCracken for their valuable comments and suggestions. I
also thank Mr. Ken Hill for his assistance in performing these
calculations. The author is also grateful to Dr. V. Ramanathan
for making his carbon dioxide radiation subroutine available
for use in the zonal model.

REFERENCES

1. Manabe, S., and Wetherald, R. T., *J. Atmos. Sci. 32,*
    3-15 (1975).

2. Luther, F. M., and MacCracken, M. C., *in* "Third
    Conference on CIAP: February 26-March 1, 1974, Report
    No. DOT-TSC-OST-74-15," pp. 437-449. Government Printing
    Office, Washington, D.C. (1974).

3. MacCracken, M. C., *in* "Proceedings of the Fourth Conference
    on CIAP: Department of Transportation, DOT-TSC-OST-75-38,"
    pp. 183-194. Government Printing Office, Washington, D.C.
    (1975).

4. MacCracken, M. C., and Luther, F. M., *in* "Proceedings of the
    International Conference on Structure, Composition and
    General Circulation of the Upper and Lower Atmosphere and
    Possible Anthropogenic Perturbations, Melbourne, Australia,
    January 14-25, 1974" (W. L. Godson, ed., V. II), pp. 1107-
    1128. Atmospheric Environment Service, Ontario, Canada
    (1974).

5. London, J., "A Study of the Atmospheric Heat Balance, Final
    Report." Contract AF-19 122 - 165, Research Division,
    College of Engineering, New York University (1957).

6. Potter, G. L., Ellsaesser, H. W., MacCracken, M. C., and
    Luther, F. M., "Atmospheric Statistical Dynamic Models.
    Model Performance: The Lawrence Livermore Laboratory Zonal
    Atmospheric Model," *in* "Proceedings of the GARP WMO-ICSU
    Joint Organizing Committee, A Review of Climate Models:
    Performance, Intercomparison and Sensitivity Studies."
    Government Printing Office, Washington, D.C. (1978).

7. Kahle, A. B., and Haurwitz, F., "The Radiation and Heat
    Budget of the Mintz-Arakawa Model." R-1318-ARPA, The
    Rand Corporation, Santa Monica, California (1973).

8. Katayama, A., *J. Meteor. Soc. Japan, Ser. 2, 45,* 1-25
    (1967).

9. ETAC (Environmental Technical Application Center),
    "U.S. Air Force, Northern Hemisphere Cloud Cover."
    Project 6168, Government Printing Office, Washington, D.C.
    (1971).

10. Miller, D. B., "Automated Production of Global Cloud
    Climatology Based on Satellite Data," *in* "Proceedings
    of the 3rd Technical Exchange Conference." Annapolis,
    Maryland (1970).

11. Van Loon, H., *Meteorology of the Southern Hemisphere,
    Meteorological Monographs, 13,* 35 (1970).

12. Holloway, J. L., Jr., and Manabe, S., *Mon. Wea. Rev. 99,*
    335-369 (1971).

13. Moller, F., *Petermann's Geographische Mitteilungen, 95(1),*
    1-7 (1951).

14. Budyko, M. I., "Atlas of the Heat Balance of the Earth."
    Gidormeteorizdat, Moscow, Russia (1973).

15. Sasamori, T., London, J., and Hoyt, D. V., *Meteorology of
    the Southern Hemisphere, Meteorological Monographs, 13,*
    35 (1972).

16. Schutz, C., and Gates, W. L., "Global Climatic Data for Surface, 800 mb, 400 mb: January." R-915-ARPA, The Rand Corporation, Santa Monica, California (1972).

17. Schutz, C., and Gates, W. L., "Supplemental Global Climatic Data: January." R-915/1-ARPA, The Rand Corporation, Santa Monica, California (1972).

18. Rashke, E., Vonder Haar, T. H., Bandeen, W. R., and Pasternak, M., *J. Atmos. Sci. 30,* 341-364 (1973).

19. Cess, R. D., *J. Atmos. Sci. 33(10),* 1831-1843 (1976).

MESOSCALE STUDIES OF THE LAKE EFFECT IN EARLY
WINTER OVER THE LAKE MICHIGAN BASIN[1]

Hsiao-ming Hsu
A. Nelson Dingle

Department of Atmospheric and Oceanic Science
University of Michigan
Ann Arbor, Michigan

A numerical model designed to simulate the development and
decay of mesoscale precipitation events in three dimensions is
presented. The model is based upon the primitive equations and
includes parameterization of three important small-scale
components: (1) the precipitation processes, (2) the boundary-
layer transfers, and (3) the ground surface energy and moisture
budgets.

In applying the three-dimensional mesoscale model to cold air
advection over the Lake Michigan basin, it is assumed that the
lake retains a constant uniform temperature whereas the air
stream has initially a temperature gradient consistent with
geostrophic flow. Two cases are considered corresponding to
early winter season situations: one with northwest winds and the
other with west winds. In both situations, two major local
maxima of the 12-hour totals of precipitation are predicted
along the eastern shore of Lake Michigan. The computed pattern
precipitation agrees very well with the climatic pattern of total
snowfall. The results in the hypothetical northwest-wind case
appear to correspond very well to observed precipitation amounts
for a real event in which the hypothetical conditions are
approximately fulfilled.

---

[1]Supported by NASA Langley Research Center through Grant
NSG 1243 and of the U.S. Department of Energy through Contract
AT (11-1)-1407.

I. INTRODUCTION

The lake effect which characterizes the late fall and early
winter seasons in the Lake Michigan Basin area is simulated by
means of a three-dimensional mesoscale numerical model. The
model is based upon the primitive equations using potential
temperature and scaled pressure (the Exner Function) as the
state variables (1).[2] The vertical velocity is calculated by
a modified Richardson's equation, and the scaled pressure is
determined by hydrostatic equilibrium. To control boundary
noise, a sponge-type zone is invoked along the top and lateral
boundaries of the model domain to damp the vertical velocity and
the local tendencies of the horizontal velocity, potential
temperature, and specific humidity. Three important physical
processes are included and parameterized in the model. They
are (1) the precipitation processes, (2) the boundary-layer
transfers, and (3) the ground surface energy and moisture
budgets.

The precipitation processes are separated into stable and
convective categories. The stable case invokes immediate precipitation when the relative humidity exceeds a critical value.
The convective case is parameterized according to Kuo's method
by means of a thermal source term which accounts for (1) the net
moisture convergence in the cloud depth, and (2) the vertical
distribution of latent energy associated with water vapor condensation and evaporation (2).

The boundary layer is composed of a constant-depth surface
boundary layer in which the vertical eddy exchange coefficient
is derived from Businger's empirical formula and a planetary

---

[2]*Hsu, H.-M., and Dingle, A. N., Numerical Simulations of
Mesoscale Precipitation Systems, I: The Model, 1979 (in preparation).*

boundary layer in which the profile of the vertical eddy exchange
coefficient is given by a cubic Hermite polynomial according to
O'Brien (3,4). The height of the planetary boundary layer is
predicted by two dynamical-empirical equations derived by
Deardorff and Smeda (5,6).

The ground temperature is predicted by means of the surface
energy budget which includes radiation, precipitation, sensible
and latent heat fluxes at the ground surface. The ground
moisture content and surface wetness are coupled by the methods
of Manabe, and Washington and Williamson (7,8).

In the present application the model domain covers a horizontal area of 480 km x 720 km and has a horizontal grid net of
17 x 25 points when a grid space of 30 km is chosen. Eighteen
vertical levels are used, their spacing being varied to permit
better resolution in the lower atmosphere.

Precipitation is notably generated in the Lake Michigan
Basin in late fall and early winter in the absence of frontal
activity as the result of strong transfers of heat and moisture
from the relatively warm lake surface to overriding cold air.

II. INITIATION

For the simulations, the initial temperature field is
chosen so that the atmospheric isotherms are parallel with the
prevailing synoptic-scale winds. In the case of northwesterly
flow, the coldest point is at the northeast corner of the
domain (θ = 260.26 K) and the warmest at the southwest corner
(θ = 273.0 K). The lake is specified to have and maintain a
constant surface potential temperature of 282 K. The horizontal
gradient of potential temperature is allowed to vary in the
vertical direction; thus, a vertical wind shear conforming with
the thermal wind relationship is allowed. At each grid point
the initial horizontal wind speed is zero at the surface and

reaches a maximum of 35 m sec^{-1} at the 8.5-km level following
a parabolic vertical profile.[3]

The initial moisture field is given by a vertical profile
of the relative humidity of 85% from the surface to 3 km, a
linear decrease from 3 to 7 km to 20%, and a further linear
decrease to 0% at 10 km. The roughness length is 10 cm over
land and is a function of the frictional wind speed over the
lake. Two cases are treated, one with a synoptic-scale northwesterly flow, and the second with westerly flow.

III. RESULTS

The simulations predict low-level convergence zones that
develop initially congruent with the pressure/density solenoid
fields along the upwind and downwind shores of the lake.
Figures 1 and 2 show results for the northwest wind case and
Figs. 3 and 4 show those for the west-wind case. After 2 hours
of simulation, the convergence fields at 50 m and the vertical
motion fields at 1250 m show the early organization of the lake-
effect system (Figs. 1a and 3a) into the two (upwind and downwind) zones of convergence, and in each of these, two upward
motion cells (north and south).

The vertical motion cells at the downwind side of the lake
move onshore and dissipate by the fifth hour of simulation
(Figs. 1b and 3b) whereas those at the upwind side move over
the lake and intensify. The most intense development is
achieved after about 8 hours of simulation as the vertical
motion centers approach the downwind shore (Figs. 1c and 3c).
Dissipation follows as these centers continue inland (Figs. 1d

---

[3]Hsu, H.-M. and Dingle, A. N., Numerical Simulations of
Mesoscale Precipitation Systems, II: Lake-effect Disturbances
over Lake Michigan, 1979 (in preparation).

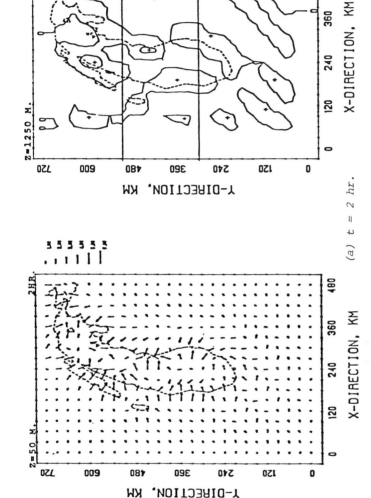

FIGURE 1. Horizontal cross sections showing the horizontal wind at z = 50 m (left) and the vertical velocity at z = 1250 m (right) for the evolution of lake-effect disturbances under prevailing NW wind. The magnitude of the wind vector is shown between the two figures in units of m-sec⁻¹. The contour interval for the vertical velocity is 10 cm-sec⁻¹.

(a) t = 2 hr.

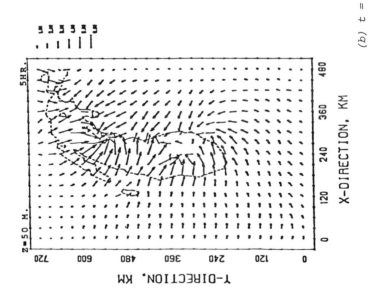

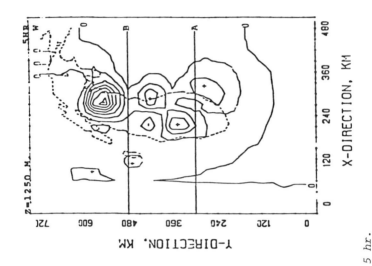

(b) t = 5 hr.

FIGURE 1--continued

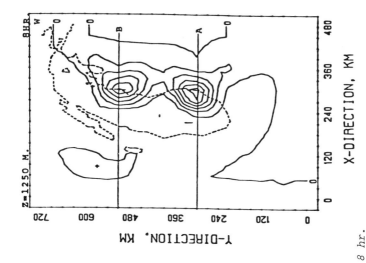

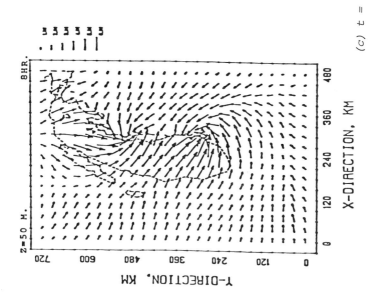

(c) t = 8 hr.

FIGURE 1--continued

461

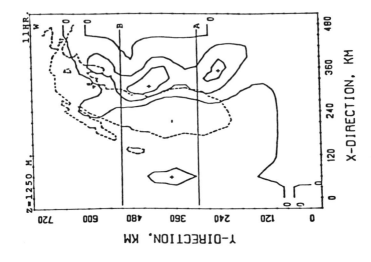

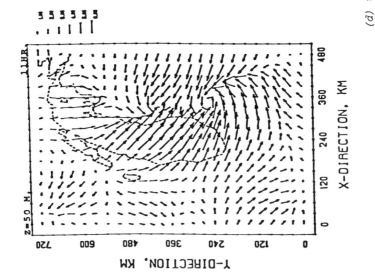

(d) t = 11 hr.

FIGURE 1--concluded

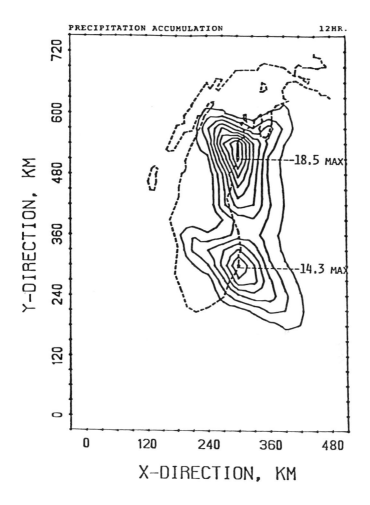

FIGURE 2. Total 12-hour precipitation for the experiment of
lake-effect disturbances under prevailing NW wind. The contour
interval is 2 mm water-equivalent.

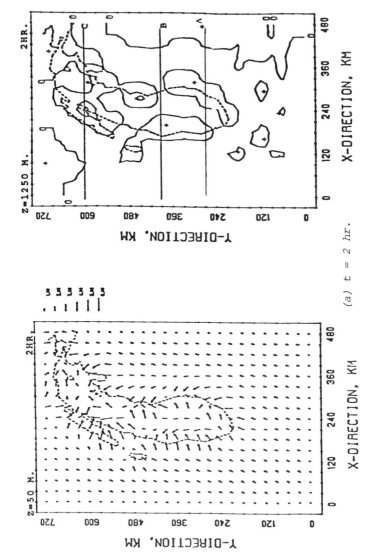

FIGURE 3. Horizontal cross sections showing the horizontal wind at z = 50 m (left) and the vertical velocity at z = 1250 m (right) for the evolution of lake-effect disturbances under prevailing W wind. The magnitude of the wind vector is shown between the two figures in units of m-sec⁻¹. The contour interval for the vertical velocity is 10 cm-sec⁻¹.

(a) t = 2 hr.

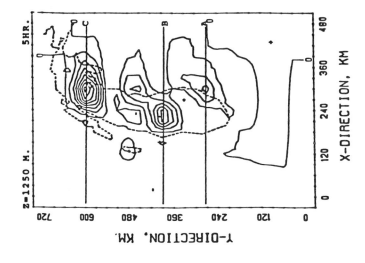

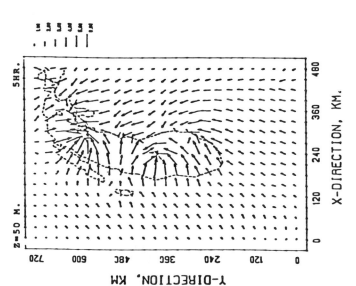

(b) t = 5 hr.

FIGURE 3 --continued

465

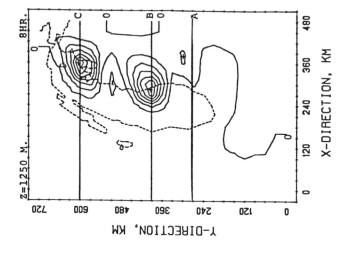

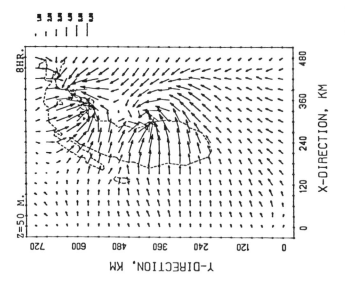

(c) t = 8 hr.

FIGURE 3 --continued

466

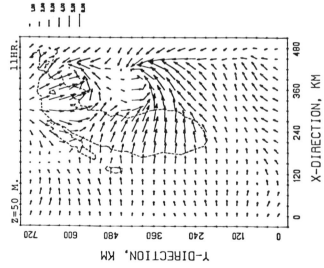

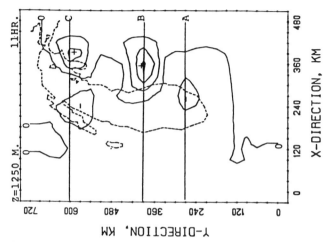

(d) $t = 11$ hr.

FIGURE 3--concluded

467

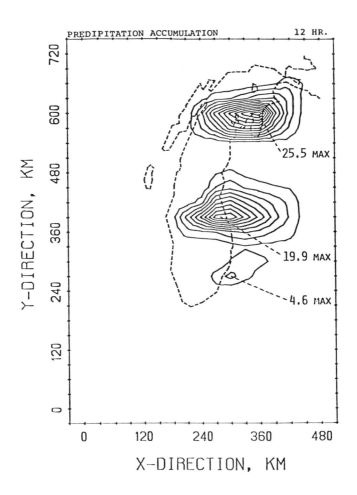

FIGURE 4. Total 12-hour precipitation for the experiment of
lake-effect disturbances under prevailing W wind. The contour
is 2 mm water-equivalent.

and 3d) and is essentially complete after 12 hours of simulation.
At this time a regeneration of the upwind-shore solenoid field
and vertical motion pattern is indicated so long as the environmental conditions (lake and air temperatures and synoptic scale
wind velocity) remain constant.

The resulting pattern of precipitation is distributed in two
prominent maxima (Figs. 2 and 4), the exact locations of which
are determined by the general direction of the synoptic-scale
flow. The intensities of these precipitation maxima appear to
depend mainly upon the lake/air temperature difference.

IV. COMPARISON AND OBSERVATIONS

The hypotheses used in the simulations were approximately
fulfilled in the weather situation of 10-11 December 1977.
Figure 5 shows the sea level weather map for 11 December at
0700 EST. Regional surface weather maps for 1200 EST on
10 December and 0300 EST on 11 December are shown in Fig. 6.
The lake-effect pressure trough is a definite feature on each
of these weather maps, and clearly indicates a low-level
convergence pattern similar to that produced by the simulations.
Precipitation patterns are inadequately resolved by the available data, but an extensive maximum south of Muskegon,
Michigan, is strongly indicated by the hourly reports from
1100 EST on the 10th through 0900 on the 11th. A second and
lesser maximum of precipitation in the Traverse City to
Cheboygan area is also indicated by the hourly data. The
correspondence of simulated and observed precipitation patterns
is by no means perfect, but considering that the real data are
only very roughly approximated by the initial and boundary
conditions for the simulations, it is remarkably close.

Satellite photographs are shown for 1231 EST on 10 December
in the visual spectrum and for 0300 EST on 11 December in the

SUNDAY, DECEMBER 11, 1977

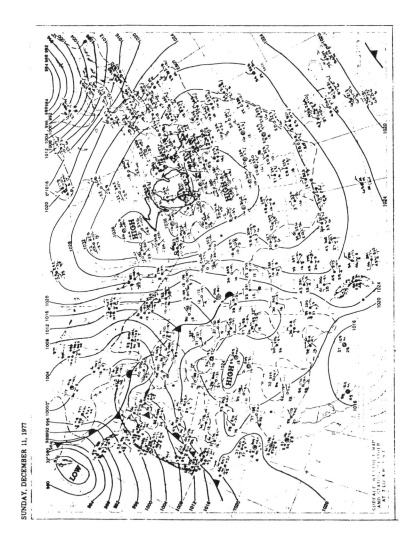

FIGURE 5. Surface weather map and station weather at 0700 EST, 11 December 1977.

470

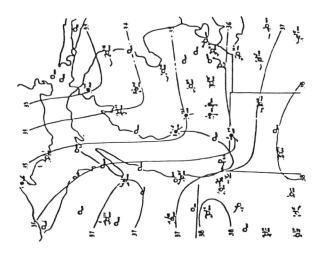

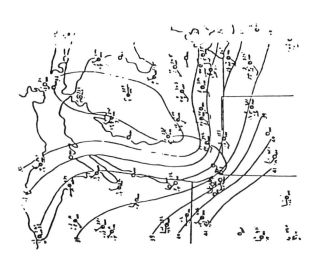

FIGURE 6. Surface maps at 1200 EST, 10 December (left) and 0300 EST, 11 December (right) 1977.

infrared in Figs. 7 and 8, respectively. Although the correspondence to the simulated precipitation patterns is only sub-
jective, the evidence of the two separated maxima appears quite
strongly in Fig. 7.

The climatic map (Fig. 9) shows the distribution of average
annual snowfall (9). Although the climatic gradient of temperature and the surface contours have the effect of intensifying
the snowfall toward the north and toward the highest land areas,
the separate maximum near and southward from Muskegon is
evidently significant over the 20-year period represented
(1940-1969).

V.   CONCLUSION AND OUTLOOK

It appears evident that the model has considerable predictive capability in these lake-effect applications. Although
the precipitation data are inadequate to resolve all details
of the pattern for any real case, it is planned to initiate the
model by means of real data for this case (10-11 December 1977)
and other cases, and to test its predictions against case-by-
case observations. Adjustment of critical parameters will be
possible by this means, and the predictive capability of the
model will be improved.

Extension of the model application to pollution removal
and deposition is contemplated as a further development. The
requisite parameterizations will account for contaminant species
differences and for the microphysical variations as the precipitation changes with latitude and season from rain to snow
(see, for example, (10)).

FIGURE 7. Satellite image (visual) at 1231 EST,
10 December 1977.

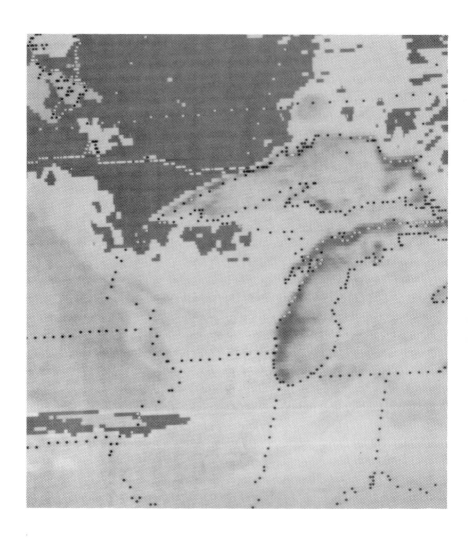

FIGURE 8.   Satellite image (infrared) at 0300 EST,
11 December 1977.

FIGURE 9.   *Average annual snowfall in inches for the period of 1940-1969 for the State of Michigan.*

ACKNOWLEDGMENT

We would like to thank Dr. Gerald L. Pellett of NASA-LRC
for his interest and support.

REFERENCES

1. Hsu, H. -M., "Numerical Simulations of Mesoscale
    Precipitations." Department of Atmospheric and Oceanic
    Science, University of Michigan, Ann Arbor, Michigan
    (1979).
2. Kuo, H. -L., J. Atmos. Sci. 31, 1232-1240 (1974).
3. Businger, J. A., in "Workshop on Micrometeorology,"
    pp. 67-100. American Meteorological Society, Boston,
    Massachusetts (1973).
4. O'Brien, J. J., J. Atmos. Sci. 27, 1213-1215 (1970).
5. Deardorff, J. W., Boundary-Layer Meteorology, 7, 81-106
    (1974).
6. Smeda, M. S., "Incorporation of Planetary Boundary Layer
    Processes into Numerical Forecasting Models." Report,
    DM-23, Dept. of Meteor., Univ. of Stockholm and Intl.
    Meteor. Inst. in Stockholm, Stockholm, Sweden (1977).
7. Manabe, S., Mon. Wea. Rev. 96, 739-774 (1969).
8. Washington, W. M., and Williamson, D. L., "General
    Circulation Model of the Atmosphere, V. 17, Methods in
    Computational Physics" (J. Chang, ed.), pp. 111-172.
    Academic Press, New York (1977).
9. Michigan Department of Agriculture, Michigan Weather
    Service, "Climate of Michigan by Stations." Michigan
    Weather Service, 1405 S. Harrison Rd., East Lansing,
    (1971).
10. Scott, B. C., J. Applied Meteorol. 17, 1375-1389 (1978).

DISCUSSION

Mitchell: These convergence centers that you show as highly
cellular in both Florida and the Lake Michigan cases, how
sensitive would the positioning of these centers be to slight
changes in the prevailing wind direction that you start with?

Hsu: We haven't had a chance to run enough new cases to answer
this question. But predominantly they depend on the synoptic
wind pattern. That is what we are trying to do next and to
include synoptic forcing through the boundary. We are trying
to study more, particularly, real cases. Did I answer your
question?

Mitchell: That's fine. I was interested for some time in this
so-called LaPorte phenomenon that the Illinois Water Survey
people have looked at. And the thought crossed my mind that
perhaps partical explaination of that might be involved in the
kind of things that you are pointing out here. So if you ever
get to study the LaPorte phenomenon, that would be an interesting
thing to look at.

Eadie: Quite often, lake effect snow bands are quite narrow, at
least if you look at them maybe when they come down a long axis
of a lake. And I notice, in your model similar to the very
primitive warm model, the conversion zones and your precipitation
zones are quite broad. I wonder if you think there is any
difference to reality of the error or what you think could be
improved in any way?

Hsu: I think that there are several elements contributing to
these wide bands in precipitation pattern. Of course, we
realize that there is one major element because of the computer
core limitation, we have to use 30 km resolution which is very
unfortunate and we would expect to use 15 km or even 10 km
resolution to resolve this situation. The strength of the total
precipitation also depends on the individual synoptic situation.
If the wind comes from north, northwest or west and it is very
cold--in other words, the temperature difference between the
water surface and the air flow above the lake is very large--
we would have very strong lake-effect storm passing over the
lake which will distribute a very wide band over the Lake
Michigan region. I indicated in one transparency compiled by
Changnon and Gatz for the Lake Michigan area that the band is
really wide. So the precipitation band depends on the weather
situation.

POLLUTION DISPERSAL MODELING AND REGIONAL
PUBLIC POLICY IMPLICATIONS

A. E. S. Green   J. M. Schwartz
E. T. Loehman[1]   R. A. Hedinger
S. V. Berg   V. E. De
R. W. Fahien   D. E. Rio
M. E. Shaw   T. J. Buckley
M. J. Jaeger   R. P. Fishe
H. Wittig   W. F. Rossley
A. A. Arroyo   D. Trimble

Interdisciplinary Center for Aeronomy and (other)
Atmospheric Sciences (ICAAS)
University of Florida
Gainesville, Florida

The overall approach and principal results of a study undertaken by members of the Interdisciplinary Center for Aeronomy
and Atmospheric Sciences (ICAAS) at the University of Florida are
presented. The study was carried out under the auspices of the
Florida Sulfur Oxide Study (FSOS) Inc., a nonprofit corporation
funded by the Electric Utilities of Florida and supervised by a
board consisting of representatives of the Electric Utilities,
environmental groups and of the Florida Department of
Environmental Regulation. ICAAS's efforts encompassed three
specific tasks: (1) modeling dose-response relationships for
sulfur oxide related pollutants, (2) risk-benefit analysis and
modeling of decision and policy alternatives, and (3) the
effect of sulfur dioxide on lung mechanics of asthmatic subjects.
In these connections, an attempt was made to carry out an integrated interdisciplinary assessment of pollution abatement
alternatives in the Tampa area, a region where coal is a major
source of electric power. Those aspects of the results that
relate to the environmental impact of coal plant emissions are
presented.

[1]Now at SRI International, Palo Alto, California.

I. PUBLIC POLICY METHODOLOGIES FOR COAL BURNING ISSUES

ICAAS's initial conceptualization of models for public policy
decisions on regulating air pollution began in 1970 in connection
with a proposed Air Quality Index (AQI) Project (1). This AQI
project was intended as a broad socio-technical research program leading to the establishment of a quantitative scale (or
scales) for air quality, the validation of this scale (or
scales), and the development of laws to implement air quality
control, particularly in Florida and the Southeast. From this
AQI program plan, the first practical cost/benefit analysis
approach was developed. Figure 1 is an adaptation of a diagram
from a proposal entitled "Preservation and Enhancement of Air
Quality (PEAQ)," August, 1971 (2). Essentially, it proposed an
economic benefit/cost analysis as a public policy decision
methodology in which pollution transport plays a major role.

These overall systems view approaches were finally imple-
mented in the FSOS study (3,4). The flow diagram illustrates a
chain of studies within the framework of two types of public
policy decision methodologies on sulfur oxide pollution. One
methodology, the Disaggregated Benefit/Cost Analysis (DB/CA), is
essentially an advanced form of economic analysis in which the
distributional aspects of benefits/costs are considered (i.e.,
the question of who gets the benefits and who pays the costs is
addressed) (5). The second methodology, the quantitative
acceptable level of risk (QALR), is a noneconomic analysis which
bypasses many of the difficult problems of translating important
decision factors into monetary terms.

The QALR methodology was in part in response to concerns
about criticisms of any procedure which places dollar values on
effects, such as human health, death, aesthetics, social good,
etc. In part, some members of ICAAS were responding to a move-
ment to quantify social questions in noneconomic terms or in

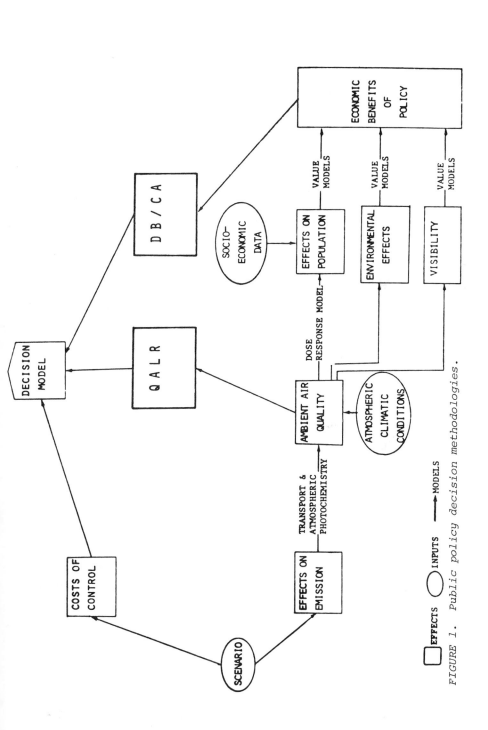

FIGURE 1. *Public policy decision methodologies.*

terms of the title of the article by Bertrand Gross entitled
"Let's Not Leave It to the Economists" (6). Consequently, other
possible public policy decision methodologies were explored.
One of these was an accounts methodology in which the components
issues involved in an overall policy decision are assembled into
various categories (7,8). Quantitative indices related to these
accounts are determined separately and a public policy board
assigns rules of combination and the weight factors to the components in some final optimization scheme.

A closely related approach was implicit in the words of
Dr. Edward D. Palmes in the NAS/NRC report on Sulfur Oxide,
1974,

It should be recognized at the outset that there is no
value other than zero that will carry with it the assurance
of absolute safety or zero risk. There are, however,
finite concentrations that, in the light of present understanding, would reasonably be expected to produce a very
small risk of adverse effects. The acceptability of a
degree of contamination depends upon the appraisal of the
risk at the present time in relation to the benefits of the
polluting process(es) (9).

Following this line of thought is in effect seeking to quantify
components of a judgmental process used in politics and
systematized extensively in World War II operations analyses.
As an illustration, consider a military decision made in March
1945 to use fire bombs on Japan in low-level night attacks
rather than explosive bombs in high-altitude daylight attacks.
The extra tonnage on target amplified by fire storms led to the
rapid destruction of Tokyo, Osaka, Nagoya, Kobe, and other
major Japanese cities. This incendiary bomb campaign was a
turning point in World War II and had already produced a
decisive result by July 1945 before the atomic bomb was
dropped (10). In essence, the decision was to go to a higher,

but still acceptable, level of risk (expected per mission losses of the B29 fleet) in order to achieve the desired result (destroy Japan's war-making potential). But for the lack of effective diplomacy the war might have ended in July 1945 without the use of atomic bombs.

Although the QALR and the accounts methodologies were discussed in the exploratory phase of ICAAS-FSOS program, most of our efforts were applied to a disaggregated benefit/cost analysis (DB/CA) which to a large extent was fostered by the development of pollution-dispersal modeling. This report describes the overall approach and principal results of the ICAAS-FSOS study--particularly as these results related to pollution dispersal modeling and the public policy implications of coal burning on a regional basis.

II. POLLUTION DISPERSAL MODELING AND DISTRIBUTIONAL ASPECTS
     OF BENEFIT COST ANALYSES

One of the unique aspects of regional analyses is the fact that air quality dispersion modeling depends intimately upon the spatial distribution of emission sources and the local meteorological conditions (11). Further regional differences in environmental control impacts arise because of physical and socioeconomic differences and the spatial distribution of the affected population relative to pollution sources.

The fact that regional B/C analyses must be carried out with detailed consideration of the spatial distribution of sources, geographic features, population groups, etc., fosters the consideration of the distributional aspects of benefits and costs. A long-standing traditional benefit-cost analysis adds benefits and costs from a policy change according to "whomsoever they may accrue" and seeks out decisions for which

ratio B/C is greater than one. This criterion is closely related
to notions of economic efficiency but ignores distributional
issues. Thus, the benefit-cost criterion might be satisfied
under conditions where costs of a policy are borne by a majority
while benefits accrue to a minority.

An alternative criterion to the benefit-cost criterion in
our society is based on majority voting. Then costs to a
majority and benefits to a minority would be unacceptable under
this criterion. However, a majority voting criterion may
result in an inefficient choice since a majority may receive
benefits while a minority bears costs greater than the benefits.
The hope that policies can be found which would pass both
benefit-cost and voting criteria was a major aim of the ICAAS-
FSOS study (5). The results suggest that efficiency and equity
criteria may indeed be combined into a more acceptable policy
choice. Thus, in addition to information on total benefits and
costs, information on disaggregation of benefits and costs is
needed for a complete evaluation of air quality control
alternatives. A regional B/C analysis provides an opportunity
for attempting such a complete evaluation. In the ICAAS-FSOS
study a specific case was used--some pollution abatement
alternatives for electric power plants in the Tampa area, a
coal burning region of Florida. Most of the methodology,
however, is general, and hence it can be applied to any region.

III. AIR QUALITY DISPERSION MODELING IN THE ICAAS-FSOS STUDY

An air quality dispersion model predicts the concentration
of pollutants that would occur at various sites within a given
study area, as a result of specified emissions from known
sources under meteorological conditions. In public policy

analyses, it is necessary to estimate the change in pollutant concentration that is likely to occur as a result of a change in policy alternatives (and its corresponding change in emissions).

In view of the complexity of the problem and the simplicity of the Gaussian model, it is perhaps surprising that the Gaussian model gives results as good as it does. However, since the model does not take into consideration surface and other conditions, it is common practice to "calibrate" it, that is, to determine a factor by which the calculated con- centrations must be multiplied. This factor must be obtained by comparing calculated concentrations at various locations in the area with those actually measured. The assumption is then made that the same factor applies to all calculated concentra- tions at all conditions.

A. Calibration of Model

Three models were considered for use in this study: The Air Quality Display Model (AQDM), the Climatological Dispersion Model (CDM), and Turner's PTMTP (more recently the PTMTP-W) (12-14). The need for annual pollutant average emissions (required inputs to the dose-response relationships developed for this study) was met by the use of the CDM model.

Tables I and II give detailed emissions for all Tampa Electric Company (TECO) sources for 1975 and 1976, respectively. These data were extracted from reports furnished by TECO to the Florida Department of Environmental Regulation (DER) in 1976 and 1977. Merging these two sources of information creates a data set which will be referred to as "Baseline 75." Coal burning plants are denoted by asterisks. Figure 2 gives the point sources of particulates and sulfur dioxide in the

TABLE I. TECO 1975 Emissions Inventory

Point	Station identification	SO_2 (g/s)	Part. (g/s)	h_s (m)	D (m)	V_s (m/s)	T_s (°C)
1	Hooker's Point #1 & 2	14.44	0.80	45.72	3.66	2.22	126.7
2	Hooker's Point #3 & 4	22.27	1.23	45.72	3.96	2.86	123.9
3	Hooker's Point #5	40.96	2.26	53.34	3.66	5.52	140.6
				(53.34)	(3.89)	(6.38)	(130.0)
4	Hooker's Point #6	16.34	0.90	85.34	2.87	11.69	162.8
5	Gannon #1	486.21	23.56	60.96	4.30	8.23	153.9
6	Gannon #2	429.08	10.47	93.27	3.05	15.76	153.9
						(18.71)	
7	Gannon #3	348.8	19.10	93.27	3.23	19.69	148.9
8	Gannon #4	63.06	3.37	93.37(2)	2.93	11.83	148.9
						(14.94)	
9	Gannon #5*	457.13	45.83	93.27	4.45	15.36	142.2
						(16.28)	
10	Gannon #6*	945.76	3.57	93.27	5.36	16.31	143.9
11	Big Bend #1* & 2*	3003.96	67.14	149.35	7.32	20.56	160.0

TABLE II. *TECO 1976 Emissions Inventory*

Point	Station identification	SO_2 (g/s)	Part. (g/s)	h_s (m)	D (m)	V_s (m/s)	T_s (°C)
1	Hooker's Point #1 & 2	4.51 ⎫	0.34 ⎫	45.72	3.66	2.50	126.7
2	Hooker's Point #3 & 4	12.82 ⎬ 37.76	1.05 ⎬ 3.49	45.72	3.96	2.69	123.9
3	Hooker's Point #5	20.43 ⎭	2.10 ⎭	53.34	3.66	4.73	140.6
4	Hooker's Point #6	18.90	1.65	85.34	2.87	14.83	162.8
5	Gannon #1	166.47	8.89	93.27	3.05	8.23/7.38	153.9
6	Gannon #2	126.49	6.13	93.27	3.05	9.08/8.23	153.9
7	Gannon #3	79.31	2.33	93.27	3.23	19.42	130.0
8	Gannon #4	102.44	7.42	93.27	2.93	12.95	141.1
9	Gannon #5*	397.09	4.60	93.27	4.45	15.67	142.2
10	Gannon #6*	360.30	3.37	93.27	5.36	15.00	144.4
11	Big Bend #1*	2564.30	35.90	149.40	7.30	21.03	131.7
12	Big Bend #3*	1081.30	5.20	149.40	7.62	9.88	137.2

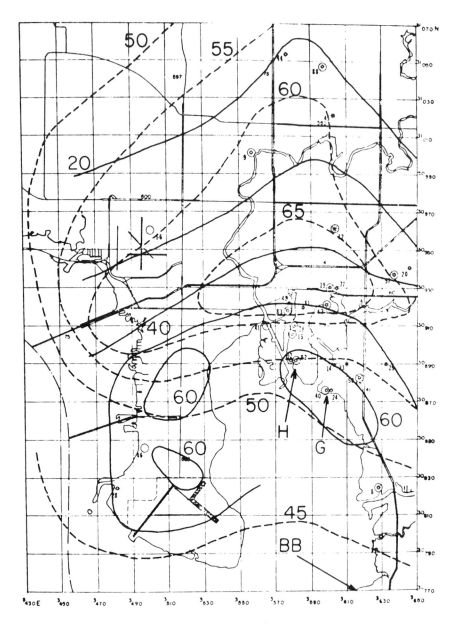

FIGURE 2.　Tampa/Hillsborough Baseline 75.　H denotes Hooker
Point, G-Gannon and BB-Big Bend.　The solid line contours give
calibrated SO_2 levels in $\mu g/m^3$.　The broken line contours give
the calibrated TSP levels in $\mu g/m^3$.

Tampa urban area in 1975. As of January 1978, the coal burning
plants at Gannon and Big Bend were the only urban coal burning
generators in the State of Florida.

B. *Average Meteorological Conditions*

ICAAS used a monthly/yearly 8 stability wind rose for the
weather station located in the Tampa Airport. This wind rose
contains observations from 1970 through 1975. By extracting
the yearly summary and combining stability classes 6, 7, and 8
into a single class (class 6), a complete 96 card stability
wind rose is obtained (16 wind direction classes, 6 stabilities
with 6 wind speed classes per card).

C. *Measured Concentrations*

A listing of all relevant SO_2 monitoring stations as well
as relevant parameters was extracted from "Air Quality
Measurements," Hillsborough County Environmental Protection
Commission, prepared by Shaw (15). By including the coordinates
of the stations as receptor points, the CDM will give values
for each station.

A plot of observed annual SO_2 concentrations against those
predicted by the CDM was made; this plot indicated considerable
data scatter. An analysis of all data led to the discovery that
the SO_2 bubblers were not temperature controlled (1975
Hillsborough Air Quality Report) and these data are not considered to be reliable. Since developing a methodology was the
main interest, only approximate ground level annual SO_2 concentration specific to a given census tract was needed. The
values for Station 63 and Station 81 were utilized to obtain
the calibration factor. Figure 2 also gives a graphical display
of the Baseline 75 results.

D. *Particulates*

Tables I and II give TSP emissions for all of the point
sources; thus, a similar strategy was pursued. In the case of
TSP, point sources contribute less than 40% of the total
emissions. Area sources and fugitive dust cannot be neglected.

The 1975 Hillsborough Air Quality Report includes are
source emissions for 1975 (includes fugitive dust and paved
roads), and TSP station data. These data were coded and
incorporated into the model (132, 5 km grid squares). SO_2
concentrations for area sources did not significantly change
the previous results, although they were also incorporated.

A plot of observed annual TSP concentrations against those
predicted by the CDM model was made. Regression analysis was
used to predict the "best linear fit." The resulting line is
plotted in Fig. 3. Note that for TSP a background level of
about 40 $\mu g/m^3$ is not unreasonable in view of the high distributed TSP emissions. The resulting TSP concentrations are
shown by the dashed contours in Fig. 2.

IV. SCENARIOS

The effects that may occur as a result of a given air pollution policy alternative are best described in terms of various
scenarios. A scenario consists of a comparison of the results
obtained after a hypothetical decision is made concerning some
type of emission regulation with those obtained from a given
base year. In this section an enumeration of the various control alternatives as applied to the existing conditions in the
Tampa area is made by isolating the effect of emission control
policy as applied to the local power industry.

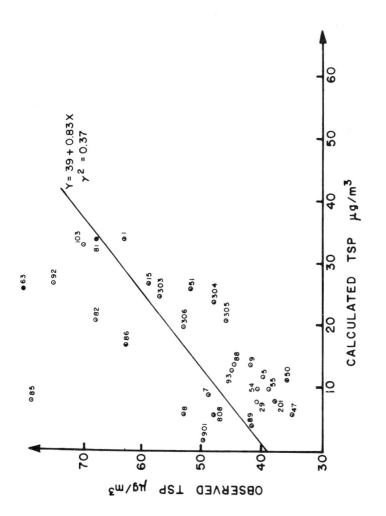

FIGURE 3. *Observed compared with calculated TSP levels for Tampa 1975.*

In order to effectively synthesize control alternatives, and for modeling purposes, an emission base year is selected. Because of the availability of various calendar year 1975 reports on Tampa Bay air quality as well as other FSOS participant reports, 1975 was selected the base year with the following modifications: "Baseline 75" consists of 1975 data supplied by PEDCO, Inc., Hillsborough County, and Tampa Electric Company. "Baseline 76" consists of data for the power generating stations supplied directly by TECO for calendar year 1976 with 1975 data for all other sources. These data consist of emission estimates, amount and type of various fuel mixes, source characteristics, ambient air quality levels, meteorology, economic analyses, socio-demographic characteristics, etc.

A major consideration in this study area is the bringing on-line of the Big Bend 3 (BB 3) generating station during April of 1976. Because of the increased capacity of the Big Bend generating station, the corresponding load at the Gannon generating station decreased. Associated with this load switching is a corresponding decrease in ambient SO_2 air quality levels in the downtown Tampa area (a high population density area), and an increase in ambient SO_2 levels in the area south of Tampa (a low population density area). As expected, total power plant SO_2 emissions increased from 1975 to 1976, yet geographical considerations accounted for a net decrease in ambient SO_2 levels around the downtown Tampa area.

New source pollution standards (NSPS) are aimed at reducing total emissions in a given area. This regulation is applicable to sources constructed after a given date, regardless of plant site or ambient meteorological conditions. In this manner, SO_2 emissions from future point sources will be strictly controlled. Scenarios were investigated in which Big Bend 3, although already operational, was required to meet the NSPS.

Based on data supplied directly from TECO, the NSPS, as applied to the three TECO generating stations, is equivalent to burning low sulfur content fuels (0.5% sulfur content oil and coal). Since the cost of using these "clean" fuels is excessive (due to scarcity and transportation costs), scrubbers and other type devices are the actual controls. Emissions from low sulfur fuels are equivalent to emissions from these "controlled" sources.

Other power plants (besides BB 3) could also burn lower sulfur content fuels. Thus, NSPS could be applied to any of the TECO stations. The effects of a strict emissions control policy can be modeled under these various conditions. Alternative "scenarios" are developed to test such policy alternatives.

One set of scenarios, S2 to S6, test the alternatives during the "post" BB 3 construction period. These conditions are more representative of the present-data pollution control policies as applied to the specifics of the Tampa Area. Another set of scenarios, S7 to S11, test the conditions during the BB 3 construction period. They illustrate alternatives that were present during the 1975-1976 period. The last set of scenarios, S12 to S16, test conditions present during the "pre" BB 3 construction period. They illustrate alternatives that could have been applied prior to April of 1976.

The next few subsections are devoted to expanding the specifics of scenario development, calculating costs associated with each alternative, deriving the various emissions under each alternative, presenting preliminary results, and commenting on the various alternatives. The 16 scenarios investigated are summarized in Table III. An example is S1.

TABLE III. Scenarios for FSOS

Code	Statement	Purpose
S 1	Load switching BB 3 comes on line	Compares levels for 1976 with levels for 1975 (baseline 76 with baseline 75)	Load switching effects plus geographical implications, no costs
S 2	Current NSPS to BB 3	Compares levels for 1976 with NSPS at BB 3 with baseline 76	Simulates BB 3 conversion to low sulfur coal at a cost of $27.9 million
S 3	Current NSPS to all TECO	Compares levels for 1976 with NSPS at all TECO plants with baseline 76	Simulates most stringent abatement policy on all TECO units at a cost of $98.3 million
S 4	Current NSPS to Gannon and Hooker's Point	Compares levels for 1976 with NSPS at Gannon and Hooker's Point with base-line 76	Simulates effect of NSPS applied to old TECO plants (where people live) at a cost of $23.193 million
S 5	Current NSPS at Gannon and Hooker's Point with tall stacks	Compares levels for 1976 with NSPS at Gannon and Hooker's Point with base-line 76 using 1974 stack heights	Simulates effect of NSPS as applied to Gannon and Hooker's Point and increased stack heights at a cost of $23.72 million

TABLE III—continued

Code	Statement	Purpose
S 6	Current increased stack heights	Compares levels for 1976 (baseline 76) with base-line 76 using 1974 stack heights	Simulates cost effectiveness of increased stack heights at a cost of $579,156
S 7	Same as S2 during construction phase	Compare levels for 1976 with NSPS at BB 3 with baseline 75	Simulates application of NSPS as soon as BB 3 comes on line; costs $27.9 million
S 8	Same as S3 during construction phase	Compare levels for 1976 with NSPS at all TECO plants with baseline 75	Simulates application of NSPS over all plants as BB 3 comes on line; costs $98.3 million
S 9	Same as S4 during construction phase	Compare levels for 1976 with NSPS at Gannon and Hooker's Point with base-line 75	Simulates application of NSPS at Gannon and Hooker's Point as BB 3 comes on line; costs $23.193 million
S10	Same as S5 during construction phase	Compare levels for 1976 with NSPS at Gannon and Hooker's Point with base-line 75 using 1974 stack heights	Simulates application of NSPS at Gannon and Hooker's Point plus increased stack height as BB 3 comes on; cost $23.7 million

495

TABLE III—continued

Code	Statement	Purpose
S11	Same as S6 during construction phase	Compare levels for 1976 with baseline 75 using 1974 stack heights	Simulates effect of stack height from one year to the next; cost, see S6
S12	No construction phase NSPS at BB 1-2	Compare levels for 1975 with NSPS at BB 1-2 with baseline 75; assume same cost as BB 3	Simulates applying NSPS to a city with no construction; cost same as S2
S13	No construction phase NSPS at all TECO plants	Compare levels for 1975 with NSPS at all TECO plants with baseline 75	Simulates NSPS applied to a city with no construction; cost similar to S3
S14	No construction phase NSPS at Gannon and Hooker's Point	Compare levels for 1975 with NSPS at Gannon and Hooker's Point with baseline 75	Simulates NSPS applied to a city with no construction or downtown units; costs like S4

TABLE III--continued

Code	Statement	Purpose
S15	No construction phase NSPS at Gannon and Hooker's Point	Compare levels for 1975 with NSPS at Gannon and Hooker's Point with baseline 75 at 1974 stack heights	Simulates NSPS applied to downtown units in a city with no construction; costs like S5
S16	No construction phase stack height increase	Compare levels for 1975 with baseline 75 using 1974 stack heights	Simulates stack height increase in 1975; costs like S6

Scenario Sl: Geographical Implication & Load Switching

Sl: Compare ambient air quality levels obtained from the

Baseline 76 data set with those obtained from the

Baseline 75 data set.

The purpose of Scenario Sl is to obtain an estimate of the

partial health benefits associated with pollutant emissions away

from population centers as compared with emissions in populated

areas.   The effects of load switching as a means for ambient

air quality level improvement is also illustrated with this

scenario.   It should be noted that total emissions increased

from 1975 to 1976, yet ambient levels in Tampa experienced a

slight decrease.

Other scenarios are described in detail in Ref. 4.

V.   HEALTH EFFECTS OF SULFUR DIOXIDE

One of the major issues in the Florida Sulfur Oxide Study

(FSOS) as a whole was the basic question of whether SO_2 has a sig-

nificant health impact at the levels of the National Ambient Air

Quality Standards (NAAQS) or the Florida Ambient Air Quality

Standards (FAAQS).   This question was exacerbated during the

course of the FSOS study when it became clear that the present

state of knowledge of health effects of low levels of pollutants

is still at a very primitive stage.   Congressional hearings held

in November 1975 on "The Costs and Effects of Chronic Exposure

to Low Level Pollutants" indicated that almost nothing is known

about the health effects of this subject (16).   Furthermore, the

aerometric methods for measuring pollutant concentrations were

also found to be of questionable validity.

The major responsibilities for clarifying the epidemiological

impact of SO_2 were given to a University of Michigan

Epidemiology Group (17).   However, responsibility for a critical

question was assigned to a University of Florida Medical team.

This study was designed as a test of the effects on pulmonary function of 0.5 ppm of SO_2 for 3 hours, the present Federal secondary standard and Florida's primary standard. Forty normal, nonsmoking subjects and 40 subjects with a history of mild to moderate asthma, but with no recent severe attacks, participated in the study. Ten lung function tests were applied, five of which being of the so-called "sensitive type." Double blind procedures were used.

The subjects were exposed for 3 hours either to air or to 0.5 ppm SO_2 gas in air. The normal subjects (16 females, 24 males) were of an average age of 25 ± 5.7 years. They were screened by history, by physical examination, and by preliminary lung function testing to establish that they were free of disease or pulmonary dysfunction. They were paid an incentive to participate, and informed consent was obtained according to a protocol established by the Human Experimentation Committee.

The 40 asthmatic subjects (19 female, 21 males) were of an average age of 27.1 ± 9.2 years. Patients known for the mild course of their disease were selected and patients suffering from acute exacerbations were excluded. Pay and informed consents were the same as in the control group. The exposures took place in a 30-foot travel trailer which was equipped with a 6 x 6 x 5 foot exposure chamber with two comfortable chairs and a table with reading material; 1% SO_2 in N_2 was delivered into the exposure chamber at a rate of approximately 0.013 liters/minute and mixed with 250 liters/minute ambient air delivered by a calibrated air pump. The control exposures were in ambient air with an average SO_2 content of 0.005 ppm, but on occasion, as high as 0.015 ppm. The exhaust was vented to the outside of the trailer. The SO_2 level was continuously sampled in the exposure chamber and monitored at 15-minute

intervals on a Meloy SO_2 meter (type 185-2A; minimum sensitivity
0.005 ppm). The meter was calibrated daily with a Meloy
calibrator (type CS10). The temperature in the chamber was
$22 \pm 2^{\circ}C$; the humidity was $60 \pm 10\%$.

Two subjects were exposed simultaneously. When possible,
a normal and an asthmatic were paired. The subjects did not
know whether they were exposed to air or SO_2. The sequence
of exposure to SO_2 or air was random, but air and SO_2 exposures
of an individual took place at the same hour of the day. Thus,
a subject exposed to SO_2 during the mid-morning would be exposed
to air during mid-morning of another day. The subjects wore a
nose clamp. All subjects underwent, before and after each
exposure, 10 lung function tests in a laboratory situated
4 walking minutes from the trailer.

The data in normals showed a pattern of progressive
bronchodilation during air exposures which, presumably, reflected
a circadian rhythm. This rhythm was not changed by the exposure
to SO_2. In asthmatics the circadian rhythm was more pronounced.
SO_2 resulted in an almost immediate decrease of mid-maximal
expiratory flow rate (MMFR) which lasted throughout the
exposure. The magnitude of this decrease of MMFR was smaller
than the amplitude of the circadian variation.

Three subjects, two in the asthmatic and one in the normal
group, complained of wheezing in the chest during the night
following exposure to SO_2. The sympton was not severe, and
certainly did not amount to an asthmatic attack. However, it
seems reasonable to conclude that some sensitive persons may
respond to concentrations of SO_2 in this range.

VI. THE FSOS-ICAAS DOSE RESPONSE APPROACH

The development of dose-response relationships for air-
pollution-induced illness presented serious obstacles during
the FSOS study (4). In an attempt to bypass these serious
roadblocks, ICAAS took an initial simplifying step by defining
an air quality index as a working measure of "dose." This
dose was given in terms of the concentrations of the major air
pollutants in a biophysically reasonable and mathematically
continuous fashion and is represented by the biquadratic
form

$$D = [\Sigma_i (C_i/S_i)^2 + b_{ij} (C_i/S_i)(C_j/S_j)]^{\frac{1}{2}} \qquad (1)$$

where C_i's are the concentrations of SO_2, O_3, NO_2, CO and TSP,
S_i's are scale factors, and the b_{ij} is an interaction term (18).
Biquadratic forms of this nature have played important roles in
many fields, such as the special theory of relativity, the
optimization of industrial processes, and nonlinear least-square
minimization. The only cross product term used was $b_{1,2} = 1.5$
to represent the synergism between TSP and SO_2. In the ICAAS-
FSOS study, scaling was chosen to place the various pollutants
on an equivalent basis in terms of health effects by using the
PSI index formulation as a guide. The approach used in the
ICAAS-FSOS study was equivalent to assuming simple exponential
frequency distribution for the daily dose characterized by
$F(D) = <D>\exp(-D/<D>)$. The conventional (air pollution) hockey
stick dose response function was also assumed

$$\left. \begin{array}{ll} I(D) = I_o & D \leq D_o \\ I(D) = I_o + \lambda(D - D_o) & D > D_o \end{array} \right\} \qquad (2)$$

With these inputs the expected number of times per year that
the threshold is exceeded can be simply calculated and the
response rate function by the frequency of any level of daily
dose weighted and the result integrated. In mathematical terms

$$I\ (<D>) \ = \ <D> \int_0^\infty \lambda(D)\ e^{-D/<D>} dD \tag{3}$$

The formulation results in an annual incidence rate having the form (see Fig. 4)

$$I(D) \ = \ I_o \ + \ De^{-D_o/D} \tag{4}$$

where now $<D> = D$ is the average annual dose.

Five health effects were modeled: asthma, chronic bronchitis, lower respiratory illness in children, chest pains, and eye irritation. Acute illness with attacks of short duration, such as asthma, were modeled differently from chronic illness with long duration, such as bronchitis. For asthma, prediction of the number of attacks per exposed population (Z) as related to age A, sex, and dose D is made. The generalized representation assumed is that

$$Z \ = \ Z(D,A) \ = \ F(A)K(D) \tag{5}$$

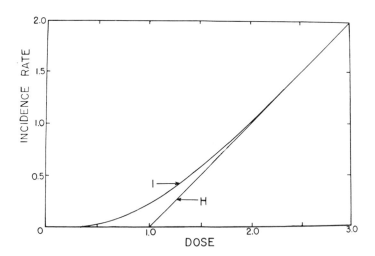

FIGURE 4. Variation of incidence rate with dose. I is the incident dose function used in this work. H is the usual hocky stock function with threshold D = 1.

where F(A) is the prevalence (number of cases of asthma per persons in each age group) and K(D) is the number of attacks per year per asthmatic. Thus, it is assumed that pollution affects the number of attacks per asthmatic but not the number of asthmatics. The form of F(A) is chosen to fit the available data. The product of these two forms results in a model which predicts a base number of asthma attacks depending on age and sex which would be present regardless of pollution plus an increment which depends on pollution.

Because of lack of data, it is not possible simultaneously to determine age and dose dependence. Parameters for F(A) and K(D) for this model were estimated separately by using data from Yoshida et $al.$, and the Chess Report (19,20). Details are found in the full report (4).

Prevalence of bronchitis P (number of cases for each age group) is modeled since it is a disease of long duration. The general form is related to a cumulative hazard model

$$P = P(A,D) = 1 = e^{-Z(A,X)} \tag{6}$$

where

$$Z(A,X) = \rho_1 A + \rho_2 AK(X) \tag{7}$$

and

$$X = D + \beta Re^{-\alpha/A} \tag{8}$$

is a generalized dose dependent on dose D, smoking rate R in packs per day, and age A.

The first term represents the prevalence of bronchitis which is independent of pollution; as age goes up, the prevalence increases. The second term is separable into age and pollution effects. The form for K(X) is chosen to simulate a curvilinear

function of the form of Eq (4). The parameters were fitted by
nonlinear least-squares techniques using the data of Lambert and
Reid and Tsunetashi *et al.* (21,22). Figure 5 shows the fits to
the data.

Similar techniques were used for the other diseases and are
reported in detail in the full report (4). Parameters were
fitted using the Chess Study for Lower Respiratory Disease (LRI)
in children and Chest Pain (20). Eye Irritation parameters
were fitted by using a study of nurses made in Los Angeles (23).

The dose index D given by Eq. (1) incorporates the
synergistic effects from more than one pollutant and puts
pollutants on an equivalent basis in terms of health effects.
For example, Larsen has shown that the product of SO_2 and TSP
may be important in predicting excess mortality associated with
air pollution (24). The weights which put pollutant levels on
an equivalent basis were first based upon the PSI index
assignments of relative effects of various pollutants (25).
The levels of SO_2 and TSP for each census tract were predicted
from the CDM model. The other pollutant levels were used as
constants throughout the study and were obtained from the 1975
Hillsborough County Air Quality Report (15).

Table IV summarizes the health effects dose response models
utilized in the ICAAS-FSOS study. The dose response functions
Z_{AA}, P_{CB}, etc., were each analytic forms incorporating the
curvilinear function given by Eq. (4) in combination with a
realistic age dependence. In the case of chronic bronchitis
the rate of cigarette smoking was taken as an important
variable. Figures 5(a) and 5(b) illustrate the P_{CB} (A,D,R)
function used in analysis in comparison with data.

One further step was needed to make health response consistent with the rest of the study. As the next section explains
dollar values were determined for symptoms (severe shortness of
breath, chest pain, severe cough, minor eye irritation) instead

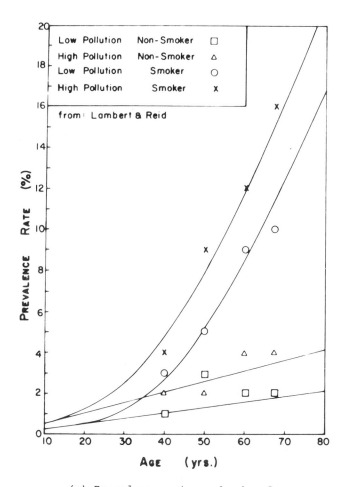

(a) Prevalence rate on basis of age

FIGURE 5.   Analytic fit on basis of age and dosage to specific
prevalence rate of chronic bronchitis for smokers and nonsmokers.
A smoking rate of 0.75 pks/day was used for smokers and D was
estimated to be 0.4 for the low pollution communities and 1.0 for
the high pollution communities.   (a).   Prevalence rate on basis
of age.   (b).   Prevalence rate on basis of dosage.

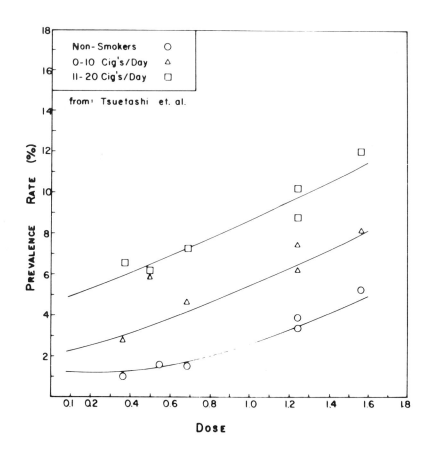

(b)　Prevalence rate on basis of dosage

FIGURE 5--concluded

TABLE IV. Summary of Health Effects Dose Response Models

Illness	Epidemiological variable	Function[a]	Sympton days per effect
Asthma	Annual attack rate[b]	$Z_{AA}(A,D)$	Severe shortness of breath, 1 day
Chronic bronchitis	Prevalence[c]	$P_{CB}(A,D,R)$	Severe coughing, 240 days
Lower respiratory illness in children	Annual attack rate[b]	$Z_{LRI}(A,D)$	Severe cough, 7 days
Eye Irritations	Annual attack rate[b]	$Z_{EI}(A,D)$	Eye irritation, 1 day
Chest pains	Worse days/year		Chest pain, 1 day
	W. Well Panel	$Z_{W}(A,D)$	
	H. Heart Panel	$Z_{H}(A,D)$	
	L. Lung Panel	$Z_{L}(A,D)$	
	HL. Heart and Lung Panel	$Z_{HL}(A,D)$	

[a] *A is age in years, D is dose of air pollution, R is rate of cigarette smoking in packs/day.*

[b] *In attacks/person-year.*

[c] *Fraction of population.*

of diseases. Thus, these disease effects had to be converted to days of each of these symptoms. Every asthma attack (Z_{AA}) was assumed equal to 1 day of severe shortness of breath (SSB) and no other source of SSB was considered. For severe cough days, bronchitis and LRI were relevant. The prevalence of chronic

bronchitis is equal to the probability that a person has 90 or
more days of severe coughing per year; the expected number of
days is calculated as 240. For LRI, the expected number of
days of severe coughing is 7 per attack (Z_{LRI}). For severe
chest pain, each incidence is assumed to last 1 day and
incidences from the various panels (W, H, L, HL) are combined.

VII. HEALTH EFFECTS VALUE MODEL

There are three alternative approaches to measure willingness to pay (26). In the first, individuals are asked directly
their willingness to pay for pollution reduction. In the
second, market transactions (wages or property values) are
used to input values of pollution reduction. The third
determines direct cost accounting of damage due to pollution,
and what individuals should be willing to pay. Each method has
its problems and weaknesses. In the ICAAS-FSOS study the
direct questionnaire approach was used.

The questionnaire design was based on an economic model of
choice. It was assumed that individuals would be willing to pay
an amount (a bid m) to avoid a given health effect (d) if they
are better off under the payment alternative. Socio-economic
variables such as income, initial health, age, sex, insurance,
and employment status affect this choice between money payment
and health status. People with similar socio-economic characteristics could be expected to make similar choices or at least
be represented by the same statistical distribution function.
Thus, a questionnaire was designed to obtain a characterization
of the distribution of individuals about a representative group
bid.

Since respondents may not be familiar with specific
diseases, such as asthma, they were asked to consider characteristics of the diseases as exemplified by the nature of its

symptoms, its temporal character (recurrence rate, number of
days of an attack), and severity or disabling effect. For
example, "Asthma" would imply recurring attacks of difficulty
in breathing, of duration of 1 day, and in many cases disabling
the individual during the attack.

The health effects in the questionnaire are given in terms
of symptoms or duration (1-, 7-, 90-days) and severity (mild and
severe). Each health effect is compared with dollar amounts
ranging from $0 to $1,000. The intent was to provide enough
"spread" for those with both low and high incomes and to
account for the diversity in seriousness of health effects.
Odor and haze were also included as aesthetic, nonhealth
effects. A typical question is: To avoid 1 day per year severe
shortness of breath/chest pains, the most I would pay is $0,
$.50, $1, $2, $10, $15, $50, $120, $250, $1,000 per year.

The biographical information requested on the questionnaire
included initial health, age, income, sex, employment status,
and insurance status. Table V gives the characteristics of the
age groups obtained from the response. The type of questions
posed in the questionnaire were a form which has given reliable
results in earlier studies (27).

The questionnaire was mailed to 1,800 people selected
randomly and 404 responses were obtained. The responses were
tested and found to be representative of the population but
with a somewhat higher proportion returned for professional,
high-income respondents than their proportion in the population.

Survey responses were tabulated according to percent of the
sample willing to pay each amount for each symptom-duration
combination. From this information averages and medians were
calculated. To avoid severe bias from "protest bids," the
medians, which are less sensitive to such problems than the
means, were used. Furthermore, the median is indicative of
majority voting since it indicates the bid which at least 50%

TABLE V. *Characteristics of Age Groups*

Age[a] group	Age range	Number in sample	Average age	Average income	Average days (1)	% with insurance	% male
2	14-24	44	20	11,180	21.56	84.1	40.9
3	25-29	56	27	14,410	21.59	87.5	62.5
4	30-34	54	32	18,090	21.27	79.6	51.9
5	35-43	49	39	22,260	21.60	89.8	66.7
6	44-50	51	47	21,740	21.49	76.5	84.3
7	51-57	48	54	19,300	21.64	95.8	64.6
8	58-68	48	63	12,830	21.06	89.4	64.6
9	69-86	46	72	10,830	20.70	97.8	60.9

[a]*Age group 0-13 not sampled but included in dose-response model.*

of the population would agree to pay to avoid an increase in illness. It was assumed that the cumulative distribution function for the probability of choosing m (money) over d (days illness) is logistic with mean and variance depending on the levels of m, d, and socio-economic factors S. In the final evaluations, bid curves were estimated separately for each age group for greater accuracy in predicting values by age. This procedure makes the form of the equation somewhat simpler. The logistic regression equation is

$$\ell n \frac{P}{1 - p} = \alpha + \gamma \, \ell n \, m + \beta \, \ell n \, d \tag{9}$$

where p is the percent favoring paying amount m over experiencing additional days of a symptom d. Take p = 1/2 to find the median bid and solve for m so that

$$m = Kd^{\delta} \tag{10}$$

where

$$K = e^{-\alpha/\gamma}$$

$$\delta = -\beta/\gamma \tag{11}$$

Bid curve equations were estimated separately for each of eight age groups and the eight symptoms. Table VI gives the values of K and δ for the various age groups and symptoms.

It is interesting that for each symptom (severe/minor shortness of breath, severe/minor coughing and sneezing, severe/minor head congestion, haze, odor) p is usually a power of $\propto 0.4$. The external coefficient also displays some shifts due to income, initial health, and sex in addition to age. The results show a statistically significant effect from these socio-economic variables; the median bid increases with income, the percent female, and initial days of illness.

The fact that the bid curve is determined for each age group is particularly useful in regional analyses since it makes it possible to match health values to available census information. The median bids by age reflects an average over other socio-economic variables except age. Since income increases with age and then decreases, the bid by age group is expected to follow this pattern. The full report gives values for each age group for each symptom (4).

VIII.   COST DISTRIBUTION

It should be obvious that costs of pollution control will either go into the rate base for capital items or into fuel adjustment charges. In either case costs of abatement are distributed to households according to consumption of electricity which is dependent upon income. In this study, the cost of

TABLE VI. Parameters for Bid Curve $m = Kd^{\delta}$

	SSB	SSC	SHC	H	O	MSB	MCS	MHC
			Values of K (in dollars)					
1	7.576	8.722	9.703	2.973	3.948	5.261	3.634	5.693
2	16.156	6.843	9.081	1.889	1.512	4.901	2.631	3.405
3	13.387	8.659	10.006	2.102	2.243	5.649	2.689	5.217
4	14.102	9.209	17.429	2.198	2.319	7.785	2.609	3.925
5	11.282	4.764	6.271	1.378	1.251	3.481	2.142	4.202
6	17.249	10.582	5.812	2.096	2.000	10.402	1.941	3.283
7	13.882	9.144	9.825	2.500	2.241	6.367	4.018	6.959
8	5.709	4.622	3.292	1.000	1.000	2.042	1.814	3.227
			Values of δ (d is in days)					
1	0.511	0.375	0.419	0.447	0.416	0.425	0.378	0.320
2	0.572	0.556	0.534	0.528	0.552	0.544	0.508	0.528
3	0.516	0.417	0.468	0.548	0.586	0.467	0.613	0.477
4	0.517	0.471	0.371	0.548	0.523	0.412	0.557	0.472
5	0.296	0.493	0.419	0.393	0.536	0.504	0.528	0.381
6	0.589	0.427	0.608	0.338	0.448	0.406	0.642	0.485
7	0.281	0.193	0.255	0.272	0.312	0.257	0.312	0.172
8	0.242	0.260	0.368	0.404	0.335	0.320	0.451	0.193

changes in fuel types which would be passed on through a fuel
adjustment change is being considered. This change would be
assumed to have no effect on the rate structure. Thus, the share
C_i of abatement costs for a household in income group i is proportional to energy consumption and is given by

$$C_i = \beta E_i \tag{12}$$

where E_i is household electricity consumption and

$$B = C/E \tag{13}$$

is the cost per kwh of electricity produced. For Tampa area
$C = \$1 \times 10^6$ and $E = 8.46 \times 10^9$ kwh; hence, it is 0.118 mills
per kwh per million dollars of cost.

Table VII gives the number of customers and average monthly
electricity costs.

IX. THE DISAGGREGATED BENEFIT/COST MODEL

A. *The Decision Model*

Sections II-VIII give the details of major components which
enter into the DB/CA overall decision model for air quality
control. It must be acknowledged that some components, e.g.,
mortality, environmental, livestock, materials, and aesthetic
damage, were not represented in the ICAAS-FSOS study. Thus,

TABLE VII. *Distribution of Abatement Costs by Household
Income Per Million Dollars of Abatement Cost*

Income class	Number of customers	Average monthly kwh consumption	Household cost per kwh per 10^6	Total costs per year income class per 10^6
0-3,000	12,700	711	1.006	12,800
3,000-6,000	20,250	816	1.155	23,400
6,000-9,000	23,230	1010	1.430	33,300
9,000-12,000	18,000	1092	1.546	27,800
12,000-15,000	11,200	1112	1.574	17,700
15,000-25,000	11,100	1187	1.681	18,700
25,000+	3,000	1187	1.681	5,000
				138,700

the attempts to put together an overall decision model based
upon the components studied must be viewed as somwehat incomplete.
Nevertheless, valuable purposes can be served by assembling an
interconnected decision model machine as long as every major
type of component is represented. Figure 6 illustrates in
greater detail the decision model machine which was assembled
from the various component models developed in the ICAAS-Florida
Sulfur Oxide Study.

The components which have been neglected can be added by
expanding the general analyses package to include tables for
each additional type of effect, e.g., mortality, materials
damage, environmental damage, livestock damage, visibility,
and other intangibles.

B. *Demographic Component*

The General Analysis Program (GAP) requires that the
Climatological Dispersion Model (CDM) grid points be contained
within demographic data areas. The current organization of the
GAP/CDM Programs utilizes grid receptors within U.S. standard
census tracts (4th count). This allows economic cross-
tabulations between these data areas.

For every census tract the GAP program groups data by age
and income characteristics. The breakdown by age groups is
utilized in the dose-response and risk-benefit models to further
illustrate the effects of atmospheric pollutants. In the
dose-response models the geometric means of each age group are
used to denote the age variable (A). In the bid curve models,
the equations have been estimated to fit the age groupings
specifically, such that medians or means are not used in the
functional forms explicitly. The family income groupings are
used in the cost analysis of the scenario in question. These
groupings are rearranged to be compatible with the data available
from utility industry reports.

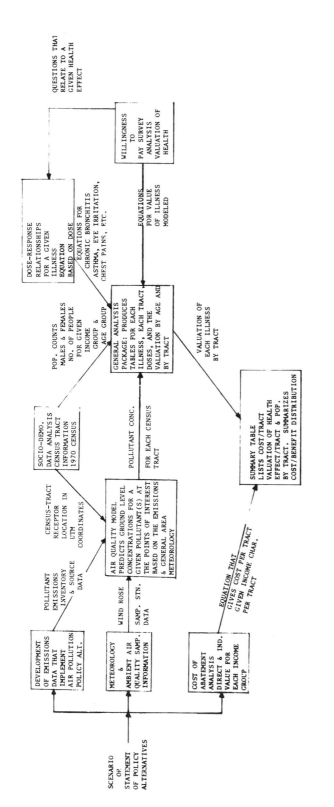

FIGURE 6. Risk-benefit analysis flow diagram.

The GAP program puts together all these components and carries out the arithmetic. After reading in the air quality parameters by census tract, the program computes health effects, value of health effects, and costs by census tract. Total benefits and costs are also given.

A summary of the health benefits/cost ratios for the 16 scenarios is listed in Table VIII. These runs grossly illustrate the relative merits of various pollution abatement scenarios. In interpreting the HB/C ratios in Table VIII it should be noted that the benefit in the ICAAS-FSOS study only includes health effects. Since mortality, environmental, agricultural, industrial, and intangible impacts are neglected, these ratios obviously are too small. However, on the basis of the National Academy of Sciences/National Research Council 1974 Study and a 1.5 indirect factor, one finds that a factor roughly of 2 corrects morbidity costs to total costs. If the HB/C numbers in Table VIII are multiplied by 2, one obtains B/C values sub- stantially in excess of 1.0 in 10 out of 16 scenarios. Furthermore, the B/C ratio is greater than 1/2 in 15 out of the 16 scenarios. From the gross average calibrated B/C numbers one would expect that the disaggregated analyses should show many cases of large DB/C values. Thus, the analyses of the Tampa/Hillsborough region suggest that benefits of pollution reduction can exceed costs of implementing such reduction for many population groups.

The ICAAS team explored several voting prescriptions in preliminary tests as to whether an informed public knowledge- able of these results would vote for various levels of control. However, this work was not completed. The illustrated cal- culations which were completed do, however, suggest that in some scenarios for the Tampa area, costs of pollution abatement

TABLE VIII.   Scenario Results Summary

Test Number	Test area cost ($)	Modeled health benefit ($)	HB/C ratio	Remarks
1	0	4,378,264	--	
	1976 post-BB 3 construction comparisons			
2	5,307,708	1,578,418	0.30	NSPS at BB 3
3	18,693,856	4,285,681	0.23	NSPS at all plants
4	4,333,079	3,571,472	0.82	NSPS at Gannon and Hookers Point
5	4,520,327	4,171,077	0.92	NSPS with stack height
6	110,023	2,307,583	20.97	Stack height increase
	1976-1975 during-BB 3 construction comparisons			
7	5,307,708	4,600,451	0.87	NSPS at BB 3
8	18,693,856	5,826,598	0.31	NSPS at all plants
9	4,333,079	5,380,697	1.24	NSPS downtown
10	4,520,327	5,751,488	1.27	NSPS with stack height
11	110,034	4,848,663	44.07	Stack height increase
	1975 pre-BB 3 construction comparisons			
12	5,307,708	2,136,855	0.40	NSPS at BB 1-2
13	13,918,030	5,696,913	0.41	NSPS at all plants
14	4,333,079	5,424,630	1.25	NSPS downtown
15	4,520,327	5,792,550	1.28	NSPS with stack height
16	110,034	2,409,095	21.89	Stack height increase

can satisfy efficiency and equity criteria. For example, in
Scenario 13 "impose NSPS on all plants" and Scenario 14
"impose NSPS on downtown plant," the methodology led to some
interesting contrasts.

For the Scenario 14 with only "downtown" plants controlled,
the total morbidity benefits exceed direct household costs by
about $2 million. Most tracts have benefits exceeding direct
costs. However, there are four tracts with costs greater than
benefits. Since these account for only about 6% of the
population, this level of control would probably be acceptable
to the majority of households. A complication occurs, however,
in that the direct costs to these census tracts was about
$3 million out of a total compliance cost of $23.2 million.
The remaining $20 million must be borne by other residential
customers, government (schools, etc.), commercial firms, and
industrial users. Some of these costs are passed directly on
to the urban census tracts, and some prices on export goods
increase.

For Scenario 13, the "all plants" policy, the annual
total household costs exceed total morbidity benefits by
nearly $4 million. However, 3% of the population in the highest
polluted areas would have morbidity benefits exceeding household
costs.

Thus, control of "downtown" emissions in the urban area
should be acceptable to households in Hillsborough County on
the basis of morbidity benefits alone. The stricter "all
plants" policy does not pass the acceptability test using only
morbidity benefits. Since mortality, aesthetic effects or
material damage have not been included in the value calculation,
it is possible, however, that these excluded effects would be
sufficient to justify this more costly policy.

To determine whether control of emissions beyond federal primary standards is justified, benefits and costs for households in affected areas besides those for the 88 tracts of this study should also be considered. To do this, three residential customer groupings may be defined as given in Table IX. It is assumed here that group 2 gets the same benefit per person (about $10) as for group 1 because of adjacent location. This may be an overestimate, given the meteorological conditions, but it illustrates how such externalities could be taken into account. For group 3 we used a health benefit of $5 per person. If one only counts total direct household costs and benefits, the control of emissions in urban areas seem justified. On the basis of electricity consumers only, there would be more persons against control (group 3) than for (group 1) although the sums of benefits and costs for groups (1) and (3) are roughly equal. The decision for a control policy thus requires

TABLE IX. *Benefits and Costs for Households in Three Groups for Control "Downtown," Scenario 14*

	Group	Population[a]	Morbidity benefits	Abatement costs
Tampa				
costs and benefits	1	369,518	5,125,555	3,098,435
S. Hillsborough,				
Manatee, Pinellas;				
no costs, benefits	2	740,182	7,401,820	0
Polk, Pasco; plant				
cost, reduced				
benefits	3	432,000	2,160,000	4,461,746
		1,541,700	14,686,375	7,560,181

[a] *Estimates based on 1970 census.*

an ethical judgment as to who should bear costs and receive
benefits. One possiblity to gain acceptance of all groups
would be to give group (3) a smaller share of abatement costs
and group (1) a larger share (by about $2 million). On the
other hand, the populations in the energy exporting areas might
object to having to pay for cleaning up pollution associated
with electricity enjoyed by the energy importing area.

These costs represent only about one-third of the total
abatement costs for this scenario; judgments also have to be
made on the allocation of costs to the commercial sector,
government, and industry. The household will bear some share
of the increase in cost to commercial and government sectors
roughly proportional to income. However, government and
commercial sectors will also receive benefits because of the
lower material damages and higher productivity. Similarly,
there will be benefits for industrial sectors with some portion
of costs passed to consumers (not all of whom will be local).
An extension of the methodology of this study would be to
consider each of these sectors separately. How such benefits
and costs are counted relative to households requires another
ethical judgment.

X. QUANTIFIED ACCEPTABLE LEVEL OF RISK (QALR) METHODOLOGY

Late in the ICAAS-FSOS study, the Clean Air Act of 1977
introduced major new elements which must be accommodated by
any new regulations at state and local levels. In particular,
the much debated concept of prevention of significant deterioration (PSD) was codified in terms of allowed incremental limits
of TSP and SO_2 over present baselines. At the regional and
state levels the PSD clauses, together with ICAAS's air quality
index or "dose," provided a simple quantified acceptable level
of risk (QALR) public policy decision methodology (PPDM) that

is very useful. For example, many regions of Florida and the
Southeast are still Class 2 regions (a Class 2 region is one with
ambient pollution levels below the National Ambient Air Quality
Standards). For such regions, Congress has specified that the
maximum allowable annual average increases over the present
baseline shall not exceed 19 $\mu g/m^3$ of TSP and 20 $\mu g/m^3$ of SO_2.
These regulations are compromise attempts at a middle course
between strict PSD and the realities of our energy and economic
crises. By specifying these incremental limits, Congress has,
in effect, given its quantitation of a maximum acceptable level
of risk for the nation as a whole. Since states are only given
the option to place more stringent incremental limits in areas
with air quality better than national standards, the range of
regional decision options has been narrowed greatly. Within
such a framework it is possible to bypass many of the complex
and more controversial steps of the DB/CA analyses. The path
through the box (QALR) illustrates the steps which are short
circuited.

In the QALR approach an air quality index is used as a
working measure of "dose" and the ICAAS dose formula which was
given in terms of the concentrations of the major air pollutants
served in this regard. In the ICAAS-FSOS study, the scaling
constants were chosen to place the various pollutants on an
equivalent basis in terms of health effects by using the
pollution standard index (PSI) formulation as a guide. This
"dose" was then used directly as an index in a quantitative
assessment of the impact of proposed PSD increments. For
example, in Table X the column labeled D indicates dose values
assigned to various regions of Florida in 1976 based upon the
concentrations given in the table. The dose values which
correspond to ambient air when all pollutant levels are at the
NAAQS is $D^{\dagger\dagger}$ = 0.67 and at the secondary standards is

TABLE X. Dose Assignments and Factors of Safety for Florida.

Pollution levels 1976	TSP μg/m³	SO₂ μg/m³	NO₂ μg/m³	O_x 1 hr max ppm	CO 1 hr max ppm	D	D*	FS	FS*
Tampa–									
Hillsborough	69.3	39.60	58.0	0.13	18.5	0.512	0.630	1.14	1.07
Miama–Dade	64.4	8.70	31.3	0.07	22.0	0.442	0.552	1.74	1.35
Duval	59.7	22.30	59.0	0.18	17.5	0.460	0.568	0.98	0.91
St. Petersburg–									
Pinellas	59.3	18.30	43.1	0.132	12.0	0.397	0.539	1.27	1.13
Gainesville–									
Alachua	47.1	5.62	26.4	0.110	5.0	0.284	0.394	1.62	1.33
Tallahassee–Leon	37.3	6.00	26.4	0.114	2.4	0.240	0.360	1.64	1.36
Orange–Orlando	49.4	6.30	2.9	0.130	5.0	0.278	0.407	1.43	1.22
Palm Beach	56.4	26.20	23.4	0.148	10.5	0.396	0.516	1.18	1.08
Ft. Lauderdale–									
Broward	59.0	5.24	35.6	0.105	23.0	0.434	0.537	1.48	1.22
NAQS: primary	75.0	80.00	100.0	0.03	3.3	$D^{††}$	0.67	1.00	
NAQS: secondary	60.0	60.00	100.0	0.03	3.3	$D^{†}$	0.58	1.13	1.00

$D^{+} = 0.58$ (18). It is reasonable to use the D^{++} level as a marker on an air quality scale corresponding to "dirty" or "unhealthy" air and the D^{+} as a marker corresponding to "marginal" or "barely acceptable" air. These markers together with the dose formula provided a simple PPDM for addressing what is now probably the most important question in air quality control--"How much of the allowed Federal PSD increments should we permit in a region?"

To illustrate this simple PPDM, the column headed D* gives the altered dose values corresponding to the full incremented limits allowed by the 1977 Clean Air Act. It would appear that several areas in Florida are already close to exhausting their accounts and the Tampa area (Location 1) would exceed its secondary account if it used the full PSD increments. The problem was not so much sulfur dioxide by itself, which took the main thrust of the FSOS project, but rather the other pollutants which are already present in some areas. Indeed, if the sulfur dioxide question were examined in isolation, one would almost inevitably come to the conclusion Florida has no problem and could be very relaxed in its regulatory measures. However, scientific evidence now suggests that pollutants cannot be examined in isolation so that in setting up the allowed local standards for one pollutant, it is essential to consider all the other pollutants already present or which must be accommodated in a specific region. Accordingly, using dose alone as a quantitative indicator, ICAAS in its final FSOS report, recommended not to utilize the full PSD incremental allowances permitted in the Clean Air Act of 1977 in areas with D values already close to D^{+}.

Subsequent to the completion of the ICAAS-FSOS study, a further and closely related study was initiated under the auspices of a State Board of Regents Grant Program and the Florida Department of Environmental Regulation (28). In this

study a broader set of air quality indices were examined for
the purposes of evaluating Florida's air quality. To place all
index systems on a common basis, a so-called Factor of Safety
(FS) has been used. To define the FS in the engineering
practice, the best technical judgment first must be used to
estimate the load limit, that is, the load level at which the
system would be expected to fail. The factor of safety is then
defined as this load limit divided by the maximum applied load
that is reported or foreseen for the structure.

The present lack of knowledge about effects of low
pollutant levels make it impossible to specify a precise load
limit for the human pulmonary system. However, for now, the
pulmonary load limit is identified as an index corresponding
to air in which all five major pollutants are present at NAAQS
simultaneously. The air quality factor of safety based upon
any index I_j is then defined as the ratio

$$FS_j = \frac{I_j(C_1^o,\ C_2^o,\ C_3^o,\ C_4^o,\ C_5^o)}{I_j(C_1,\ C_2,\ C_3,\ C_4,\ C_4)} \tag{14}$$

where C_1^o, C_2^o, C_3^o, C_4^o, and C_5^o denote the NAAQS concentration for
the five major pollutants, TSP, SO_2, NO_2, O_x, and CO, and the
C_1, C_2, C_3, C_4, and C_5 denote the acttural reported concentrations. The column of Table X headed by FS gives the ICAAS
factors of safety for each of the locations relative to this
chosen standard. The last column headed FS* gives values
obtained when the PSD increments are added to the baseline
levels.

Normally, engineering factors of safety range from 1.5 to
10, with larger FS values being used when the uncertainties
are greater. From the SO_2 standpoint, Florida is generally in
good shape since most areas are considerably below federal or
state standards. Unfortunately, Florida is generally fairly
high in TSP and ozone which makes its air quality worse

than it would be based upon the SO_2 levels alone. If Florida were to utilize the full PSD increments, the air quality of Florida would begin to fall into rather poor categories. It would appear, therefore, that in these instances the Clean Air Act of 1977 would actually permit significant degradation. Again it was recommended that Florida should be more restrictive than Federal law permits.

The fundamental problem in judging air quality at present is lack of knowledge about the effects of chronic exposure to low levels of pollutants, particularly upon the young, the elderly, and the infirm (16). It is unlikely that the standard methodologies used in estimating air pollution health effects will resolve these problems in the near future. Under such circumstances it is usually helpful if a discipline (e.g., air pollution science and engineering) reaches beyond its normal boundaries for new ideas and approaches. The Factor of Safety concept is a time-tested approach used in many other fields of science and engineering; hence, its application as an index of air quality should be fruitful. It might be noted that in a subsequent study using a national data base, it was found that the factors of safety computed with the ICAAS index correlate very well with FS values based upon other combinative indices including the ORNL index, the Ottawa index, the ACGIH index, and the PSI index (29-32). Thus, the overall FS implications are not expected to be very sensitive to a specific choice of an air quality index.

In Hillsborough County, site specific data for the year 1977 was used to generate ICAAS FS values. The data used was taken from the Environmental Quality 1977 Report which is assembled by the Hillsborough County Environmental Protection Commission (33). For the purpose of generating site specific FS values, the pollutant data set was often incomplete,

especially with respect to ozone and carbon monoxide. Therefore,
certain assumptions were made to provide complete sets of
pollutant concentrations (TSP, SO_2, NO_2, O_x, CO) at each site.

All sites chosen had readings for TSP and SO_2. Most sites
had readings for NO_2, the exceptions being the outlying regions
from the city of Tampa. By recognizing that the major contributors to NO_2 pollution are the power plants, it is felt that an
estimate for the rural pollution sites which are not near
power plants could be made by using the closest available
city readings. This procedure amounted to using a level of
12 ppb for rural sites where data were not available.
Regardless of the pollution levels extrapolated for NO_2 in the
rural section, these levels should be less than those observed
in the city or near power plants and are observed to be low
compared with the NAAQS.

Only three sites monitored ozone; these ranged from the
heart of the city (Davis Island) to a residential area. Ozone
is a secondary product associated with the photochemistry of
hydrocarbons and nitrogen dioxide. It is probably high in high
density traffic areas and in the neighborhood of power plants.
In particular, ozone levels are assumed to be high and constant
over the city and then drop off linearly toward the rural
areas. The formulation used is

$$\text{Max. 1 hour ozone level (during a year)} = \begin{cases} 150 \text{ ppb} & 0 \leq r < 2.2 \text{ mi.} \\ -20.5r + 195 \text{ ppb} & 2.2 \leq r \leq 7.7 \text{ mi.} \end{cases}$$

where r is measured radially from site 63 (see Fig. 7) located
on Davis Island.

CO data is also severely limited, in fact, only one site
(81) near the inner city monitors this pollutant. Comparison
with other sites for other years established that CO is

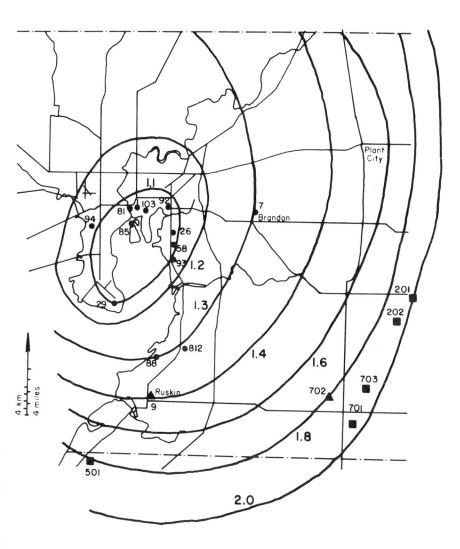

FIGURE 7.   Factor of safety contour map for Hillsborough
County.   Contours near numbered locations are firm; those at
a distance are guestimates.

typically high near major roadways, as would be expected, but
still relatively low if compared with the NAAQS. After noting
that the 1977 averages for site 81 were a little lower than those
of past years a constant background value of 11 ppb (1 hour maximum for 1 year) CO over the entire county was assumed.

With these assumptions, the ICAAS FS was then calculated
for 22 sites in Hillsborough County. The TSP, SO_2, and NO_2
levels of pollution used were annual arithmetic averages.

Since TSP was only given in terms of a geometric average,
this quantity was converted to an arithmetic average by
assuming a log normal distribution. Ozone and CO were used
in terms of their maximum 1 hour averaged reading for 1 year.
The pollution values used for the 22 sites are given in
Table XI. The FS values based upon various AQI's systems are
given including the Toronto index which is based only upon
SO_2 and TSP (34). It is seen that apart from the Toronto
factor of safety (TOR-FS) values, the other FS values are all
moderately comparable. The dominant effect of ozone is
illustrated in the PSI index FS values which are only sensitive
to the highest component sub index (4 corresponds to ozone).
These index values are computed on the basis of the 1977 standard
for a maximum for ozone which was 80 ppb or 118 $\mu g/m^3$. If the
new EPA standard of the 120 ppb is used, the FS values for Tampa
will be larger.

Figure 7 presents an interpolated FS contour map based upon
the 22 ICAAS index points in Table XI. The results are somewhat
confused by the dominance of the ozone component. However, they
display a reasonable sensitivity to urban air pollution and
suggest that the simple FS methodology in conjunction with a
pollution dispersion model can be used to obtain spatial displays
of a health indicator for any regulatory scenario. Thus, the
simple QALR decision methodology can also be utilized for a
detailed regional decision.

TABLE XI. Factors of Safety for Various Sites in Hillsborough County

Site	Pollutants					Factor of Safety						
	TSP	SO₂	NO₂	Oₓ	CO	ICAAS	Ottawa	ORNL	TOR	ACGIH	PSI	
Standards	75	80	100	157	40.0							
1	63	9	46	320	12.0	1.10	0.98	1.47	2.70	1.05	1.29	4
7	46	3	22	289	12.0	1.28	1.13	1.99	4.61	1.23	1.40	4
9	38	7	18	256	12.0	1.44	1.28	2.30	5.15	1.37	1.54	4
15	49	14	39	320	12.0	1.14	1.01	1.59	3.31	1.08	1.29	4
29	38	12	16	320	12.0	1.18	1.05	1.87	4.51	1.15	1.29	4
55	40	11	21	284	12.0	1.30	1.16	2.00	4.31	1.24	1.42	4
58	55	6	26	320	12.0	1.14	1.01	1.68	3.38	1.10	1.29	4
81	48	5	39	320	12.0	1.15	1.02	1.66	4.13	1.09	1.29	4
85	69	14	34	320	12.0	1.09	0.98	1.46	2.24	1.05	1.29	4
86	60	10	38	320	12.0	1.12	0.99	1.54	2.85	1.07	1.29	4
88	40	10	17	287	12.0	1.30	1.15	2.05	4.50	1.25	1.41	4
92	69	8	51	320	12.0	1.08	0.96	1.41	2.49	1.03	1.29	4
93	40	9	29	320	12.0	1.17	1.04	1.77	4.51	1.12	1.29	4
103	69	13	47	320	12.0	1.08	0.97	1.41	2.32	1.03	1.29	4
201	35	9	18	155	12.0	2.14	1.93	3.35	5.28	1.94	2.22	4
202	49	15	18	160	12.0	1.89	1.75	2.80	3.21	1.81	2.15	4

TABLE XI. Continued

	Pollutants				Factor of Safety					
	TSP	SO_2	NO_2	O_x	CO	ICAAS	Ottawa	ORNL	TOR	ACGIH	PSI
401	56	6	28	251	12.0	1.39	1.23	2.00	3.35	1.31	1.56	4
501	35	4	18	199	12.0	1.80	1.60	2.91	6.06	1.66	1.85	4
701	62	2	18	153	12.0	1.90	1.68	2.85	3.16	1.84	2.25	4
702	60	2	18	181	12.0	1.74	1.54	2.61	3.28	1.67	1.98	4
703	62	2	18	162	12.0	1.84	1.63	2.76	3.17	1.78	2.13	4
812	40	7	18	287	12.0	1.30	1.15	2.06	4.75	1.25	1.41	4

XI.   CONCLUDING REMARKS

Two quantitative public policy decision methodologies for
regional air pollution decisions which are dependent in detail
upon pollution dispersal modeling (PDM) have been illustrated.
The first, the DB/CA, is an economic methodology in which the
PDM makes possible detailed consideration of the distributional
aspects of B/C analysis, one of the frontier problems of
economic analysis.   In the Tampa Bay region which was examined
in detail in the ICAAS-FSOS study, it turned out that coal
burning power plants are major sources of SO_2, TSP, and NO_2
emission; hence the intensive work on this project is very
relevant to coal burning issues in general.

The results of the distributional analyses of regional
benefits and costs implies that the Scenario S14 "use NSPS
downtown" gives net household benefits to the majority (both
by age and location).   Thus, both efficiency and equity tests
appear to be satisfied by this policy for broad groupings of
households.   On the other hand, the Scenario S13 "use NSPS for
all plants" was not cost effective.   These results were based
upon the use of a standard CDM air quality model, a "smooth
hockey stick" dose response model, and a value model based upon
a "willingness to pay" survey.   It must be acknowledged that each
of these models have uncertainties characteristic of the various
frontiers of knowledge involved.   In particular, the fact that
cost or the benefits were not considered in detail nor were the
indirect costs due to electricity price impacts upon commercial
and industrial goods and services subjected the results to
uncertainties depending upon the import and export characteristics of the region.

Despite these caveats, the overall policy DB/CA machinery
developed is quite tractable, efficient, and should certainly
be useful in examining the *relative* merits of various possible

policy scenarios. It is also useful in exploring the sensitivity
of the disaggregated results to various assumptions used in the
inputs, models, and calculational methods. Furthermore, each
component of the machinery could at a later date be replaced by
an improved component which should preserve or actually improve
the decision machinery as a whole.

In earlier studies the usefulness of the QALR methodology as
a rapid quantitative screening procedure for identifying "hot"
urban areas and as a guide for the allocation of PSD increments
has been demonstrated (18,28). In section X previous QALR
methodology has been extended so that in conjunction with PDM,
it can be used to generate Factor of Safety (FS) contour maps
for any regulatory scenario. The results for the Tampa Bay
area gave only marginal FS values for the city center and
suggest that in this case the details of source distributions,
meteorological characteristics, photochemical transformations,
the character of the AQI used, the choice of standards or scale
parameters, and the choice of PDM should all be reconsidered
very carefully. The overall QALR methodology should be applied
to a number of cities so that judgments can be assembled as to
FS levels that currently exist. Then one can quickly examine
the FS levels that can be attained with current control
technology for any particular region in the context of National
experience.

A public policy board could probably make judgmental
decisions in such a context. Alternatively, a fast economic
analysis could be developed in which an expert group prices out
the costs of various feasible FS contours and lets the public
policy panel make choices among those options that appear to
involve an "acceptable level of risk."

Finally, it should be mentioned that DB/CA and QALR
methodologies examined during the ICAAS-FSOS study of policy
alternatives independently pointed to the conclusion that the

air in some regions of Florida cannot absorb much additional
degradation without bringing Florida's air to marginal levels
from the health standpoint. Estimates by two routes of the value
of visibility deterioration gives bids of $12 to $15 per person
per year for the avoidance of haze or the maintenance of visibility or clear sky. These bids increase the benefit of
pollution abatement substantially. Accordingly, some of the
topics discussed earlier in this conference are highly relevant
to public policy decisions on abatement measures for coal
burning.

In view of the foregoing considerations, the ICAAS team as
its general conclusion cautioned the FSOS Board not to recommend
a great relaxation of control lest air pollution in Florida go
too far, possibly beyond repair. As to its specific recommendations in view of Florida's important role as a retirement state,
its large tourist industry and in consideration of the health
and welfare of all of Florida's citizens, ICAAS recommended that
the State of Florida maintain its historic practice of using
more stringent regulations than the Federal regulations.
Finally, ICAAS recommended that its two Public Policy Decision
Methodologies (DB/CA) and (QALR) be used routinely to assess air
pollution problems in each Florida region and to determine the
optimum local regulatory response.

LIST OF SYMBOLS

A. *Dose Response*

    A   age of population

    b_{ij} pollutants interaction term

    C_i concentrations of SO_2, O_3, NO_2, CO, and TSP

    D   dose index of air pollutants

    <D>   arithmetic mean of dose

$F(A)$	prevalence of attacks asthma/person/age group
$F(D)$	daily dose function
H	hockey stick function with threshold $D = 1$
$I(D)$	incidence-dose function
$K(D)$	number of attacks/year/asthmatic
$K(X)$	generalized dose threshold dependence
M	one parameter function for hockey stick function
P_{CB}	prevalence of bronchitis/age group
R	smoking rate in packs/day
S_i	scale factors for pollutants
Z	number of attacks per exposed population
Z_{AA}	asthma annual attack rate
Z_{EI}	eye irritation annual attack rate
Z_H	chest pain--worst days/year, heart panel
Z_{HL}	chest pain--worst days/year, heart and lung panel
Z_L	chest pain--worst days/year, lung panel
Z_{LRI}	lower respiratory illness annual attack rate
Z_W	chest pain--worst days/year, well panel
$\lambda(D)$	incidence rate as function of dose

B. *General Analysis Program*

A	age
C_i	abatement costs for household in income group i
d	days illness for bid
E_i	household electricity consumption
H	haze effect
K	coefficient parameter for bid
m	money for bid
MCS	minor coughing and sneezing
MHC	minor head congestion
MSB	minor shortness of breath
O	odor effect

P   percent favoring paying m to avoid d

S   socio-economic factors

SHC   severe head congestion

SSB   severe shortness of breath

SSC   severe coughing and sneezing

S1   Scenario Number 1 (S1 through S16 for 16 scenarios)

β cost per kwh of electricity produced

δ exponent coefficient for bid

C. Quantified Acceptable Level of Risk

D^* altered dose values at incremented limits

$D^\dagger$ dose at National Ambient Air Quality Standards

$D^{\dagger\dagger}$ dose at Secondary AAQ standards

FS   factory of safety

FS*   factor of safety with PSD increments

I_j any air quality index

ACKNOWLEDGMENTS

In such a broad interdisciplinary effort many persons besides those listed as authors have made important direct or indirect contributions. First it is a pleasure to thank some of the key University and State administrators who gave early support and encouragement to our endeavors on interdisciplinary analyses of public policy issues. These include H. P. Hanson, E. T. York, G. K. Davis, Pat Rambo, R. E. Uhrig, and H. E. Spivey of the University of Florida, and V. D. Patten, D. Levin, P. P. Baljet, Alice C. Wainwright, formerly with the Florida Department of Pollution Control Board. We would also like to thank some of the faculty founders of ICAAS, who, although no longer with the University of Florida, imparted to us some of their knowledge. Leaping in time from those days when academicians looked out at

the outside world with rose-colored glasses, we want to thank
the contributors to this report, the faculty, research staff
and graduate students whose names are on the title page, most
of whom worked far beyond the call of duty.   In addition we
would like particularly to thank Professor E. D. Palmes,
Ms. L. L. Gibb, Ms. G. P. Penland and Ms. W. A. Wallace whose
professional efforts influenced the technical evolution of this
work.   We also would like to thank Dr. A. Randazzo, Mrs. C.
Kennedy, and Dr. L. Polopolus, who contributed facilities during
parts of this effort and M. Gray, O. Berger, L. Smith, K. Cross,
C. Jackman, F. Riewe, L. Rippetoe, and F. Green who bore the
brunt of many of our reports.

    This work was supported directly by the Florida Sulfur
Oxide Study, Inc., the U.S. Department of Energy, The State
University System Star Grant Program, and the Division of
Sponsored Research of the University of Florida and indirectly
by the Office of Academic Affairs and the Colleges and
Departments of the listed authors.

REFERENCES

1. ICAAS, "An Air Quality Index (AQI) Project," a proposal by
    the Interdisciplinary Center for Aeronomy and (other)
    Atmospheric Sciences (ICAAS), University of Florida,
    Gainesville, Florida (November 1970).

2. ICAAS, "Preservation and Enhancement of Air Quality (PEAQ)
    in Florida and the Southeast," proposal dated August 1971.

3. Wilson, S. U., Anthony, D. S., Hendrickson, E. R., Jordan,
    C. L., and Urone, P, "Final Report of Florida Sulfur Oxide
    Study," Post, Buckley, Schuh & Jernigan, Inc., Orlando,
    Florida (1978).

4. Green, A. E. S., Berg, S. V., Loehman, E. T., Shaw, M. E., Fahien, R. W., Hedinger, R. A., Arroyo, A. A., De, V. H., Fishe, R. P., Gibbs, L. L. Penland, G. P., Rio, D. E., Rossley, W. F., Wallace, W. A., and Schwartz, J. M. An Interdisciplinary Study of the Health, Social and Environmental Economics of Sulfur Oxide Pollution in Florida. Interdisciplinary Center for Aeronomy and Other Atmospheric Sciences, University of Florida, Gainesville, Florida (February 1978).

5. Loehman, E. T., Berg, A. V., Arroyo, A. A., Hedinger, R. A., Schwartz, J. M., Shaw, M. E., Fahien, R. W., De, V. H., Fishe, R. P., Rio, D. E., Rossley, W. F., and Green, A. E. S., *J. Environ. Economics and Management, 6,* n.p. (1979).

6. Gross, B., *Challenge, September,* pp. 30-33 (1965).

7. Green, A. E. S., *Physics Today, June,* pp. 32-38 (1965).

8. Green, A. E. S., and Loehman, E. T. On Preserving Decision-Making Options in Complex Public Interest Studies. ICAAS, University of Florida, Gainesville, Florida (1974).

9. National Academy of Science/National Research Council, Committee on Sulfur Oxides. "Air Quality and Automobile Emission Control," Vols. I and II. Government Printing Office, Washington, D.C. (1974).

10. "Encyclopedia Britannica," 14 Ed., Vol. I, p. 470 (1970).

11. Robson, A. J., *Urban Studies, 14,* 89-93 (1977).

12. U.S. Department of Health, Education and Welfare. Air Quality Display Model. PB-189-194, National Air Pollution Control Administration, Washington, D. C. (1969).

13. Busse, A. D., and Zimmerman, J. R. User's Guide for the Climatological Dispersion Model. EPA-R4-73-024, EPA, Research Triangle Park, North Carolina (1973).

14. Turner, D. B. User's Guide to UNAMAP, Environment Applications Branch, Meteorology Laboratory. EPA, Research Triangle Park, North Carolina (1975).

15. Shaw, A. J. Air Quality Measurements 1975. Hillsborough County Environmental Protection Commission, Tampa, Florida (1975).

16. Brown, G. E., Jr., Chairman, Subcommittee on the Environment and the Atmosphere of the Committee on Science and Technology, U.S. House of Representatives, 94th Cong. "The Costs and Effects of Chronic Exposure to Low-Level Pollutants in the Environment." No. 49, Government Printing Office, Washington, D. C. (1975).

17. Welch, K. B., Haring, M. K., Morris, L. E., and Higgins, I. T. T. Health Effects of Sulfur Oxides, Literature Review. Report for Florida Sulfur Oxides Study, Inc., University of Michigan, Ann Arbor, Michigan (February 1978).

18. Green, A. E. S., Rio, D. E., and Hedinger, R. A., *The Florida Scientist, 41,* 182-190 (1978).

19. Yoshida, K., Oshima, H., and Imai, N., *Arch. Env. Health, 13,* 763-768 (1966).

20. U.S. EPA. Health Consequences of Sulfur Oxides: A Report from CHESS, 1970-71. EPA-640/1-74-004, EPA, Research Triangle Park, North Carolina (1974).

21. Lambert, P. M., and Reid, D. D., *The Lancet, April,* 853-857 (1970).

22. Tsunetashi, Y., *et al., Intern. Arch. Arbeits. Med. 29,* 1 (1967).

23. Harnmo, D. E., *Arch. Env. Health, 28,* 255-260 (1974).

24. Larsen, R. I., *J. Air Poll. Control Assoc. 20,* 214 (1970).

25. Ott, W. R., and Thom, G. C., *J. Air Poll. Control Assoc. 26,* 460-470 (1976).

26. Zeckhauser, R. Willingness to Pay as an Efficiency Guide for the Regulation of Noise Disturbance. Economic Welfare Impacts of Urban Noise. EPA-600/S-76-002, EPA, Research Triangle Park, North Carolina (1976).

27. Shaw, M. E., and Wright, J. M., "Scales for Measurement of Attitudes." McGraw-Hill, New York (1967).

28. Green, A. E. S., Buckley, T. J., MacEachern, A., Makarewicz, R., and Rio, D., *Atmos. Environ.* (1979).

29. Babcock, L. R., Jr., *J. Air Poll. Contr. Assoc. 20,* 653-659 (1974).

30. Inhaber, H., *Science, 186,* 798-805 (1974).

31. ACGIH TLVS: Threshold Limit Values for Chemical Substances and Physical Agents in the Workroom Environment with Intended Changes for 1977. American Council of Governmental Industrial Hygienists, c/o Secretary, P. O. Box 1937, Cincinnati, Ohio 45201 (1977).

32. Ott, W. R., and Hunt, W. F., Jr., *J. Air Poll. Contr. Assoc. 26,* 1050-1054 (1976).

33. Shaw, A. J., and Wilkins, R. G., "Environmental Quality 1977." Hillsborough County Environmental Protection Commission, Tampa, Florida (1977).

34. Shenfield, L., *J. Air Poll. Contr. Assoc. 20,* 612-614 (1970).

DISCUSSION

Kramer: When you went to national source performance standards, did you consider the change in emission rates in terms of velocity and temperature? Did you allow re-heat, etc., with scrubbers?

Green: We didn't go into that degree of detail. Here we were guided by TECO (Tampa Electric Power Company) and their engineers. Most of the cases examined were as if we were simply switching to 0.5 percent coal.

Kramer: So in other words, you just reduced the emissions without changing the temperature or the velocity?

Green: Yes. These are fairly gross changes because TECO was not only burning coal but very low-grade coal. Thus, switching to the best available coal involved quite a cost.

Kramer: I understand that. The reduction in your exit conditions, though, will have a greater effect. If you are burning low-sulfur coal already you may have a greater effect on your calculated ground-level concentrations in terms of annual distributions than you might see. It's something to be considered.

Green: Yes, I am sure we weren't using the state-of-the-art of climatalogical dispersion modeling and that uncertainty would be there but I don't think it would be as great as some of the others.

Shaw: It looks like you restricted all your studies to the existing new source performance and to the ambient clean air standards. I gather you didn't really attempt to look at carcinogens or trace metals that come from coal?

Green: Not in this study. I might say one of the big issues involved in this study was that Florida's standards are a little more stringent than the federal standards. We use as our primary standards the secondary federal standards. And a big issue was whether Florida should switch over to the federal primary standards. Our report, ICAAS study, suggests that we sould maintain our more restrictive standards. Stanford Research Institute's conclusions were the opposite. They said we could do anything in the State of Florida and that the benefit would be substantially less than the cost. So we somewhat canceled each other in our impacts, and maintained the status quo, particularly the more restrictive standards on TSP and SO_2.

ASSESSING THE AIR QUALITY RELATED IMPACTS OF
COAL CONVERSION FACILITIES[1]

F. C. Kornegay

Oak Ridge National Laboratory
Oak Ridge, ·Tennessee

In an effort to accurately assess the impacts of coal conversion facilities, Oak Ridge National Laboratory has assembled a
multi-disciplinary team of experts. One facet of this effort has
been to assess the air-quality-related effects of the Department
of Energy demonstration facilities. The goal of the study is to
quantify the emissions of the various processes, estimate the
average and worst-case ground-level concentrations resulting from
plant operations, determine the potential effects of such concentrations upon the environment, and design a preoperational and
operational monitoring program to determine the actual impact of
the facility. As no commercial-scale conversion ventures are
presently in operation, the team relies upon laboratory and pilot-
plant experiences to determine probable emissions, emission rates,
and potential mitigative measures. Given realistic estimates of
pollutant emission rates, the heat, velocity, and physical
characteristics of the emission streams, and the meteorological
features of the proposed facility, ground-level concentrations of
various pollutants are estimated. These estimates of ground-level
concentrations of the chemicals emitted by a proposed facility are
then used to determine the probable effects upon soils, flora,
fauna, and future development of the area due to the operation of
that facility.

[1]Research sponsored by the Div. of Fossil Fuel Processing,
U.S. Department of Energy under contract W-7405-Eng-26 with the
Union Carbide Corporation.

541

I. INTRODUCTION

The United States Department of Energy (DOE) has initiated a
program to demonstrate the technical feasibility, economic
viability, and environmental acceptability of various coal conversion facilities. The Environmental Impact Section of the Oak
Ridge National Laboratory (ORNL) is providing assistance to DOE
in determining the environmental acceptability of the facilities
proposed in this program. Such assistance includes, but is not
limited to, efforts in the National Environmental Policy Act (NEPA)
process culminating in preparing Environmental Impact Statements
(EIS) for the proposed facilities.

The demonstration program presently includes two high-Btu
plants (CONOCO and Illinois Coal Gasification Group (ICGG)), two
fuel gas plants (W. R. Grace and the Memphis, Light, Gas and Water
(MLGW)), and the Solvent Refined Coal (SRC) plants, producing both
solid (SRC-I) and liquid (SRC-II) products. Plant sites being
considered are located in Noble County, Ohio (CONOCO); Perry
County, Illinois (ICGG); Henderson County, Kentucky (W. R. Grace);
Shelby County, Tennessee (MLGW); Davies County, Kentucky (SRC-I);
and Monongahela County, West Virginia (SRC-II). The approximate
locations of these sites are shown in Fig. 1.

Before funds for the construction of any of these facilities
are released by the DOE, a project review will be held. These
reviews will determine the technical feasibility, economic viability, and environmental acceptability of the proposed designs.
To ensure that the environmental impacts associated with the construction and operation of these proposed facilities can be
included in the funding decision process, EISs must be completed
before project reviews are held.

The primary input to the EIS is the Environmental Report (ER)
prepared by the industrial partner. To ensure that the monitoring
data provided in the ER is sufficient to allow a complete environmental impact assessment, ORNL maintains, through DOE, a close

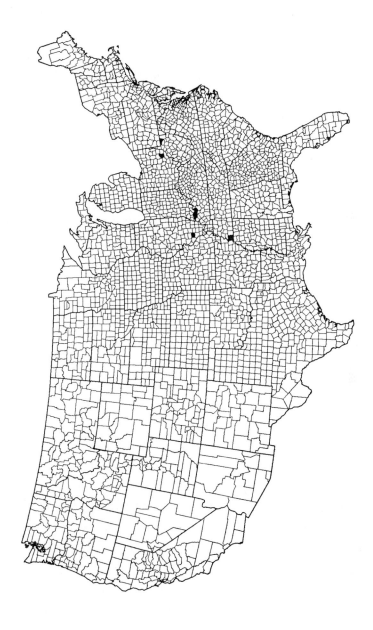

FIGURE 1. Approximate locations of demonstration plant sites presently under consideration.

interaction with the environmental contractors on each of the proposed facilities. ORNL has developed a guidance document,
"Environmental Monitoring Handbook for Coal Conversion Facilities,"
to assist the environmental contractors (1). This early interaction provides ORNL with the opportunity to recommend a preoperational monitoring program suitable for each facility.

II. IMPACT ASSESSMENT

Because the proposed facilities are demonstration projects,
actual operational characteristics are not well defined. To determine the probable source terms for the various processes, two
paths are taken. First, since all these demonstration plants are
scale-ups of pilot operations, some data from these facilities
are available (2-7). Unfortunately, the proprietary nature of
the processes prohibits complete disclosure of all acquired data.
Second, to estimate unknown emissions, similar processes in other
industries, such as petroleum refineries, are evaluated (8-10).
The success of pollution abatement devices in these related industries is studied to determine the probable emission rates from the
proposed coal conversion facilities. The preoperational monitoring
program guidelines incorporate the knowledge available from pilot-
scale facilities as well as from the literature. Careful
attention is paid to determine ambient concentrations of those
materials most likely to be emitted by the proposed facility,
particularly polycyclic aromatic hydrocarbon (PAH) compounds and
trace elements, as well as the Environmental Protection Agency
(EPA) pollutants.

In determining the air quality impacts of the proposed facilities, any plant process that emits a gaseous, liquid, or solid
pollutant to the atmosphere is a source. The primary sources
include the coal handling and storage facilities, steam generators,
vented process gases, acid gas removal systems, and the sulfur
recovery units as shown in Fig. 2. Cooling towers associated with

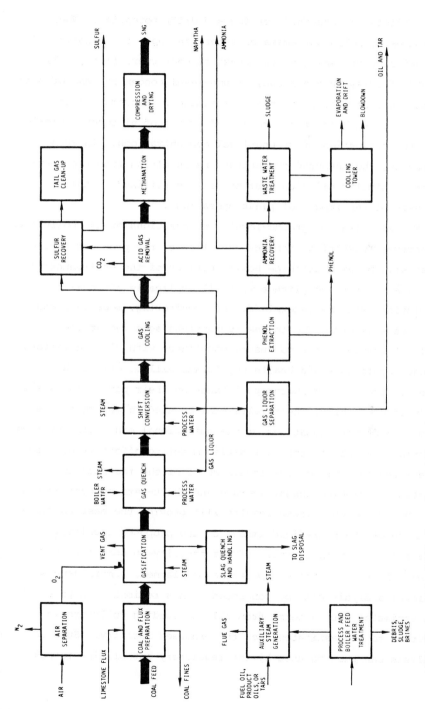

FIGURE 2. Simplified coal gasification system.

the projects are examined for fog and drift potential. When
estimates of pollutant emission rates, the heat and velocity of
the emission streams, and the physical characteristics of all
points of emission are known, ground-level concentrations of many
of the various pollutants can be estimated. Extremely reactive
pollutants, such as hydrocarbons and oxides of nitrogen, compli-
cate the estimation of ground-level concentrations. In such
instances, the measured concentrations resulting from the operation of similar emission sources will be utilized in the
analysis of the impact of plant operations.

The estimates of ground-level concentrations of the pollutants
of interest resulting from the operation of the proposed facilities will be used in three basic sections of the environmental
analysis. First, compliance with all applicable air quality
regulations will be determined. All monitoring, modeling, and
analyses were presented to the appropriate agencies at the
earliest possible time to ensure compliance with all regulations.
Second, ecologists and health physicists will determine the
probable impact of airborne pollutants on their particular areas
of interest. Third, the results will be used to determine the
scope and primary locations of the operational monitoring systems.
To produce these results, a two-tier modeling effort is planned.
A detailed analysis of the local impact of the facility in
question, including annual average and short-term concentrations,
will be conducted. These results will provide the best possible
estimate of the near-field impact of the proposed facility. To
determine the overall impact of the facility in conjunction with
other pollution emitters, operating or planned, a more general
regional analysis will be conducted. This regional analysis will
use estimated source terms from all surrounding polluters in
addition to the estimates from the proposed facility. Such an
analysis can then be used by the regulatory authorities to make

decisions concerning the more desirable use of the air resources of the region in which the proposed facility may be located.

III. PROGRAM UTILIZATION

This detailed modeling program allows ORNL to best estimate the potential impacts of the proposed facilities. Well-designed monitoring programs, both before and after plant operation, allow a determination of the true impact of the facility and a judgment of the applicability of the techniques used to predict impacts. In this manner, the environmental acceptability of the proposed facilities can be estimated before construction decisions are made, and engineering design changes can be made to mitigate the predicted impacts. The monitoring programs will provide a means for determining the actual impact of the demonstration facilities and an invaluable data base for estimating the environmental effects of commercial-scale coal conversion facilities.

REFERENCES

1. Salk, M. S., and DeCicco, S. G., "Environmental Monitoring Handbook for Coal Conversion Facilities," ORNL-5319, Oak Ridge National Laboratory, Oak Ridge, Tennessee (1978).

2. Elgin, D. C., and Parks, H. R., "Gasification of U.S. Coals at Westfield Scotland," Institution of Gas Engineers Communication 946 (November 1974).

3. Swanson, V. E., et al., "Collection Chemical Analysis, and Evaluation of Coal Samples in 1975," USGS Open File Report 76-468 (1976).

4. "Evaluation of Background Data Relating to New Source Performance Standards for Lurgi Gasification, EPA-600/ 7-77-057," U.S. Environmental Protection Agency, Research Triangle Park, North Carolina (June 1977).

5. "Emissions from Processes Producing Clean Fuels," Contract Number 68-02-1358, U.S. Environmental Protection Agency, Research Triangle Park, North Carolina, pp. 111-151 (March 1974).

6. Ralph M. Parsons Company, "Coal Conversion and Utilization-Liquefaction: Solvent Refined Coal," ERDA, FE-1234-6 (May 1976).

7. Hittman Associates, Inc., "Standards of Practice Manual for the Solvent Refined Coal Liquefaction Process," U.S. Environmental Protection Agency, EPA-600/7-78-091, Research Triangle Park, North Carolina (June 1978).

8. Hammer, M. J., "Water and Wastewater Technology," John Wiley and Sons, Inc., New York (1975).

9. "Air Pollution Control and Design Handbook, Part 2" (P. N. Cheremisinoff and R. A. Young, eds.), Marcel Dekker, Inc., New York (1977).

10. Carson, J. E., "Atmospheric Impacts of Evaporative Cooling Systems," Argonne National Laboratory ANL/ES-53, Argonne, Illinois (1976).

DISCUSSION

Green: I would like to make just one comment bearing on your talk. We did make a cost estimate of the value of the visibility degradation. It was also done by Stanford Research Institute. And we did find that this was a rather large figure for the state as a whole, that the willingness of the public to pay to avoid haze was fairly substantial compared to the health benefits, and that this was a case where the two independent studies tended to agree. So the dollar value of visibility is very important to economic analyses.

TRACE ELEMENT REDUCTION
BY THE JPL COAL CHLORINOLYSIS PROCESS

E. R. du Fresne
John J. Kalvinskas

Jet Propulsion Laboratory
California Institute of Technology
Pasadena, California

Removal of sulfur from coal by low-temperature chlorinolysis offers a dividend: substantial removal of trace elements, including toxic heavy metals and other elements contributing to aerosol and gaseous emissions. Boiler corrosion and slagging problems should also be alleviated. Preliminary analytical data show good removal of lead, arsenic, vanadium, phosphorus, titanium, and beryllium.

Many of the nuisance elements, including the radioactives, are concentrated in coal accessory minerals such as pyrite, sphalerite, and apatite. Chalcophile elements are released in soluble form when the sulfide minerals are attacked by chlorine and by the HCl formed in chlorinolysis. HCl also attacks the phosphate minerals, which ordinarily contain much of the uranium and its daughter elements, as well as some vanadium.

"Recapture" of some of these elements by thiol groups in the coal would be possible if only pyrite sulfur was removed. But thiol groups are highly vulnerable to chlorinolysis, and there is no evidence for "recapture."

The importance of trace element removal by chlorinolysis compels the collection of more extensive analytical data, including the use, for elements such as mercury, of more sensitive or accurate methods. Mineralogical and microscopical studies are also advisable.

I. INTRODUCTION

A bewildering variety of trace elements can occur in coal,
trapped either in the organic portion or in the inorganic
accessory minerals. Whether a given trace element can be found
in a given coal sample is a function of the past history of the
coal; for example, the element-bearing ground waters that the
coal may have been exposed to during the various stages of
coalification. The subject is too large to treat in detail here;
those that wish an introduction to the literature can consult
references 1-3. But as an introductory overview, one can mention
these processes:

First, during the growth and partial decay of the original
vegetation in the peat-bog, certain elements may be biologically
concentrated. These elements typically are Cu, Zn, Ni, V, Ti,
and Fe, when they are available.

Second, sediments, both wind-borne and water-borne, may make
a contribution, notably uranium-bearing volcanic ash, and so
forth.

Fluorine, a typically volcanic element, has been found in
some coals at levels as high as 100 g/ton (100 ppm). Zircons
containing uranium and its daughter elements are by no means rare
as coal accessory minerals, and presumably were deposited in the
coal as dust or silt as the peat was accumulating.

(Before proceeding, it is well to remember that peat bogs
contain not only plants and sediment, but animals as well:
insects, amphibians, etc., which tend to concentrate phosphorus
to a greater degree than plants. Even if no fossil bones are to
be found in a given sample of coal, there yet is likely to be
some apatite, precipitated at a later stage of coalification.)

Third, the peat bog is in due course covered by an overburden, which eventually develops a soil horizon and vegetation
of its own. Soil acids from rotting vegetation will leach Fe,
Al, Ti, V, and U, and other heavy metals from the new soil

horizon, and these elements, carried by ground water, will be percolated through the peat at various stages of coalification. The humic acids of the peat, of course, constitute a natural ion-exchange resin. When U, Ge, and V are available in the ground water, they are found to accumulate early in diagenesis, while the coal is still in the lignite stage.

Fourth, it should be remembered that many of the toxic heavy elements owe their toxicity to the fact that they denature protein by binding strongly to it, often to the sulfur atoms in such amino acids as methionine and cysteine. They tend to concentrate in biomass for this reason and, as that biomass decays in a reducing environment, it produces H_2S. These elements will be trapped in the sulfide accessory minerals.

Thus, many of the toxic heavy elements can be trapped both in the early stages of diagenesis, when there are still some protein fragments left, and at any later stage, if sulfur is present. These elements can become incorporated in the sulfide accessory minerals, which is the usual case, or become bound to the organic sulfur of the coal.

Such sulfur-loving elements are often called "chalcophiles," presumably because they are associated with copper in the sulfide ores. The group includes, for present purposes, besides Cu itself, Fe, Ni, Zn, As, Se, Ag, Cd, Sb, Te, Hg, and Pb, etc. Note that many of them, besides forming insoluble sulfides, share the property of mercury of binding tightly to thiol groups or mercaptans. It is important to realize that they can be associated with organic sulfur as well as with the inorganic sulfide phases. This property will be discussed later.

The sulfur content of coal often increases during diagenesis. One mechanism for the increase is the transport of sulfate by ground water. Once in place, the sulfate is subject to reduction to a sulfide.

II.   THE ACCESSORY MINERALS OF COAL

Geologists quite properly distinguish between the "synge-
netic" minerals, formed at the peat stage or earlier, and the
"epigenetic" minerals, which are deposited from solution at a
later stage of coalification. But for the present purposes,
it makes sense to divide the minerals as follows: First, those
that usually occur in fissures and cavities in the coal and are
therefore rather coarsely intergrown with it; thus, a simple
gravity separation would remove most of them. Second, those
that are often so finely intergrown with the coal that gravity
separation is either not effective in removing most of them or
results in the discard of substantial amounts of the coal in the
effort to get rid of them.

Those clay minerals which are actually in the coal, rather
than an accidentally added overburden, are usually finely intergrown. The importance of the clays lies in their ion-exchange
capacity, which provides another accommodation for trace elements.

In contrast, quartz and chalcedony occur in both forms, but
this is of little interest as silica is not efficient in trapping
the toxic trace elements.

Hematite and other iron ores occur in both forms also, and
again this is of little interest.

Zircon and rutile, and such other minerals as tourmaline,
orthoclase, and biotite, are always finely divided. The first
two are of importance because of the effect of titanium, etc.,
on boiler slags, and, of course, the burden of radioactives
in zircon.

Evaporites, that is, chlorides, sulfates, and nitrates, are
almost always coarse and, in any event, are easily leached.

The remaining minerals, which are of the greatest interest,
are summarized in Table I.

TABLE I.   *Characteristic Accessory Minerals of Coal*

	Fine	Coarse
SULFIDES	FeS_2	
	Pyrite	Pyrite
	Melnikowite	
	Marcasite	Marcasite
	$CuFeS_2$	$CuFeS_2$
		(Chalcopyrite)
	ZnS	ZnS
		(Sphalerite)
		PbS
		(Galena)
PHOSPHATES	Apatite/	
	Phosphorite	
CARBONATES	Ca (Fe, Mg, Mn)	
	$(CO_3)_2$	
	(Ankerite)	Ankerite
	Calcite	Calcite
	Dolomite	Dolomite

The chalcophile elements, many of them quite toxic heavy metals, tend to concentrate in the sulfide phases, as indicated.

The carbonates are not nearly so important in this study. They have little effect on slag formation and hence are not a nuisance in that respect. One might expect barium to concentrate in one or the other carbonate minerals, but it is more likely to be present as sulfate. Beryllium, again, is more like aluminum than an alkaline earth, in spite of its valency, and would be expected to be trapped by clay or phosphate.

The later-arriving portion of the uranium, epigenetic uranium, tends to concentrate in the phosphate minerals. (Dinosaur bones are often found to have abnormal, extremely high uranium contents.) Naturally, the daughter elements, radium, polonium, and so forth, are likely to remain trapped in the mineral where they were formed, whether it be apatite or zircon.

Vanadium tends to follow uranium and, of course, fluoride is readily trapped by apatite.

Do not get the impression that *all* the trace elements are trapped in the accessory minerals--far from it. The coal itself not only contains thiol and thiophenol groups capable of trapping chalcophile metals, but carboxylic and phenolic functionalities, pyridine and amino nitrogen, and other groups capable of sequestering trace elements.

Vanadium, for example, is held in Venezuelan crude oil as complexes by such ligand groups alone, without benefit of an accessory mineral to trap it. One may suppose that the organic portion of the coal, like crude oil, is also capable of holding vanadium, as well as other metals. It has indeed been suggested that most of the sodium in coal is held in the organic phase. Yet it is undeniable that many trace elements tend to concentrate in the accessory minerals, so that these, although making up only a small percentage of the total mass, will contain the preponderant part of many important trace elements. Thus, the removal of these minerals from coal can achieve a very substantial reduction of these elements in the treated coal.

III. CHLORIDE CHEMISTRY OF THE TRACE ELEMENTS

In the current version of the JPL low temperature chlorinolysis process, the coal is initially exposed to chlorine in the presence of water and an immiscible chlorinated solvent. HCl and sulfuric acids are produced during the attack of chlorine on the sulfide minerals. Later on, the coal is subjected to hydrolysis by steam to remove any chlorine that may have become bound in the organic phase; this procedure again produces HCl.

Now, it is not always realized, outside the field of inorganic chemistry, just how many of the heavier elements have chlorides that are soluble in organic solvents, or volatile, or both. One may make several bold generalizations: first, the chlorides of the metals and metalloids are almost always more volatile than the corresponding oxides; second, the volatility of the chloride tends to increase with the valence of the metal; and third, those chlorides of polyvalent metals which are not soluble in water will often be soluble in one or more organic solvents.

The simple explanation for these generalizations is that oxygen, being divalent, is much more apt to form links between the metal atoms than is monovalent chlorine. A familiar comparison would be that between aluminum oxide, which boils around 3000° C, and aluminum trichloride, which sublimes as Al_2Cl_6 at only 180° C. Less familiar, perhaps, is the fact that ferric chloride is very similar to aluminum chloride in this respect, being nearly as volatile and as soluble in organic solvents.

Many of the heavy metal chlorides are subject to hydrolysis, of course, but often, as in the cases of arsenic and antimony, the hydrolysis is reversible in the presence of excess chloride ion. For practical purposes, the hydrolysis of a given chloride would be of concern only if it resulted in the metal being redeposited in the coal. Whether the trace element in question remains soluble in the water phase in some other form, or whether it precipitates as a colloidal hydroxide that leaves the coal upon filtration, is really only of academic interest at this point, so long as it is removed from the coal one way or another.

Table II shows the boiling points of some chlorides of the chalcophile elements.

Since most of these elements occur in coal to the extent of a few tens of ppm or less, not much solubility is required, either in water or in the organic solvent, to remove them completely.

TABLE II.   Volatilities of the Chlorides of Some Metals
Found in Sulfide Phases

	Boiling point	Solubility in water
Antimony		
$SbCl_5$	$79^{\circ}C$ at 22 Torr (decomposes)	
$SbCl_3$	283	
Arsenic		
$AsCl_3$	130.2	
Cadmium		
$CdCl_2$	960	v.s. water
Iron		
$FeCl_2$	1030	v.s. water
$FeCl_3$	300 (sublimes)	
Lead		
$PbCl_2$	950	s.s. dil HCl
Mercury		
Hg_2Cl_2	400 (sublimes)	
$HgCl_2$	302 (sublimes)	
Nickel		
$NiCl_2$	993 (sublimes)	v.s. water
Polonium		
$PoCl_2$	190 (sublimes)	
Selenium		
$SeOCl_2$	176.4	
Zinc		
$ZnCl_2$	732	

TABLE III. *Volatilities of the Chlorides of Other*
Trace Elements of Interest

		Boiling point	Solubility in water
Barium	$BaCl_2$	1560	v.s. water (but $BaSO_4$ = 0.006/ 100 ml 3% HCl)
Beryllium	$BeCl_2$	492	sim. to $AlCl_3$
Lithium	$LiCl$	1360	v.s. water
Phosphorus	$POCl_2$	105.3	v.s. water
Radium	$RaCl_2$	1000	s. water (but sulfate insoluble)
Thorium	$ThCl_3$	829 (sublimes)	v.s. water
Titanium	$TiCl_3$	660 (108 Torr)	v.s. water
Uranium	UO_2Cl_2	Melts 578 (decomposes)	s. water
Vanadium	$VOCl_3$	126.7	s. water
	VCl_4	148.5	s. water

As mentioned earlier, polonium, as a radioactive daughter of uranium, will be found in apatite or in zircon rather than in the sulfide minerals. It is included in this table because it is a chalcophile by chemistry, if not by provenience.

Table III covers the remaining trace elements of interest.

Note that radium, being so very similar to barium, is likely to be trapped as sulfate after chlorinolysis.

The chlorides and oxychlorides of phosphorus, thorium, titanium, uranium, and vanadium are all susceptible to hydrolysis, but, as will be shown, this property need not interfere with the removal of such elements from the coal.

This short review of the chemistry of these trace elements
shows that there is something distinctively different between
chlorinolysis and those oxidation methods of sulfur removal
which do not employ chlorine or chlorides. It is anticipated,
in short, that chlorinolysis will be much the more effective
method for removing most of the trace elements of interest.

IV. RESULTS AND DISCUSSION

Chlorinolysis was originally conceived of simply as a
method of desulfurization, without regard to other possible
benefits. But a few analyses for trace elements were made
during Phase I of the project, and these turn out to be very
suggestive.

There are two types of analyses: those done on the hydrolysis water, and the before-and-after analyses done on coal
samples. Phosphorus was the only element determined by a
colorimetric method; all the rest were done by atomic absorption,
and in every case the methods were those prescribed by the EPA
for the element in question.

Less sample preparation is required in analyzing the
hydrolysis water, and thus minimizes the chances of analytical
error; hence, these analyses are probably the most accurate.
Yet some of the trace elements may be taken up in the organic
solvent as chlorides, as was shown, and the solvent was not
analyzed. Hence, the figures in Table IV may not tell the whole
tale: for some elements, they may indeed be minimum figures.

The hydrolysis water from run 138 has picked up significant
amounts of arsenic and lead; this result is confirmed by the coal
analyses that are given shortly. The removal of alkaline earths
is no great surprise, for the carbonates are attacked. The
results for the alkalies are more surprising and leave one with
the question whether they are being removed from the organic

TABLE IV. *Trace Elements in Hydrolysis Water After*
Each of Two Chlorinolysis Runs on PSOC-219
Coal. Water/Coal Ratio Varied as Shown.

	Run 118 1000 cc water/100 g coal	Run 138 600 cc/100 g
Fe	800 mg/l	1,600.0 mg/l
Ca	140	55.0
Al	50	125.0
Na	160	no data
Pb	no data	12.5
As	no data	3.8
Mg	no data	41.0
K	no data	22.5
Ti	no data	8.5
P	no data	6.6

phase or whether clay minerals are being attacked; either way,
large amounts are being removed from the coal. The figures
for aluminum suggest that clay *is* being destroyed. It is worthwhile to translate these raw data to the basis of mg/kg of
coal and to compare them with the analysis of untreated PSOC-219

TABLE V. The Preceding Hydrolysis Water Analyses
          Reduced to an On-the-coal Basis,
          With the Raw Coal for Comparison

	Run 118	Run 138	Raw PSOC-219
Fe	8,000 mg/kg	9,600.0 mg/kg	
Ca	1,400	330.0	
Al	500	750.0	
Na	1,600		
Pb		75.0	46 mg/kg
As		22.8	73
Mg		246.0	
K		135.0	
Ti		51.0	1086
P		39.6	131

in Table V. The large discrepancy in the figures for lead
suggests either a substantial sampling error, or that the analytical procedure does not recover all the lead in converting the
coal to a form suitable for atomic absorption. High sampling
error is, of course, characteristic of minor minerals distributed
throughout the matrix, and elaborate precautions are required to
minimize such error. It is not always practical to observe these
precautions in a desulfurization test, and future studies of the
trace elements must provide figures that are not beset with this
error.

The same sampling error shows itself in the Ca figures and
probably has a strong effect on the As and P values as well.
It is in this light that the more abundant data on the treated
and untreated coal must be examined.

A toxic trace element may be known to occur in some coals
in significant amounts without being found in other coals or
the amount found may be so small that the analytical method

fails to give usable numbers. This last result is the case for mercury and selenium in the coals for which analyses were obtained.

Fortunately, cadmium, which behaves much like mercury in many respects, is considerably easier to determine accurately by atomic absorption. It is felt that when the analyst reports a greater than 10:1 reduction of cadmium between the untreated and treated coal, that, in fact, some substantial reduction did take place, even though the analyses are in 1 ppm range, and even though a sampling error is acknowledged.

A group of four elements, which are represented in the coals at the 4 to 20 ppm level, were determined with considerably less analytical precision than cadmium. These were vanadium, lithium, beryllium, and barium. The figures were often reported qualitatively, for instance, as "less than 10," leaving one to wonder whether that is greater or less than the simple "five" reported in the comparison analysis.

In the specific case of barium, the insolubility of the sulfate leads one to believe that removal of the element will be next to nothing, simply because chlorinolysis produces a substantial concentration of sulfate by oxidation of coal sulfur. The imprecision of the analytical data for barium does not permit even *that* hypothesis to be verified or contradicted.

In short, after cadmium, three elements for which the data seem to be meaningful (Table VI):

TABLE VI. Analysis of PSOC-219 Coal

	Raw	Run 107		Run 120	
	(ppm)	(ppm)	(Change)	(ppm)	(Change)
Arsenic	73	25	-66%	49	-32%
Lead	46	4	-91%	5	-89%
Titanium	1,086	510	-53%	680	-38%

Another coal, for which there were fewer determinations made, gave similar results (Table VII):

TABLE VII. Analysis of PHS-398 Coal

	Physically cleaned (ppm)	After run 140 (ppm)	(Change)
Arsenic	85	9	-89%
Titanium	1400	700	-50%

It would be unsound to make deductions about the relative efficiency of the three chlorinolysis runs from these data: sampling error is undoubtedly contributing a major fraction of the variability.

These data should be supplemented with new studies, using the best possible analytical methods for each element of interest, and taking all possible precautions to hold down sampling error. The labor and expense of such studies appear to be justified by the very positive results that have been obtained for the few elements presented here.

Lead is being removed to the extent of about 90%. This result is precisely what could be expected if most of the lead were in a sulfide phase that is destroyed by chlorinolysis. But there is more; lead forms mercaptides quite readily at the temperature of the process, and yet it is not being captured by the coal. Cadmium forms mercaptides even more readily than lead, and yet it is removed with about the same efficiency. The complexes of these metals with chloride are not very strong compared with their readiness to form sulfides and mercaptides. One explanation that springs readily to mind is that thiol

groups in the coal have been substantially removed by chlorinolysis: - indeed, thiol sulfur should be the organic form most
vulnerable to attack by chlorine. Future studies should test
this working hypothesis.

Other elements are removed from the coal in surprisingly
large amounts. The reduction of alkalies and of titanium should
have a favorable *economic* effect; slagging and fouling problems
in the furnace can be expected to be drastically reduced. An
empirical relation used in boiler design is

Slagging index = (B/A) × S

= (Basic oxides/acid oxides) × pyritic sulfur

If this formula were to be taken literally, then the removal
of pyrite by chlorinolysis would suffice to solve all problems
with furnace slag and corrosion. It can also be imagined that the
removal of an acid oxide necessarily *worsens* slagging. The formula
does make sense, however, in that certain slagging problems *have*
been ameliorated by the addition of silica, as sand, to the coal,
to reduce the relative proportion of alkalies. Alkalies foster the
formation of fume, and make the slag stickier and more fusible.

Chlorinolysis, as shown, does remove substantial amounts of
sodium and potassium. Within that context, removal of titanium
and aluminum is of benefit also, even if it appears to work
against the formula.

Thus, even though reliable analytical data are available for
only a few trace elements in a couple of coal types, it is clear
that the benefits of chlorinolysis in trace element removal are
worthy of serious investigation.

V. RECOMMENDATIONS FOR FURTHER WORK

It is obvious that the trace element question deserves
chlorinolysis experiments in which every effort has been made to
avoid sampling error. Such experiments should include not only

before-and-after analyses of the coal itself, and coordinated
analyses of the hydrolysis water, but also analyses of the solvent and water in contact with the coal during chlorination.

With sampling error at a minimum, it becomes worthwhile to
scrutinize the analytical methods for the individual trace
elements. Consider mercury. Past methods for mercury were so
insensitive at the ppm level that atomic absorption was rightly
hailed as a tremendous advance in the art. But the blunt truth
is that atomic absorption is simply not good enough for our purposes. The importance of mercury might well justify the expense
of neutron activation analysis, if no other method can be found
that will quantify mercury in the 10 to 1000 parts-per-billion
range in coal.

Vanadium is a particularly important element in that its
presence in slags leads to rapid rates of corrosion of the boiler
tubes. For what the present vanadium data may be worth, it
seems to occur in the coal samples at about the 25 ppm level.
How this is partitioned between the accessory phosphate minerals
and the organic phase of the coal itself is not known. One way
to settle the vanadium question would be the use of the scanning
electron microscope to excite the X-ray emission of vanadium, so
that both its total amount and its distribution in the raw coal
and in the treated coal can be determined. The same technique
can obviously apply to a number of the heavy elements.

Lithium and beryllium are clearly not going to be susceptible
to this technique. But the analytical difficulties that have
been had with these elements may not be due to atomic absorption,
but to the techniques used to work up the coal samples. Unlike
the normal alkalies and alkaline earths, these elements have a
strong tendency to precipitate with silica or to remain attached
to siliceous residues. Better methods of working up the samples
may solve the problem.

Finally, the radioactives can be identified and quantified by their own gamma-ray emissions. The techniques have long existed; they just have not been applied as yet to the chlorinolysis of coal.

It is not the first time that coal samples are observed to be distinct individuals, and that rules based on one coal do not necessarily apply to another. Much of this "individuality," as far as trace elements are concerned, is likely to be due to the suite of accessory minerals present in each coal. It will be desirable, therefore, to supplement these proposed analytical studies with identification of the accessory minerals in each different type of coal that is subjected to chlorinolysis. This can readily be done by an X-ray diffraction microtechnique and by conventional coal petrography.

REFERENCES

1. Nicholls, G. D., *in* "Coal and Coal-Bearing Strata" (Murchison and Westoll, eds.), pp. 269-337. American Elsevier, New York (1967).

2. Mackowsky, M. -Th., *in* "Coal and Coal-Bearing Strata, (Murchison and Westoll, eds.), pp. 309-312. American Elsevier, New York (1967).

3. Manskaya, S. M., and Drozdova, T. V., "Geochemistry of Organic Substances" (L. Shapiro and I. A. Breger, eds.), Pergamon Press, New York (1968).

DISCUSSION

Beadle: What is the disposal problem of the wash water containing the trace elements?

duFresne: Well, at this point, we haven't given much thought to it. It is not a question of having a sludge; we have a solution without much suspended matter in it. So we do not have the option of, say, concentrating these elements by ion exchange or we could reuse the hydrolysis water by bringing it up to steam temperatures again, which is, in effect, distilling it. So we should be able to get the "nasties" out in fairly concentrated form; and even if they are not economically recoverable, at least we don't have to spread them out all over the environment.

Singh: What about the organic solvent?

duFresne: The organic solvent is distilled for reuse and consequently recovery from that material should be possible also. Again, as I say, we are looking at data that were taken more or less as an afterthought and have not fully worked out what we are going to do next. But it does seem to us, biased as we perhaps are, that keeping these materials from going into the environment at large is going to be much, much easier than in the case where you are lime-scrubbing stack gas, for example, and where the toxic trace elements wind up in the gypsum sludge.

Unidentified Speaker: What size are the samples?

duFresne: Well, these are from bench scale runs which are perhaps of the order of a kilogram--600 grams down to 100 grams in some cases. Quite a few bench scal runs made under different conditions, with different sizes of samples. Right now we are in the throes of getting the pilot plant operating and I could tell you some stories about that, but I am afraid my time is running very short.

SIZE/COMPOSITION PROFILES OF
RESUSPENDED FLY ASH

T. A. Cahill and L. L. Ashbaugh

Crocker Nuclear Laboratory
University of California
Davis, California

Characterization of fly ash has been made in the laboratory by a resuspension system. Samples of fly ash from four different power plants were resuspended in an air jet system, followed by transport of one meter in an airstream. The particles were sized with a Lundgren rotating drum impactor with coated Mylar stages. The five size fractions were subsequently analyzed for elemental content using particle induced X-ray emission (PIXE) and X-ray fluorescence systems. The chemical flash size characterization of the resuspended fly ash matched semi-quantitatively the results taken in the stack by previous studies, even for particles below 0.5 μm diameter. Other studies have characterized fly ash particles collected on site at power plants. The fine particle modes in this fly ash were likewise recovered via resuspension by a Wright dust feeder and analyzed by a spiral centrifuge impactor.

I. INTRODUCTION

Considerable effort is being expended to provide accurate characterizations of fly ash from coal combustion. The data are needed to provide estimates of impacts of coal combustion on health and welfare, including optical extinction. The problem has been compounded by the discovery of strong chemical composition shifts in submicron components of the fly ash, so that simultaneous measurements of size and composition are important. Recent studies

of stack emissions have become quite complex and expensive for
this reason, since in-stack impactors with post collection elemental and chemical analyses are difficult to accomplish quanti-
tatively. For this reason, only a limited number of relatively
complete data sets are available, and routine detailed evaluations
of combustion particulate effluents are not being done.

II. PROCEDURES

One possible method for increasing the information on particulates from coal combustion involves detailed analysis of the fly
ash itself as collected by electrostatic precipitators or other
devices. The key question here revolves around the changes that
occur in the fly ash once it has been collected in bulk, and, in
particular, whether the important and chemically unique submicron
components can still be isolated. Resuspension of fly ash in an
air stream could be a relatively simple method for characterizing
stack effluents if this were so, and this method would allow
routine evaluations of emissions at numerous sites and over
extended periods of time. A further reason for evaluating the
size-composition profile of resuspended fly ash and the fly ash
present in stack effluents is that major studies of health impacts
are being made with resuspended ash. Fly ash, sometimes presized
on site at a power plant, is resuspended in exposure chambers by a
dust feeder. The resuspended ash is then characterized by
numerous methods to guarantee that the size-composition profiles
are representative of the effluent. These comparisons would be
made more easily, however, if the same types of instrumentation
that have been used on stack analyses could be used on the
resuspended ash. An opportunity to do this arose following an
Environmental Protection Agency (EPA)-financed study of a power
plant burning low-sulfur western coal (1). The study involved
extensive stack monitoring for fly ash composition as a function

of size using an MRI impactor with grease-coated stages. Fly ash
was also removed from the hopper on the same day and sealed in
jars, and it was this ash that was used in the laboratory studies.

The fly ash was placed in a chamber and resuspended with a
flow of air through a sintered glass disk. The particles were
then accelerated and injected into clean air in order to achieve
about a 100 : 1 dilution prior to sizing. (See Fig. 1.) A
standard Lundgren impactor sized the ash into diameter size
ranges: above 15 microns, 2 to 5 microns, 0.5 to 2 microns, and
less than 0.5 micron. Analysis was subsequently performed by
particle-induced X-ray emission (PIXE) for the elements sodium and
those heavier.

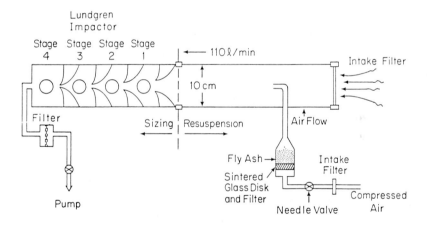

FIGURE 1. Fly ash resuspension and sizing system, University
of California, Davis, California.

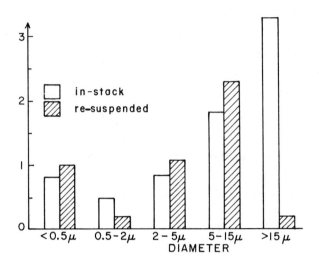

FIGURE 2.   Silicon: Relative mass compared with size.

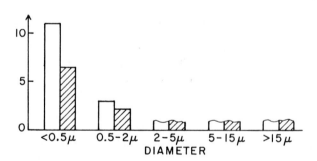

FIGURE 3.   Ratio S/Si as a function of size.

III.   RESULTS AND DISCUSSION

The results of this test are shown in Figs. 2 and 3.   The
fine particle silicon mode seen in the stack was recovered in the
resuspended ash, whereas the strong submicron sulfur peak was also
seen in both analyses.   Other crustal elements (Al, K, Ca, Ti, Fe)
showed only minor changes in ratios to Si on the basis of size.
This behavior was subsequently seen in fresh fly ash samples from
three other power plants, although comparison with stack samples
was not possible in these latter cases.

There are numerous arguments one can marshall as to why the
resuspended ash should not have the submicron modes seen in the
stacks.   The collected ash tends to be the coarser fraction, and
fine particles should not be seen in it, whereas any moisture
should, for agglomerates, further suppress submicron particle production.   The fact that they are present may be due, in part, to
the separation of small particles stuck to large particles in the
resuspension process, since only large particles are efficiently
collected by the precipitator.   Whatever the mechanism, the result
opens a possibility of more extensive studies of plant emissions
at moderate cost, while supporting inhalation studies based on
resuspended fly ash.

REFERENCES

1.   Ensor, D. S., Cahill, T. A., and Sparks, L. E., Elemental
     Analysis of Fly Ash from Combustion of a Low Sulfur Coal,
     Proceedings of the 68th Annual Meeting--Air Pollution Control
     Association, Boston, Massachusetts, 75-33.7, pp. 1-18 (1975).

ACID PRECIPITATION--A PROBLEM IN METEOROLOGICAL
PHYSICS/CHEMISTRY

Donald H. Pack

Consulting Meteorologist
McLean, Virginia

Examination of observations on the chemistry of precipitation in the United States and Canada, obtained in the last three years, is used to describe differences in the behavior of various chemical compounds and to outline some of the problems that must be addressed to relate emissions of acid-forming materials to the changing precipitation chemistry.

I.   INTRODUCTION

The purpose of this paper is to describe, briefly, some of
the issues that have developed as more knowledge is obtained
about acid precipitation in the United States. There is no doubt
that precipitation in the eastern United States is much more
acidic than the value that would exist if the water were just in
equilibrium with natural carbon dioxide (a pH of about 5.6).
Annual pH values (weighted by precipitation amounts) near 4.0 are
common. Individual rains with pH values of 3.2 have been measured.
Although there is controversy regarding the timing of the trend,
the evidence that acidity has increased over the last two decades
is considerable. What is not known is the quantitative relations
between the amount and character of anthropogenic emissions and
the acidity of the collected precipitation.

In a large part, this is the result of a poor record of
interest in the United States as reflected in the sporadic and
discontinuous observational record in this country. In Europe,
following the lead of Carl Gustav Rossby, atmospheric chemistry,
and rain chemistry in particular, was felt to be an integral part
of meteorology and an element requiring continuous surveillance.
Observations of the chemistry of precipitation over most of
Europe now have a history of about 25 years in duration. In contrast, in the United States, only 1 year of data is available in
the mid-1950s from an Air Force program. In the mid-1960s, 2
years of data were collected by a PHS-NCAR effort. A small NOAA-
EPA network of only 10 stations for the entire United States
began in the mid-1970s. There have been a few other programs for
collection and chemical analyses at a few locations but none that
would provide the spatial coverage for the many climatic and
land-use zones of the United States.

Interest in this topic has, however, burgeoned in the last
few years and modern measurement and analyses programs have been
put into place. Unfortunately, these programs are not uniform in
character, nor are they conceived, sponsored, and funded as long-
term continuing efforts, at least not yet.

Without attempting to be comprehensive, it is useful to list
the present major programs for the collection of precipitation
for subsequent chemical analyses.

NOAA-EPA	10 sites	Monthly composited samples
CANSAP	50 sites	Monthly accumulated samples (Canada)
TVA	5 sites	Bi-weekly composited samples
NADP	30 sites	Weekly composited samples
MAP3S	8 sites	"Event" sampling
EPRI	9 sites	"Event" sampling (24 hr accumulation)

Research efforts--Usually single site collections, but with
                   a few small-area networks, e.g.,
                   Missouri, Illinois, New York, etc.

Figure 1 shows the distribution of these networks in the United
States and southern Canada.

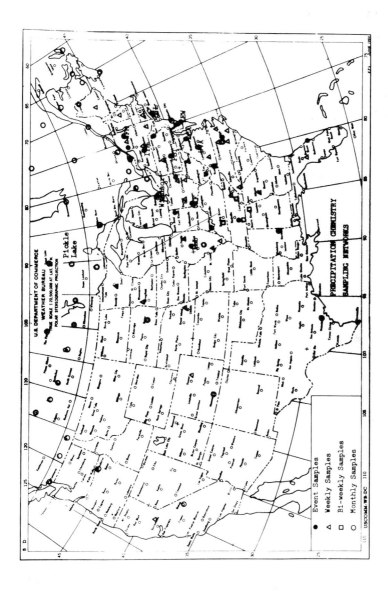

FIGURE 1. Continuing programs for collection and chemical analyses of precipitation.

Each of these programs uses a different laboratory to perform
the chemistry of the samples. However, a program for inter-
laboratory comparisons has been established. Different collectors have been used to sample the precipitation, although there
is now convergence on one type of design. The major difference
between the networks is in the sampling interval. A recent paper
suggests that depending on the particulate loading of the collected samples, dramatic alterations may occur in the measured ion
concentrations if the analyses are delayed for any significant
amount of time (1). There is also evidence that the integrity of
sample concentrations may be a function of the acidity, the more
acid samples changing less with time than those with a pH above
about 5.0.

II. DATA SOURCES

The data presented here are analyses of the Multi-State
Atmospheric Power Production Pollution Study (MAP3S) published by
Battelle's Pacific Northwest Laboratory, from the periodic publi-
cations of the Canadian Atmospheric Environment Service (CANSAP)
and from the Air Resources Laboratories, NOAA Washington Area
network (2,3). MAP3S and the Washington Area data are all "event"
samples; that is, the precipitation was collected for each individual rain or snow. The Canadian data are monthly accumulations
of precipitation. The longest record is 4 years for the
Washington data; the MAP3S record covers 2 years and is slightly
longer than the Canadian data. No other information of comparable
length and quality were available.

Chemical analyses for pH, conductivity, and at least 10 ions
have been published for the MAP3S and CANSAP data but only pH
analyses for the ARL network. Monthly weighted ion concentrations
have been calculated for all the United States data to reduce the
large variations that exist between individual precipitation

events and to provide a more stable statistic. (See Reference 4
for a description of the procedure.) This study also permits
direct comparison with the Canadian data which is, de facto, a
monthly average. Attention has been primarily confined to pH
(expressed as H^+), $SO_4^=$, NO_3^-, NH_4^+, Cl^-, and Na^+. Only a fraction
of the analyses can be presented here. A fuller discussion by
D. H. Pack and D. W. Pack will become a part of the Proceedings
of the World Meteorological Organization Technical Conference on
Regional and Global Observations of Atmospheric Pollution
Relative to Climate, August 1979.

There has not yet been a controlled comparison of the differences that might be occasioned by the different sampling intervals. This lack suggests caution in comparing the results of
different networks. Indeed, one of the purposes of these analyses
was to see whether the data could reveal something about these
differences.

III. DISCUSSION

The analyses of Pack and subsequent extension of this work
indicate seasonal and spatial variations in the various ions and
different types of sites (4). There is a very strong annual cycle
in the $SO_4^=$ ion concentration, even when weighted by the observed
precipitation amounts. Sulfate concentrations peak very strongly
during the warm season in the United States. This is in direct
contrast to the European data where the $SO_4^=$ is at a maximum in
winter. It is not now possible to determine whether this effect
is due to a different pattern in fuel usage (i.e., emissions) or
to a different climatology created primarily by latitudinal differences. Hales and Dana have suggested that warm, moist weather
favors the rapid oxidation of SO_2 to $SO_4^=$ in aqueous solutions (5).

This annual cycle is most clearly shown in the pH record in
the Washington Area. Figure 2 is the time history for this area
and has a number of interesting aspects. First is that the annual
peak does *not* occur in the same month each year. In 1975, the
maximum was in May, in August in 1976, in July in 1977, whereas in
1978, the pattern was quite different with a double, and lower,
peaks in July and September. It can also be seen that although
the patterns vary from year to year, the annual weighted average
value's range is small and actually decreased (less acidity) in
1978. This result should not, however, be interpreted as any kind
of trend yet.

The correlations between the concentrations of various ions
with the distance separating the sites was computed for all possible combinations of the 8 sites in the MAP3S network and similarly for 11 sites in the CANSAP network. These calculations
show that in the United States, the $SO_4^=$ ion concentrations are
much more closely related; that is, they vary in parallel much
more than do the Canadian sites. This different behavior might
be brought about by a less pronounced cycle in Canada, by different source-receptor relations, by the different sampling
schemes, or by a different overall climate. It is not yet possible
to separate out the various factors.

"Internal" ion correlations were also calculated. These
were between the weighted monthly ion concentrations of pH and
5 different ion concentrations at each of the 19 locations. For
example, $SO_4^=$ was correlated against pH, NO_3^-, and NH_4^+, etc. The
Canadian data show very low correlations, site by site, between
pH and either the sulfate or nitrate concentrations. In contrast,
the MAP3S sites with 2 years of data show high correlations, near
0.9, between pH and sulfate. Even the lowest value was 0.70. In
contrast, the Canadian data have correlations ranging from -0.36
to 0.76. Sulfate and nitrate concentrations are poorly correlated
with each other in the United States network, but are much higher

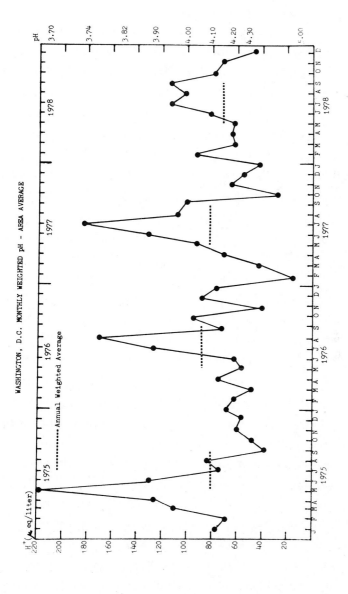

FIGURE 2. Time history of precipitation pH--Washington, D.C.

for the Canadian collections. Both data sets show high correlations between $SO_4^=$ and NH_4^+, the correlations being generally
above 0.75.

Separation of the data into logical subdivisions shows further
contrasts. The Canadian data were selected to represent the
St. Lawrence Valley and, in contrast, remote locations. The remote
sites showed, as might be expected, lower acidity and lower concentrations for all the ion species than was the case for the more
developed region. However, the intra-ion relations (i.e.,
correlations) were about the same for both regions. The United
States data indicated that the two coastal sites (Brookhaven,
Long Island, New York and Lewes, Deleware) are different, in the
mean, from the inland locations. This is most evident in the
Na^+ - Cl^- correlations which were 0.71 and 0.98 at the coastal
sites. In contrast, these same correlations inland ranged from
-0.11 to 0.24. (One must be cautious, however, since both of the
coastal locations had less than 1 year of reported data.) Another
interesting observation is that when the Lewes collector site was
changed, the Na^+ and Cl^- concentration averages were 228 and
197 μmol/liter, respectively. After the move, the averages at the
second site were 21 and 31 μmol/liter.

One of the most interesting aspects of the United States
record is the variation in the ions that dominate the observed
acidity. As stated previously, the sulfate concentrations show a
large annual amplitude. On the other hand, the nitrate ion concentrations, although variable from month to month, do not show
this organized pattern. To examine the seasonal control of acidity,
it was necessary to determine, for each site and for each month,
the ratios $SO_4^=/H^+$ and NO_3^-/H^+. Figure 3 shows these data plotted
for each site in terms of whether the NO_3^-/H^+ ratio was larger,
or smaller, than its counterpart, the $SO_4^=/H^+$ ratio. A very clear
pattern emerges. For all sites, the sulfate ion dominates the
acidity in the warmer months. In winter, the nitrate contributes

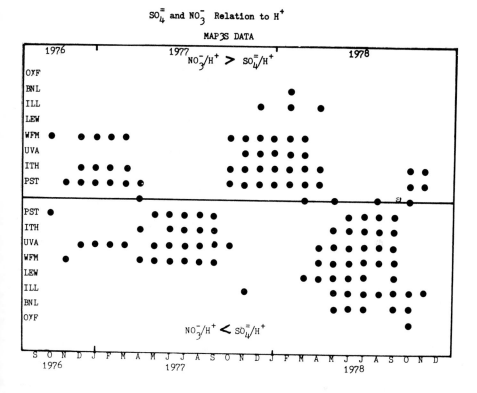

FIGURE 3.   *Seasonal variability in the dominant ion controlling acidity.*

most to the observed acidity. There is a very abrupt shift in these relationships with the transition to sulfate control of acidity occurring in April or May and nitrate becoming predominant in October. The causes of this sequence are not known, but one can speculate that the seasonal meteorological changes--less solar radiation, lower temperatures, and water vapor, etc., play a role in reducing the $SO_4^=$ levels.

IV.   CONCLUDING REMARKS

We do not now have a complete model or theory that can explain
these observations nor does a complete "end-to-end" simulation
seem near at hand.   Such a complete simulation must include, at
least:

(1) Natural and anthropogenic emission factors as a function
of time-space (A point of special concern is NO_2. Recent work
(6,7) suggests that lightning-produced NO_2 may be a significant
input to atmospheric levels and, hence, to precipitation NO_3^-.)

(2) Homogeneous and heterogeneous conversion of pollutants:

(a) in dry air

(b) in clouds

(c) in air after cloud evaporation, as a function of
atmospheric variables, e.g., temperature, relative humidity,
radiation, etc.

(3) Three-dimensional transport of pollutants

(4) Presence or absence of clouds

(5) Quantitative occurrence of precipitation

(6) Pollutant removal rates for gaseous and aerosol pollutants,
etc.

All these conditions presuppose that the observed precipitation
chemistry is adequate in spatial, temporal, and chemical docu-
mentation to test and improve the model(s).   Thus, whether
simulation capability or not is possible, improvement in the
observational coverage of the United States and exploration of the
observations for guidance in the transformations of the various
ions, regional differences, source-receptor relation, and trends
in acidity and ion composition must continue.   The variability in
the data thus obtained even over the short 2-year period suggests
that this study will take time, probably many years of dedicated
observations.

Finally, one of the more interesting unknowns and one on which data might shorten this time scale as it improves one's understanding of acid precipitation is the chemical content of cloud water. Does it differ as a function of season, cloud type, storm system, altitude, etc.? All the ground-based sampling can only hint at these vital data.

REFERENCES

1. Peden, M. E., and Skowron, L. M., *Atmos. Environ. 12,* 2343 (1978).
2. Dana, M. Terry, The MAP3S Precipitation Chemistry Network: First Periodic Summary Report, PNL-2402, Battelle Pacific Northwest Laboratory (1977).
3. Atmospheric Environment Service, Canada, CANSAP Data Summary, Quarterly Summaries (January 1977 through June 1978)(1978).
4. Pack, Dee W., *Geophy. Res. Letters, 5,* 673 (1978).
5. Hales, J. M., and Dana, M. Terry, *J. Appl. Meteor. 18,* 3 (1979).
6. Noxon, J. F., *J. Geophy. Res. 83,* 3051 (1978).
7. Yung, Y. L., and McElroy, M. B., *Science, 203,* 1002 (1977).

DISCUSSION

Harriss: Don, I was surprised there wasn't more of a maritime
effect on the BNL and the Lewes stations. Could you comment on
that?

Pack: I did the Na/Cl ratios and at the inland stations, they
run about 0.07 to 0.21. The Brookhaven value is 0.71 and the
Lewes is 0.98. Now, let me say something else. At Lewes,
Delaware, there were sodium chloride concentrations of around
200- to 300-micromoles per liter in the first 3 months of
their data. The reason that ratio is in brackets in Lewes,
they moved it off the roof, 1/4 mile away, to a permanent site.
The values dropped by an order magnitude. They tried to correct
for sea-salt sulfate for the rooftop data. So there is a very
definite bias. Brookhaven is a little different. They are
sitting where they are getting a very large New York City, New
York State component. They are much further from the beach--of
the order of 10 miles at least--but Lewes is within about 300
meters of the ocean itself. But we have got to have locations
based on land use, on meteorological patterns, and on source
patterns. It's a very complex problem.

Singh: Are chlorides totally insignificant with precipitation?

Pack: No. It represents about 10% of the acidity. It is
highly variable, but about 10%.

Beadle: There is a second station in Delaware--it's an EPRI
event sampling station colocated in the vicinity of their class-
one station at Indian River which is about 20 miles inland.
It would be interesting to make comparisons between the MAP3S
site which is, as Don said, just over the dunes, and essentially
within spitting distance of the ocean versus the other station
which is further inland.

Pack: We are getting the EPRI data Monday night. Unfortunately
I didn't have it for this meeting.

Mattson: I am fascinated by this presentation. It would seem to
me, though I don't have the technical capability to really
comprehend most of this, that what I am concerned about is just
the natural environment itself as a baseline. You just keep
talking about sources that are man-made sources, and I keep
thinking about the way things are.

Pack: My collaborator, Dr. Miller, has been doing similar work
at Mauna Loa Observatory, Point Barrow, Alaska, Samoa, Hawaii,
and then I have some data from Tazmania. These show, of course,
very much less acidity than we see in our part of the country.
But Mauna Loa still runs about 4.8 in terms of pH; Samoa is about
5.1; and Tazmania, a very remote location, comes the closest to
the carbon dioxide equilibrium--it is 5.3. The subject is so
complex, we could talk all afternoon. But we think we are
beginning to vector in on the natural component. So they know
how much additional acidity is added by man.

TRACE METAL SOLUBILITY IN AEROSOLS
PRODUCED BY COAL COMBUSTION

Steven E. Lindberg

Oak Ridge National Laboratory
Oak Ridge, Tennessee

Robert C. Harriss

NASA Langley Research Center
Hampton, Virginia

The results of a combustion plume aerosol solubility experiment and geochemical applications of the measurements are presented. The particle size distribution and relative solubilities of aerosol-associated Cd, Mn, Pb, and Zn were measured in plume samples taken 0.25 km and 7 km from a Tennessee Valley Authority 2600 MW coal-fired power plant. Also measured for comparison are relative solubilities of the same elements in stack ash and in ambient aerosols collected 33 km from the plant. When the total plume aerosol is compared with the stack ash, each of the elements increases in relative solubility with increasing distance from the power plant. The trend of increasing solubility with decreasing particle size, which is typical for ambient aerosols, was not observed in plume aerosols. The range in relative solubility of the elements studied is sufficiently large to be an important factor in environmental assessment. Inclusion of the aerosol solubility concept provides improved prediction of both potential pollutant dose and pathway in areas subject to atmospheric deposition from coal combustion sources.

589

I.   INTRODUCTION

The critical links between coal combustion emissions and
their effects on ecosystems are the mechanisms and rates of
atmospheric deposition and the subsequent bioavailability of the
deposited pollutants.   A key factor determining bioavailability
is the water solubility of aerosols.   This point has been considered for the leachability of $SO_4^=$, NH_4^+, and Cl^- in marine
aerosols, and recently for trace metal leachability in the total
size fraction of ambient aerosols collected near the Pacific
Coast, and for urban aerosols collected in New York City (1-3).
However, because of sampling and analytical difficulties, the
solubility of aerosol-associated trace metals in various size
fractions of coal-combustion emissions has not been studied in
sufficient detail for evaluation as a potential indicator of
ecological impacts.   Reported herein are the results of an
aerosol solubility experiment and some biogeochemical implications of the measurements.

II.   METHODS

A.   Aerosol Collections

Aerosol samples were collected in the plume of the
Cumberland power plant during a cooperative study with the Air
Quality Branch of the Tennessee Valley Authority.   The
Cumberland Power Plant is located on the Cumberland River
approximately 80 km northwest of Nashville, Tennessee.   The
plant includes two horizontally opposed boilers with a total
electric power generation capacity of 2600 MW.   Each unit is
equipped with a 305-m stack and an electrostatic precipitator

with a design efficiency of 99%. The units burn pulverized coal
from western Kentucky with a sulfur content of $\sim$ 4%. During the
studies reported here (26-27 October 1976) only one of the two
units was operating.

A helicopter was used to collect samples during flights
across the path of plume flow during early morning hours while
the plume was visually well defined. Plume center was determined
by continuous readout of SO_2, NO, NO_x, Aitken and condensation
nuclei levels while samples were drawn through a teflon intake
probe which extended beyond the prop wash. Further details
relating to plume identification, local weather conditions, and
airborne sampling methods have been described in Meagher *et al.*
and Lindberg *et al.* (4,5). Atmospheric stability during the
period of sample collection was determined to be class D
(J. Meagher, personal communication). Samples were collected
at altitudes ranging from 250 m to 450 m and distances from
0.25 km to 7 km downwind of the source stack during horizontal
passes perpendicular to the wind direction in an attempt to
sample similarly aged aerosols at each downwind distance.

Particulate samples were collected by use of both total
filters and Anderson impactors modified for isokinetic flow
under the experimental conditions. The impactor modification
involved removal of the lower five stages based on calculation
of the critical flow velocities and impaction parameters for
the conditions of sampling needed to maintain isokinetic flow
through the intake probe and the impactor inlet. The aerodynamic particle diameters for each stage, assuming an impaction
efficiency of 50%, were calculated from relationships between
flow rate, impaction efficiency, and particle diameter supplied
by C. H. Erickson, Director of Engineering, Anderson 2000 Inc.
These values are as follows: particle size class 1 = 1.3 μm
aerodynamic diameter, class 2 = 0.56 μm, class 3 = 0.14 μm,
class B (backup filter) = <0.14 μm. For comparison purposes

ambient aerosols were collected at Walker Branch Watershed,
a location not directly influenced by the combustion plume.
The watershed is located $\sim$ 240 km east of Nashville. In
addition, samples of particulate matter were collected in the
stack and from the electrostatic precipitator hopper.

To facilitate the aerosol leaching experiments, the standard
glass impaction plates supplied with the Anderson impactor
were coated with nuclepore polycarbonate film. This material
was chosen as the impaction surface as well as for the backup
and total filters (0.4 μm pore diameter) because of its low
trace constituent concentrations, its nonhygroscopicity, its
hydrophyllic nature, and its low absorption and adsorption
losses, which make it an ideal substrate for aerosol leaching
experiments (6,5).

B. *Solubility Experiments*

The laboratory leaching experiments were designed to result
in the complete extraction of the water soluble component of
the ambient aerosols in an initial step and in the extraction
of the dilute acid leachable component in a following step.
For selected samples, the residue remaining after the water
and dilute acid extraction was subject to an additional acid
leaching at 10 times the original acid concentration, or the
residue was carried through a complete digestion procedure to
determine the residual metal concentrations.

Briefly, these methods involved the following: (1) water
extraction--2 hour agitation of aerosols in distilled water
using a reciprocating shaker, (2) dilute acid extraction--similar
agitation of residue from step 1 in 0.08N Ultrex ultra-pure
HNO_3 (pH = 1.1) followed by a 7-day nonagitated leach in the
same acid. All samples were filtered through 0.4 μm nuclepore
membranes following extraction and stabilized with Ultrex HNO_3.

The additional acid extractions follow: (3) 1N HNO_3 extraction--
step 2 was repeated using 1N Ultrex HNO_3, (4) acid digest--heat
treatment of residue with Ultrex HNO_3, Ultrex HCl (conc.), and
distilled water (7). All handling of aerosol samples was performed in a laminar flow clean bench.

Leachates were analyzed for Cd, Pb, Zn, and Mn by using
graphite furnace atomic absorption spectrophotometry (GFAAS)
and by employing the standard additions method on each sample.
Sulfur, extracted as sulfate, primarily in the distilled water
leach, was determined by using an automated colorimetric procedure employing the improved method of McSwain et al. (8).
The GFAAS method resulted in mean accuracies (based on analysis
of NBS Standard Fly Ash) on the order of ± 5% with precision
ranging from 5% to 10% for leachate concentrations in the
10 µg/l range. The accuracies of the sulfate analyses were
generally < ± 2% whereas precision was < ± 5%.

Since our interest was in the environmentally mobile forms
of aerosol-associated elements, the work was concentrated on
the initial two leachate fractions. The sum of the air concentrations in these two fractions was termed the available metal
concentration, and the ratio of the water soluble fraction to
the available concentration was defined as the relative
solubility. The purpose of the additional acid extractions
(steps 3 and 4) was to compare the available with the residual
concentrations. Briefly, these experiments indicated the
following: continued extraction using 1N HNO_3 released minor
amounts of trace elements (< 5% additional Cd and Mn, $\sim$ 10%
additional Pb and Zn) and no further $SO_4^=$, whereas the acid
digest released larger amounts of Pb and Zn. The ratio of
the available to the total (available plus digestible) metal
concentrations was 0.9 to 1.0 for Cd and Mn but 0.4 to 0.5 for
Pb and Zn. However, since the environment rarely approaches
the conditions of the acid digest, the consideration of available

concentrations and relative solubilities is valid, particularly
with respect to intersample comparisons and biological effects.
Further details of the extraction, digestion, and analytical
methods have been described in Lindberg *et al.* (5,9).

III. RESULTS

A. *Aerosol Composition and Particle Size*

The Cd, Mn, Pb, Zn, and S (as $SO_4^=$) concentrations in the
total aerosol from the combustion plume and in the regional
ambient aerosol are presented in Table 1. The relative decrease
in the plume concentrations from 0.25 km to 7 km is similar for
all of the elements studied, values ranging from a 2.8-fold
decrease in Pb to a 4.7-fold decrease in Mn. Comparing these
values with mean elemental concentrations in the ambient aerosol
indicates that at 7 km, elements considered as having significant

*TABLE 1. Concentrations of Trace Elements (Available
Fraction, ng/m^3) and Sulfate (μg/m^3) in Total Aerosol Samples
from a Coal-fired Power Plant Plume and in the Regional Total
Ambient Aerosol*

Element	Distance from emission source		
	0.25 km	*7.0 km*	*Ambient*
Cd	27	8.2	0.17
Mn	47	10.0	9.40
Pb	820	290.0	110.00
Zn	990	270.0	35.00
$SO_4^=$	400	115.0	9.00

coal combustion sources (Cd, Zn, $SO_4^=$) are present at concentrations of $\approx$ 10 to 50 times those in ambient air. Concentrations
of constituents considered to be primarily derived from soil
(Mn) and automotive (Pb) sources are more comparable to ambient
aerosols by the time the plume has reached 7 km.

The particle size distributions of each element were as
expected for emissions from combustion sources with Cd, Pb, Zn,
and $SO_4^=$ being concentrated on fine particles, whereas Mn was
somewhat more uniformly distributed. (See Fig. 1.) At the
0.25 km distance, the geometric mass weighted mean diameters
(Table 2) are similar to the geometric mean diameters (GMD)
measured by Whitby *et al.* for plume samples collected at 2 km
to 4 km from the stack of a coal-fired plant (GMD = 0.20 ±
0.01 μm) (10). As the Cumberland plume aged during travel from
0.25 km to 7 km, there was an increase in the geometric mass
weighted mean diameter for Cd whereas this parameter decreased
somewhat for the other elements. Whitby *et al.* reported a
slight increase in the GMD of aerosols during plume travel from
1 km to 4 km with no subsequent change during travel from 4 km
to 8 km (10).

B. Aerosol Solubility Experiments

The results of solubility experiments are presented in
Figs. 2 and 3 which describe the relative solubility for each
trace element in four particle size classes, in the total
plume aerosol, in the total stack ash, and for the total ambient
aerosol ($\overline{X}$ ± σ). The relative solubility of $SO_4^=$ ranged from
98% to 100% in all samples and is not included in this discussion.

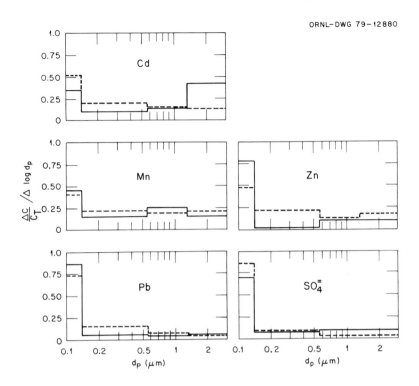

FIGURE 1. Normalized size distributions for Cd, Mn, Pb,
Zn, and S (as $SO_4^=$) in aerosols from the plume of a coal-fired
power plant. Size distributions of samples collected at 0.25 km
downwind are designated by a dashed line, while samples
collected at 7 km downwind are designated by a solid line.
$\Delta C/C_T$ = relative concentration of the constituent in the atmosphere, where ΔC = air concentration within a given size class
(ng/m^3) and C_T = total air concentration in all size classes
(ng/m^3); d_p = particle aerodynamic diameter (μm).

TABLE 2. *Geometric Mass Weighted Mean Diameters for Particulate Trace Elements and Sulfate from a Coal-fired Power Plant Plume and Ambient Air Samples (All Values in μm).*

Element	Distance from emission source		
	0.25 km	7.0 km	Ambient
Cd	0.16	0.34	1.50
Mn	0.23	0.18	3.40
Pb	0.14	0.10	0.50
Zn	0.17	0.13	0.89
$SO_4^=$	0.12	0.12	0.60

ORNL-DWG 78-18667

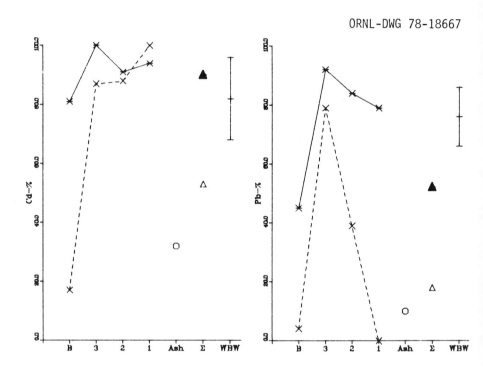

FIGURE 2. *Relative solubilities of aerosol-associated Cd and Pb in the plume of a major coal-fired generating plant at two distances downwind of the stack, 0.25 km (---) and 7 km (——). Also shown for comparison are relative solubilities of the stack ash, the total (Σ) aerosol fraction in the plume (Δ = 0.25 km, $\blacktriangle$ = 7 km), and ambient aerosols (WBW, $\overline{X} \pm \sigma$).*

ORNL-DWG 78-18667

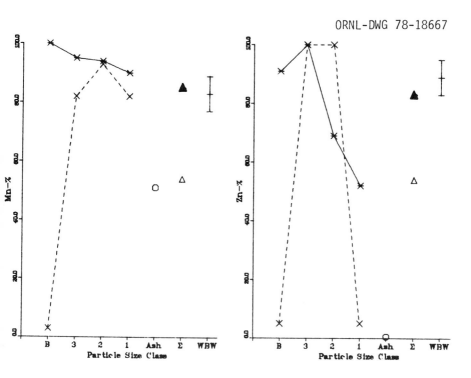

FIGURE 3.   Relative solubilities of aerosol-associated Mn
and Zn in the plume of a major coal-fired generating plant at
two distances downwind of the stack, 0.25 km (---) and 7 km
(———).   Also shown for comparison are relative solubilities of
the stack ash, the total (Σ) aerosol fraction in the plume
(Δ = 0.25 km, ▲ = 7km), and ambient aerosols (WBW, X̄ ± σ).

1. *Solubility of the Total Aerosol*. The relative
solubility of each of the trace elements exhibits the same
trend, increasing with increasing distance from the emission
source. The plume aerosols always exhibit a higher relative
solubility than the stack ash. This difference is supported
by the work of Gordon who made a similar observation during
studies of element solubility in precipitator ash collected
at the Chalk Point Power Plant (11). Detailed in-plant
studies led to the conclusion that the general nature of
in-plume aerosols (chemical composition and particle size
distribution) may be quite different from that of fly ash
collected in the plant. This condition is apparently also true
for aerosol solubility. The similarity of the relative
solubility of the plume aerosol at 7 km to the regional ambient
aerosol is an interesting and unexpected finding. The relative
solubilities of Cd, Mn, and Zn at the 7 km distance are within
1 σ of values measured for the ambient aerosol, whereas that
for Pb is a factor of $\approx$ 1.5 lower.

Although there are no other in-plume aerosol solubility
studies by particle size class with which to compare these
data, the recent papers by Hodge *et al.* and Eisenbud and Kneip
do provide data on total aerosol solubility of ambient
aerosols (2,3). These authors employed water leaching plus
concentrated acid wet digestion to determine water solubility
defined as: (water leachable trace element concentration) ·
100/(total leachable plus digestible concentration). Hence,
their results are not directly comparable with the relative
solubility data. However, as previously discussed, water
solubility for the total size fraction of one ambient aerosol
sample was determined with the following results: Mn (75%) >
Cd (68%) > Pb (28%) > Zn (19%). For comparison, the marine
aerosols collected by Hodge *et al.* exhibited the following
water solubilities (mean and range): Cd (83%, 63 to 91) >

Zn (53%, 10 to 82) > Mn (43%, 23 to 58) > Pb (30%, 6 to 54);
the values reported for the urban aerosols collected by Eisenbud
and Kneip were as follows: Cd (85%, 68 to 100) > Zn (64%,
41 to 100) > Mn (62%, 55 to 68) > Pb (9%, 5 to 13) (2,3). In
general, the values reported herein show somewhat lower
solubilities for Zn, but higher for Mn, the values for Cd and
Pb being roughly comparable. However, because of the variability
in reported solubilities, the solubility ranges of the marine
and urban aerosols encompass the values measured during this
study, with the exception of Mn. It is this wide range in
solubilities which suggests that this aspect of aerosol
chemistry requires further study to determine the solubility
controlling parameters.

   2. *Solubility as a Function of Particle Size in Plume
Aerosols.* The relative solubility behavior of trace elements
in different particle size fractions of the plume aerosols
provides important information for the interpretation of the
total aerosol data. Generally, the most significant change
in relative solubility occurs in the finest particle size
class (class B). Increases in relative solubility of class B
particles range from $\sim$ 85% to 98% (Zn, Mn) to $\sim$ 45% to 65%
(Pb, Cd) when samples collected at 0.25 km are compared with
those collected at 7 km. Lead and Zn exhibit particle-size-
related changes in solubility in the larger size classes, the
most pronounced being an $\sim$ 80% increase in the solubility of
lead on class 1 particles at 7 km relative to 0.25 km.

   The expected trend of increasing relative solubility with
decreasing particle size reported for ambient aerosols by
Lindberg *et al.* does not exist for the 0.25 km plume aerosols
and is exhibited only marginally by Mn and Zn in the 7-km plume
samples (5). The decreased relative solubilities of the finest
particle size class are particularly pronounced for Cd and Mn
at 0.25 km.

3. Factors Influencing Aerosol Solubility. The general
characteristics of the plume aerosols described in the preceding
paragraphs include relatively small changes in particle size
distribution from 0.25 km to 7 km, similar dilution rates during
dispersion for all elements studied, increases in relative
solubility of particles in the finest size class from 0.25 km
to 7 km, and constant (Cd, Mn) or slightly increased solubility
(Zn, Pb) of largest particles from 0.25 km to 7 km. With one
exception (Zn in class 2 particles), the relative solubilities
of aerosol particles at 7 km always exceeded the values measured
at 0.25 km.

The primary factors which are known to influence the physical
and chemical properties of plume aerosols during travel from the
emission source include gas-particle interactions, dilution with
ambient air, and the formation of aggregates. The data suggest
the hypothesis that vapor condensation on the fine aerosols,
which are characterized by relatively high surface area to
volume ratios, leads to the formation of thin, highly soluble
coatings, and is the major factor resulting in the increased
relative solubility of the total aerosol at 7 km. This condition
is discussed in further detail in Lindberg *et al.* (5). This
hypothesis is supported by studies which have characterized the
surfaces of aerosols by direct chemical analysis, demonstrating
thin (< 100 $\overset{\circ}{A}$) film enrichments of trace elements on coal
combustion aerosols (12).

IV.   IMPLICATIONS FOR ENVIRONMENTAL ASSESSMENT

The results of this study have interesting implications for
understanding processes related to the chemistry of precipitation, biogeochemical pathways of trace elements following dry
deposition of the landscape, and potential human health hazards.
Aerosol-associated trace elements with relatively high

solubilities may have a higher potential for immediate impacts on biogeochemical cycles. Any future attempts to develop dose-response models for predicting ecological impacts of coal combustion should include consideration of variations in both the total flux of a trace element and the water soluble fraction of the total flux.

A particularly important example of the potential significance of water soluble aerosol material occurs when aerosols deposited on leaf surfaces are exposed to localized increases in moisture (i.e., dew on the leaf surface). Droplets of dew can mobilize and concentrate potentially hazardous trace elements on leaf surfaces leading to very high exposure levels at localized positions on the leaf (5). However, the less soluble trace elements will more generally follow geological and detrital pathways in soil and stream environments.

V. SUMMARY

Few generalizations should be drawn from the limited data reported in this paper. However, because data of this type are difficult to obtain, are largely absent from the literature to date, and have interesting implications for understanding potential impacts of coal combustion, it is believed that the data should receive careful study. The following observations are of particular significance to the problem of assessing potential environmental impacts:

(1) The concentrations of aerosol associated Cd, Mn, Pb, Zn, and $SO_4^=$ in the emissions from a major coal-fired power plant exhibit similar dilution rates for distances up to 7 km from the source stack.

(2) During stable atmospheric conditions the power plant
plume can increase the levels of Cd, Zn, and $SO_4^=$ in the atmosphere 10 to 50 times above ambient levels at distances up to
7 km from the source stack.

(3) The elements Cd, Mn, Pb, and Zn in the total aerosol
from the power plant plume become increasingly soluble as the
distance from the source stack increases from in-stack to
0.25 km to 7 km. The relative solubilities of Cd, Mn, and Zn
in the total aerosol samples 7 km from the stack are within 1 σ
of values measured for the regional ambient aerosol. These
results suggest that coal combustion sources may be a major
contributor of soluble trace elements to the regional tropospheric aerosol.

(4) The finest particles in the plume (< 0.14 μm) exhibit
the most significant changes in relative solubility when aerosols
sampled at 0.25 km are compared with 7-km plume samples. Zn and
Mn exhibit ∿ 85% to 98% increase; Cd and Pb exhibit ∿ 45% to 65%
increase in relative solubility. Pb and Zn also exhibit
significant increases in relative solubility in larger particle
size classes. These results support the concept of surface
enrichment of trace elements by condensation of plume gases on
aerosol surfaces during the initial cooling of emissions close
to the stack. These thin surface coatings are a key factor in
trace element transport and fate in the environment.

(5) Aerosol solubility is an important parameter for understanding processes related to the chemistry of precipitation,
biogeochemical pathways of trace elements following dry
deposition to the landscape, and potential human health
hazards.

ACKNOWLEDGMENTS

The authors wish to acknowledge the assistance given by J. Meagher and staff of the Air Quality Branch of the Tennessee Valley Authority and staff of the Cumberland Power Plant during the in-plume sampling experiments and for the assistance in collection of precipitator and stack ash. Thanks also goes to J. Story who provided field assistance during this work, and to D. Shriner and D. Huff who provided useful comments on the manuscript.

This research was sponsored by the Office of Health and Environmental Research, U.S. Department of Energy, under contract W-7405-eng-26 with Union Carbide Corporation and has been presented in Publication No. 1368, Environmental Sciences Division, ORNL.

REFERENCES

1. Meszaros, E., *Tellus, 20,* 443 (1968).
2. Hodge, V., Johnson, S., and Goldberg, E., *Geochem. J., 12,* 7 (1978).
3. Eisenbud, M., and Kneip, T., Trace Metals in Urban Aerosols. EPRI-117, Electric Power Research Institute, La Jolla, California, 418 pp (1975).
4. Meagher, J., Stockburger, L., Bailey, E. M., and Huff, O., *Atmos. Environ. 12,* 2197 (1978).
5. Lindberg, S. E., Harriss, R. C., Turner, R. R., Shriner, D. S., and Huff, D. D., Mechanisms and Rates of Atmospheric Deposition of Selected Trace Elements and Sulfate to a Deciduous Forest Watershed. ORNL/TM-6674, Oak Ridge National Laboratory, Oak Ridge, Tennessee (1979).

6. Andren, A. W., Lindberg, S. E., and Bate, L., Atmospheric
    Input and Geochemical Cycling of Selected Trace Elements in
    Walker Branch Watershed. ORNL/NSF/EATC-13, Oak Ridge
    National Laboratory, Oak Ridge, Tennessee (1975).

7. Anderson, J., *At. Abs. News, 13,* 31 (1974).

8. McSwain, M. R., Watrous, R. J., and Douglass, J. E., *Anal.
    Chem. 46,* 1329-1331 (1974).

9. Lindberg, S. E., Turner, R. R., Ferguson, N. M., and
    Matt, D., *in* "Watershed Research in Eastern N. America"
    (D. Correll, ed.), Smithsonian Institute Press,
    Washington, D.C. (1977).

10. Whitby, K., Cantrell, B. K., Husar, R. B., Gallani, N. V.,
    Anderson, J., Blumenthal, D., and Wilson, W., Jr.,
    Aerosol Formation in a Coal Fired Power Plant Plume.
    Extended abstract, presented before the Division of
    Environmental Chemistry, American Chemical Society (New
    York) April (1976).

11. Gordon, G., Study of the Emissions From Major Air Pollution
    Sources and Their Atmospheric Interactions. Progress
    Report-75, University of Maryland, College Park, Maryland
    (1975).

12. Keyser, T. R., Natusch, D., Evans, C., and Linton, R.,
    Environ. Sci. Technol. 12, 768 (1978).

DISCUSSION

Whitby: What does the fine particle mean?

Harriss: The aerodynamic diameters of those fine particles
were less than 0.14 microns. This corresponds to the backup
filter for the modified Anderson impactor.

Whitby: Next question. This is all relative. Were you able
to compare the absolute in water solubility relative to the
total content?

Harriss: Yes. Basically the acid soluble material and the
total material differ for certain elements. But, in some
cases, there are considerable differences between the acid
soluble and the total digestible material.

Yue: Trace metal ions in water droplets can act as a catalyst
for the oxidation of sulfur dioxide. Did you analyze the
sulfur in that aerosol to find out whether it related with the
trace metal ions inside the aerosol?

Harriss: Yes, we have. We have data available on the total
composition. In case of solubility studies, the sulfate is
always 100 percent soluble--it always dissolves in all particle
size ranges.

Yue: Did you find which metal ions are more efficient in act-
ing as catalysts for the oxidation of sulfur dioxide?

Harriss: No, I have not seen any consistent relationships
which would suggest that, but we did not design our experiments
to study the problem.

Beadle: It would be interesting to compare solubility data
from the use of distilled water for your solubility experiments,
with so-called "acid rainwater"--one could use real rainwater,
for that matter--and also to compare these solubilities with
the solubilities that Ondov and Jerry Fischer are obtaining in
a biological fluid, namely, lung fluid. I also have a comment
to Dr. Glenn Yue. I was curious about this myself, and I
inquired of Don Gatz the other day whether he sees significant
amounts of the transition metals, iron and manganese, in rain
which they have analyzed. He indicated there is very little
iron in solution in a rain drop or in rain, most of it is
insoluble. More manganese is in solution though; this might

indicate, depending on what he means by very little compared to the manganese level, that the mechanisms that have been proposed for solution catalysts of SO_2 by iron may not be too important, and we can probably look more at a manganese catalyzed reaction--if it is indeed working.

THE EFFECTS OF ELEVATED CO$_2$ CONCENTRATIONS
ON GROWTH, PHOTOSYNTHESIS, TRANSPIRATION,
AND WATER USE EFFICIENCY OF PLANTS

R. W. Carlson and F. A. Bazzaz

University of Illinois
Urbana, Illinois

*Photosynthesis, transpiration, and growth were measured at
300, 600, and 1200 ppm CO$_2$ for selected crop, agricultural weed,
and tree species after up to 35 days of growth in 300, 600, and
1200 ppm CO$_2$. Growth and photosynthetic rate increased by varying
amounts with increased CO$_2$. Transpiration decreased with
increasing CO$_2$ for all species. The amount of biomass supported
by a unit of transpired water increased by an average of 52% for
crop species, 76 to 128% for weed species, and 38 to 71% for tree
species with a doubling of the CO$_2$ concentration within the range
of 300 to 1200 ppm. The magnitude of the increase and variation
between species in water use efficiency and growth response suggest a change in the competitive relationships between crop and
weed species and between species within a community as a function
of an increase in the ambient CO$_2$ concentration.*

I. INTRODUCTION

The input of CO$_2$ into the atmosphere from anthropogenic

sources is presently increasing at a rate that may cause a doub-

ling of the atmospheric CO$_2$ concentration by the year 2020 (1).

These authors suggest that a peak in CO$_2$ production will occur

around the year 2060. This peak will result in a maximum atmo-

spheric CO$_2$ concentration of 5 to 10 times the preindustrial

value. There is increasing evidence to suggest that because of forest fires, terrestrial plants are now a net source of CO_2 into the atmosphere (2-4). In fact, the input of CO_2 from the biota may be as large as the fossil fuel source which is estimated as $1 \pm 0.6 \times 10^{15}$ g carbon per year. By comparison, the amount of carbon in the present atmosphere is of the order of 600 to 700 $\times$ 10^{15} g. It is well known that for many species, growth and photosynthesis increase with increasing CO_2 concentration. Attempts have been made to estimate the effect of elevated CO_2 concentrations on crop (5,6) and forest (7) productivity. The results suggest that CO_2 uptake may be increased by up to 45 percent in a corn field when the concentration of CO_2 within the crop canopy is doubled. As the concentration of CO_2 in the air increases, there is a substantial decrease in the rate of transpiration (8-12). Thus, for plants grown in high CO_2 concentrations, the amount of CO_2 taken up in photosynthesis per unit of water expended in transpiration (water use efficiency) may be higher than that of plants grown at normal CO_2 concentrations.

Thus, under elevated CO_2 concentrations not only are plant growth and productivity likely to increase but the amount of water required to support plant growth may decrease. Reported herein are the results of an experiment in which growth, photosynthesis, transpiration, and water use efficiency were measured for several crop, weed, and tree species grown in a controlled environment of either 300, 600, or 1200 ppm CO_2. In most species, growth and photosynthesis increased with CO_2 concentration and transpiration declined for all species. The increase in water use efficiency with a doubling of the ambient CO_2 concentration varied between species from 38 to 128 percent. The magnitude of this increase and the variation between species suggest a potential for (1) plant species to expand into drier habitats, and (2) altered competitive relations between species in agricultural and natural habitats should the concentration of CO_2 in the atmosphere rise significantly above present levels.

II. METHODS

Crop and weed species were started from seed and grown, respectively, for 20 and 28 days in a common environment within a glasshouse before CO_2 treatment. The tree species were obtained as dormant saplings 0.3 to 0.5 m tall, pruned to 0.2 m and placed in a glasshouse with CO_2 treatment beginning 14 to 28 days after bud break. For CO_2 treatment, the plants were transferred to 1 m^3 glass-sided growth chambers maintained at CO_2 concentrations of 300, 600, or 1200 ± 30 ppm. The environment within each growth chamber was maintained at 27 ± 3° C day temperature, 18 ± 3° C night temperature, and 75 ± 5% relative humidity with 15-hour days during which the natural sunlight of the glasshouse was supplemented with illumination from metal halide lamps. The light intensity at plant height varied from 2000 $\mu E\ m^{-2}\ s^{-1}$ on clear days to 800 $\mu E\ m^{-2}\ s^{-1}$ of photosynthetically active radiation (400 to 700 nm) on overcast days. Plant biomass of crop and weed species was measured at 14 and 28 days and of tree species at 21 and 35 days after the initiation of the CO_2 treatments. Each plant was fertilized with 2.2 g of 13:13:13 NPK twice during the first and third weeks of CO_2 treatment. Soil moisture was maintained near field capacity by watering the free-draining pots daily to slightly above saturation. In a typical experiment, 6 plants of 3 different species were simultaneously grown at each CO_2 concentration with 3 plants being measured at each harvest. Additional replication and treatment of other species were accomplished by a series of successive experiments over time. Photosynthesis and transpiration were measured just prior to each harvest with a semi-closed infrared gas analysis system (13). This system consists of an airtight environmentally controlled chamber connected by glass tubing to a nondispersive infrared gas analyzer. A portion of a plant, usually a single leaf, was placed in the chamber and the change in concentration of CO_2 was measured over time. Transpiration was

measured simultaneously by measuring the change in relative
humidity within the chamber. Measurements of gas exchange were
made at 300, 600, and 1200 ppm CO_2, 25° C air temperature, 50%
relative humidity, 1.5 m s^{-1} wind speed, and a light intensity of
2200 μE m^{-2} s^{-1} (400 to 700 nm).

III. RESULTS AND DISCUSSION

A. *Growth*

At the end of 28 days of CO_2 treatment, growth was significantly greater for sunflower (*Helianthus annuus* L.) and soybean
(*Glycine max* L.) at 1200 ppm than at 300 ppm. Total plant weight
increased by 38% for sunflower and by 75% for soybean (Table I).
at 600 ppm the increase in total plant weight was 20% and 58%,
respectively, for sunflower and soybean. Total plant weight of
maize (*Zea mays* L.) was not different between treatments. For
all three species, the amount of leaf area per plant was not different between treatments. Dividing leaf area by plant weight
gives rise to a term called leaf area ratio (LAR) which, in a
broad sense, represents the ratio of photosynthetic to respiring
material within a plant. LAR declined with increasing CO_2 concentration for both sunflower and soybean but was the same at all
CO_2 treatments for maize.

At the end of 28 days of CO_2 treatment at 1200 ppm, there was
a significant increase in growth for two of the weed species with
velvet leaf (*Abutilon theophrasti* Medic.) attaining a 75% and
jimson weed (*Datura stramonium* L.) a 96% increase over growth at
300 ppm (Table I). Increased growth was also observed at 1200 ppm
for pigweed (*Amaranthus retroflexus* L.) and common ragweed
(*Ambrosia artemisiifolia* L.) but to a lesser degree. Leaf area
per plant changed only a small amount across treatment concentrations. Velvet leaf showed a slight increase, ragweed a slight

TABLE I. *Change in Total Weight for Plants Grown at 600 and 1200 ppm CO_2 Expressed as a Percent of Plant Weight for Plants Grown at 300 ppm CO_2*

Species	Common name	600 ppm (%)	1200 ppm (%)
Glycine max	*Soybean*	58	75
Helianthus annuus	*Sunflower*	20	38
Zea mays	*Maize*	24	-7
Datura stramonium	*Jimson weed*	74	96
Abutilon theophrasti	*Velvet leaf*	44	75
Amaranthus retroflexus	*Pigweed*	41	21
Ambrosia artemisiifolia	*Common ragweed*	10	24
Acer saccharinum	*Silver maple*	61	89
Populus deltoides	*Cottonwood*	65	74
Platanus occidentalis	*Sycamore*	13	30

decrease, and pigweed and jimson weed no change. LAR declined with increased CO_2 for all species, jimson weed exhibiting the largest decline and pigweed, the smallest.

Significant increases in growth of 89 and 74 percent, respectively, were observed for silver maple (*Acer saccharinum* L.) and cottonwood (*Populus deltoides* Bartr.) at 1200 ppm over that at 300 ppm (Table I). Large increases in growth were also obtained at 600 ppm. Growth of sycamore (*Platanus occidentalis* L.) increased by 30% at 1200 ppm and 13% at 600 ppm. Leaf area per plant increased with increasing CO_2 concentration for silver maple and sycamore. For cottonwood, the amount of leaf area per plant was the same at 300 and 1200 ppm but slightly higher at 600 ppm. LAR remained constant across treatment concentrations for sycamore but declined slightly with increasing CO_2 for silver maple and cottonwood.

Leaf, stem, and root weight varied in proportion to total plant weight for all species. Thus, the allocation of structural carbohydrate to various plant parts was not altered by elevated CO_2 during vegetative growth. Experiments are needed to see whether reproductive allocation is altered by elevated CO_2

concentration. Presently, there is some evidence to suggest that
reproductive allocation varies in proportion to total plant weight
but that flowering might be initiated at an earlier date for
plants grown at high CO_2.

These results suggest that for many species, an increase in
the atmospheric CO_2 concentration may lead to an increase in
plant growth. The response appears to be curvilinear with an
upper asymptote that is probably not much above 1200 ppm. This
result has been observed for other species (15). For maize and
perhaps pigweed, there may be an optimum CO_2 concentration for
growth, that is, less than 1200 ppm. The response of plant
growth to increased CO_2 concentration may be limited by other
factors in the environment, such as light intensity and soil
nutrition. Large differences in growth between low and high CO_2
concentrations are not exhibited unless the plants are adequately
fertilized. The interaction between elevated CO_2 and other
environmental factors must be known to determine the true impact
of high atmospheric CO_2 on plant growth.

The differences in growth between species in response to
increased ambient CO_2 concentration make generalizations difficult but the data presented in Table I do suggest the possibility
of some interactions that should not be overlooked. The crop
group includes soybean which responds to increased CO_2 with a
large increase in growth, maize which shows a small or even a
negative effect at high CO_2 levels, and sunflower which is intermediate. The weed group also contains species with a broad range
of responses to CO_2.

The weed species are common potential inhabitants of agricultural fields and, as such, are of great concern to the agri-
culturalist. Given a doubling of the present ambient CO_2 concentration, it would appear that jimson weed, velvet leaf, and
pigweed would increase their competitive ability and possibly
cause reduced crop growth and yield, especially in maize fields,

whereas ragweed would become less of a problem. In soybean
fields, jimson weed would have an advantage at higher CO_2 concentrations while the other three weed species would either pose the
same or less of a problem than at present. These species are
also components of a natural community that forms during the
first year after agricultural land is abandoned. Differential
response to elevated CO_2 concentrations may alter competitive
relationships and could lead to significant changes in community
structure and dynamics. Weed communities are well suited to the
study of community dynamics and the information derived from them
can often be used to suggest the response of later successional
communities to perturbation.

B. Photosynthesis

The rate of photosynthesis increased with increasing CO_2 concentration for sunflower, soybean, and the four weed species up
to 1200 ppm (Figs. 1 and 2). Photosynthesis became saturated
with respect to CO_2 at $\approx$ 600 ppm for maize and the tree species
(Fig. 3). The photosynthetic response to CO_2 concentration was
the same for plants grown at different concentrations. This
result suggests that the controls on net photosynthesis are not
altered by external CO_2 even when the level of CO_2 during the
growth period is increased by a factor of 4 above atmospheric
concentration. This result is in sharp contrast to the effect
of light intensity on photosynthesis where maximum photosynthetic
rates are lower and occur at lower light intensity for plants
grown under less than an optimal light (14).

C. Transpiration and Water Use Efficiency

Transpiration decreased with increasing CO_2 concentration for
all species. The difference between growth concentrations was

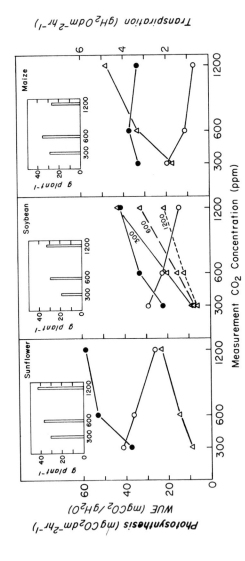

FIGURE 1. Response of photosynthesis (●), transpiration (0), and water use efficiency (Δ) to CO_2 concentration and response of growth (inset) for sunflower, soybean, and maize grown at 300, 600, or 1200 ppm CO_2. The trends in photosynthesis and transpiration were the same for plants grown at different CO_2 concentrations. Therefore, a composite curve was calculated for each species. The same was true for water use efficiency of sunflower and maize. Water use efficiency of soybean varied as a function of growth concentration.

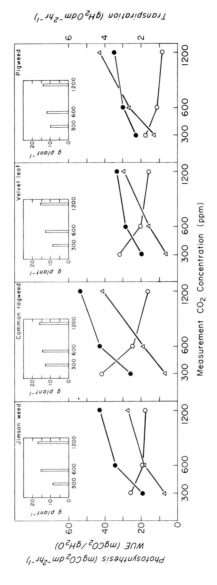

FIGURE 2. Response of photosynthesis (●), transpiration (0), and water use efficiency (△) to CO_2 concentration and response of growth for jimson weed (Datura stramonium), common ragweed (Ambrosia artemisiifolia), velvet leaf (Abutilon theophrasti), and pigweed (Amaranthus retroflexus) grown at 300, 600, or 1200 ppm CO_2.

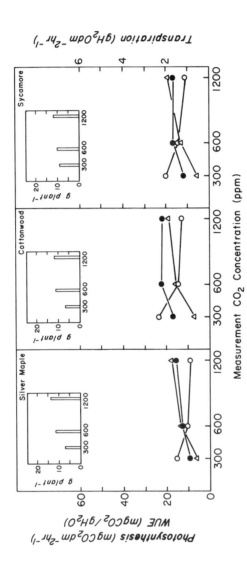

FIGURE 3. Response of photosynthesis (●), transpiration (O), and water use efficiency (△) to CO_2 concentration and response of growth for silver maple (Acer saccharinum), cottonwood (Populus deltoides), and sycamore (Platanus occidentalis) grown at 300, 600, or 1200 ppm CO_2.

not statistically significant but there was a trend in soybean toward greater transpiration at high CO_2 concentrations for plants grown at 600 and 1200 ppm.

Water use efficiency of individual leaves was calculated by dividing the photosynthetic rate by the rate of transpiration. The resulting curve for leaf water use efficiency increases with increasing CO_2 concentration for all species. For all species except soybean, this curve was not significantly different between plants grown at different CO_2 concentrations; therefore, data for each species were combined into a single curve. However, for soybean, leaf water use efficiency measured at 600 and 1200 ppm progressively decreased from high values for plants grown at 300 ppm to lower values for plants grown at 1200 ppm (see Fig. 1). This is an important aspect of the present study. Others have observed a decrease in transpiration and stomatal aperture with increased CO_2. However, the results may not be applicable to a prediction of the effect of elevated ambient CO_2 on the overall water economy of a plant unless the leaves are allowed to form and mature in the concentration in which the measurements are made.

An estimate of the water use efficiency of the whole plant (WUEp) was calculated from the per hour rate of transpiration of individual leaves multiplied by the total green leaf area of each plant, divided into total plant weight, and expressed as a fraction of the water transpired by plants grown at 300 ppm. For each species, the values of WUEp were neither significantly different between harvests nor between replicate experiments and were therefore combined into a single data set for curve fitting. The data for each species were fit by least squares analysis to a linear form of $y = ax^b$ where y = WUEp, x = the CO_2 treatment concentration, and a and b are constants. The resultant

coefficients of determination (r^2) are listed in Table II. The
exponent b in this expression can be used to calculate the percent change in WUEp (Δ_2) that occurs with a doubling of the CO_2
concentration by solving $\Delta_2 = (2^b - 1)/100$ for Δ_2. Change in
water use efficiency (Δ_2) varied from 38.0 for silver maple to
127.7 for common ragweed (Table 2). The values of Δ_2 for crop
plants varied over a smaller range (47.5 to 54.5) even though
these species are very different morphologically and biosyn-
thetically. Soybean and sunflower are dicotyledonous plants
which fix CO_2 by the ribulose diphosphate-pentose (C_3) pathway
characteristic of the majority of plant species whereas maize is
a monocot which fixes CO_2 by the tricarboxylic acid pathway (C_4).
Moreover, there is a large difference in the rate of transpiration
at 300 ppm CO_2 between these three species with sunflower at 3.5,
soybean at 2.5, and maize at 1.7 g H_2O dm^{-2} hr^{-1}.

The data on water use efficiency suggest that with an increase
in the CO_2 concentration, it may be possible to expand crop ranges
into geographical areas which are presently too arid to support
them. Too much haste in this direction would be premature for
at least two reasons. First, similar experiments must be done
at relative humidities lower than that used in this study since
atmospheric aridity is usually concomitant with soil aridity.
Second, the weed species examined in this study uniformly had
higher Δ_2 values than the crop species. This suggests that an
elevated CO_2 concentration in combination with decreased moisture
availability might confer a competitive advantage on weed species
in an agricultural field. Another aspect of the water use efficiency data is the effect of elevated CO_2 combined with moisture
stress on competitive relationships within a natural community.
Although three of the weed species have nearly the same Δ_2, the
value for ragweed is 50 percent higher and there is a wide range
for the tree species. Again, as with the influence of CO_2 on

Table II. Water Use Efficiency of Whole Plants $(WUEp)^a$

Species	Common name	r^2 (%)	Δ_2 (%)
Helianthus annuus	Sunflower	0.772	54.5
Zea mays	Maize	0.693	53.9
Glycine max	Soybean	0.767	47.5
Ambrosia artemisiifolia	Common ragweed	0.644	127.7
Abutilon theophrasti	Velvet leaf	0.587	87.1
Datura stramonium	Jimson weed	0.834	83.5
Amaranthus retroflexus	Pigweed	0.877	76.3
Platanus occidentalis	Sycamore	0.887	70.9
Populus deltoides	Cottonwood	0.663	55.7
Acer saccharinum	Silver maple	0.935	38.0

a WUEp was calculated from the per hour rate of transpiration of individual leaves multiplied by the total
green leaf area of each plant, divided into total plant
weight, and expressed as a fraction of the water transpired by plants grown at 300 ppm. The data were fit by
least squares analysis to a linear form of $y = a\ x^b$ where
$y = WUEp$, $x = CO_2$ treatment concentration and a and b are
constants. The exponent b was used to calculate the percent change in WUEp (Δ_2) that occurs with a doubling of
the CO_2 concentration by solving $\Delta_2 = (2^b - 1)/100$.

growth, it is necessary to examine the importance of these Δ_2
values in a situation where the plants are allowed to compete
with each other for moisture.

IV. CONCLUDING REMARKS

Photosynthesis, transpiration, and growth were measured at
300, 600, and 1200 ppm CO_2 for selected crop, agricultural weed,
and tree species after up to 35 days of growth in 300, 600, and
1200 ppm CO_2. Growth and photosynthetic rates increased by
varying amounts with increased CO_2. Transpiration decreased with
increasing CO_2 for all species. The amount of biomass supported
by a unit of transpired water increased by an average of 52% for

crop species, 76 to 128% for weed species, and 38 to 71% for tree species with a doubling of the CO_2 concentration within the range of 300 to 1200 ppm. The magnitude of the increase and the variation between species in water use efficiency and growth response suggest a change in the competitive relationships between crop and weed species and between species within a community as a function of ambient CO_2 concentration.

REFERENCES

1. Siegenthaler, U., and Oeschger, H., *Science 199*, 388 (1978).
2. Adams, J. A. S., Mantovani, M. S. M., and Lundell, L. L., *Science 200,* 54 (1977).
3. Wong, C. S., *Science 200,* 197 (1978).
4. Woodwell, G. M., Whittaker, R. H., Reiners, W. A., Delwiche, C. C., and Botkin, D. B., *Science 199,* 141 (1978).
5. Allen, L. H., Jensen, S. E., and Lemon, E. R., *Science 173,* 256 (1971).
6. Lemon, E. R., Stewart, D. W., and Shawcroft, R. W., *Science 174,* 371 (1971).
7. Botkin, D. B., Janak, J. F., and Wallis, J. K., in "Carbon and the Biosphere" (G. M. Woodwell and E. V. Pecan, eds.), p. 328. (Available NTIS, Springfield, Virginia.)
8. Moss, D. N., Musgrave, R. B., and Lemon, E. R., *Crop Sci. 1,* 83 (1961).
9. Egli, D. B., Pendleton, J. W., and Peters, D. B., *Agron. J. 62,* 411 (1970).
10. Akita, S., and Moss, D. N., *Agron. J. 12,* 789 (1972).
11. Van Bavel, C. H. M., *Agron. J. 14,* 208 (1974).
12. Regehr, D. L., Bazzaz, F. A., and Boggess, W. R., *Photosynthetica 9,* 52 (1975).
13. Bazzaz, F. A., and Boyer, J. S., *Ecology 53*, 343 (1972).
14. Boardman, N. K., *Ann. Rev. Plant Physiol. 28,* 355 (1977).
15. Wittwer, S. H., and Robb, W. M., *Economic Botany 18,* 34 (1979

DISCUSSION

Green: I wonder if any of the climate modelers have considered this as a feedback mechanism seriously?

Carlson: I have no knowledge of them having done so but I think it is something that should definitely be considered.

Shaw: In addition to the doubling the CO_2, modelers have also told us that we will have a temperature increase. Would you comment on what you think a 5 or 6 degree temperature increase for the zones where these types of plants grow would do to photosynthesis?

Carlson: If you are talking about as much as a 5 or 6 degree increase, you probably would have some influence just on growth per se. If you are only talking about a couple of degrees, it 'vould be less so as far as just growht per se. But, of course, with increasing the temperature, you might be increasing the length of the growing season, and if so, I think this could have a considerable effect on total productivity. With an increase in the air temperature, you also get a retreating of permanent snow fields and an advance of hospitable growing con- ditions northward, and this will of course increase the amount of land available. I think those two aspects are probably stronger than just an increase per se in the growth due to 2 or 3 degrees rise in temperature.

Shaw: For example, you mentioned that in the presence of CO_2 water is used more efficiently because the stomatas tend to contract. What would temperature increase do?

Carlson: Very little. But I should caution that the experiments that we have done have been at a fairly high relative humidity, and I think obviously the idea of trying to, perhaps, introduce or expand crop ranges into more arid areas, obviously the atmosphere is going to be more arid, and the experiments will have to be done at lower humidity before we run off half-cocked in that direction.

THE FORMATION AND REGIONAL ACCUMULATION
OF SULFATE CONCENTRATIONS IN THE
NORTHEASTERN UNITED STATES[1]

T. F. Lavery, G. M. Hidy, R. L. Baskett
and P. K. Mueller

Environmental Research & Technology, Inc.
Westlake Village, California

The Sulfate Regional Experiment (SURE) is being conducted to
provide information concerning the regional behavior of sulfur
oxides throughout an area that includes most of the United States
east of the Mississippi River. Ground-station and aircraft
measurements of various air quality and meteorological variables
were taken during the period August 1977 through October 1978.
The SURE also includes the assembly of ground-level and upper air
meteorological observations taken by the National Weather Service,
the compilation of seasonal, regional and special hourly (for
selected electric utility plants) emissions inventories, and the
development, refinement and verification of an Eulerian grid
model for the simulation of sulfate formation and evolution.
This paper is intended to provide a preview of some early results
of the SURE, especially the observed variability in particulate
sulfate data. The reported data come primarily from the six
intensive study months, when the ground network of 54 rural
stations was operating and the aircrafts were sampling. Monthly
average sulfate, sulfure dioxide, nitrate, and total suspended
particulate concentrations and dosages and their seasonal and
monthly variability are presented. A brief discussion of
occurrences of regional sulfate accumulation and the two types of
meteorological regimes that contribute to this accumulation is
also given.

[1]The SURE was sponsored by the Electric Power Research
Institute under Contract No. RP 862

I.   INTRODUCTION

The Sulfate Regional Experiment (SURE) is being conducted to
provide information concerning the regional behavior of sulfur
oxides throughout an area that includes most of the United States
east of the Mississippi River.[2] The SURE program includes:
(1) measurements to establish the extent and intensity of
regional scale, rural sulfur oxide levels and other pollutants in
the northeastern United States; (2) estimation of seasonal
average emissions of SO_x, NO_x, particulates and hydrocarbons, and
hourly data for major utility SO_2 sources during certain study
periods; (c) establishment of relationships between sources and
ambient air quality; and (4) analysis and interpretation of the
results to extend knowledge of sulfur oxide and nitrogen oxide
behavior in the atmosphere. To complete this program, ground-
station and aircraft measurements of various air quality and
meteorological variables were taken during the period August 1977
through October 1978. Ground-level and upper air meteorological
observations taken by the National Weather Service were assembled
and seasonal, regional and special hourly (for selected electric
utility plants) emissions inventories were computed. The
program also includes the development, refinement and verification of an Eulerian grid model for the simulation of sulfate
formation and evolution. The design of SURE and other program
details are given in the papers by Perhac, Hidy *et al.*, and
Mueller *et al.* (1-3).

[2]*The regional scale for pollution is analogous to the scale
for synoptic meteorological processes, involving an area of a
million square kilometers and time periods up to a week.*

This paper is intended to provide a preview of some early results of the SURE, especially the observed variability in particulate sulfate data. The reported data come primarily from the six intensive study months, when the ground network of 54 rural stations was operating daily and the aircrafts were sampling for 5- to 10-days each month. Monthly average sulfate, sulfur dioxide and nitrate concentrations and dosages, sulfur oxide emissions and their seasonal variability are presented. A brief discussion of occurrences of regional sulfate accumulation and the two types of meteorological regimes that contribute to this accumulation is also given. Finally, the regional grid model components and a simulation from 5 August 1977 are briefly summarized.

II. SEASONAL VARIABILITY OF AIR QUALITY LEVELS AND SO_2 EMISSIONS

The observations from the ground network illustrate the distribution of rural air quality over the northeastern United States. Average sulfate concentrations for each of six intensive measurement periods (August 1977, October 1977, 9 January to 10 February 1978, 3 April to 2 May 1978, July 1978, and October 1978) have been determined from particulate sulfate levels analyzed from samples collected on Teflon-coated glass fiber filters at 54 stations and are depicted in Fig. 1. Sulfur dioxide and nitrate averages are shown in Figs. 2 and 3, respectively. As another way of illustrating the sulfate concentrations and for modeling purposes, the region of the United States shown in Fig. 1 has been divided into 690 grid squares, each with 80 km x 80 km dimensions, 23 from south to north and 30 from west to east. The data measured at the 54 stations were than extrapolated using an inverse distance squared exterpolation procedure to specify sulfate concentrations for each grid square.

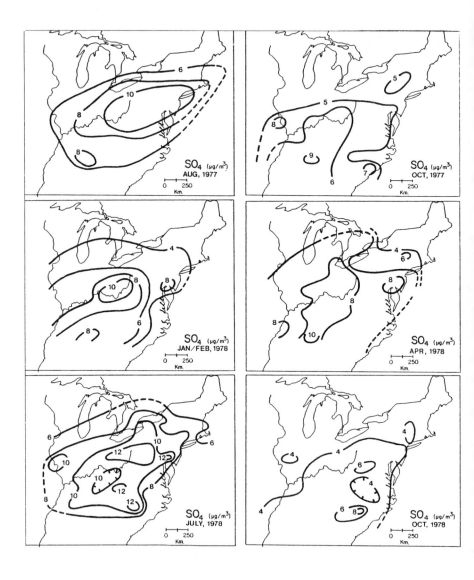

FIGURE 1. Geographic distribution of monthly mean 24-hour
sulfate values.

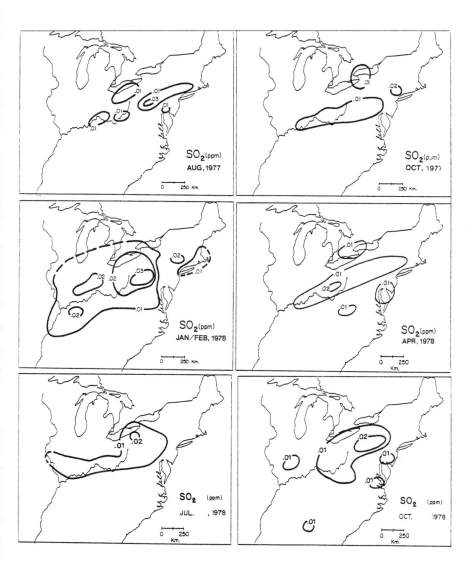

FIGURE 2. Geographic distribution of monthly mean 24-hour sulfur dioxide values.

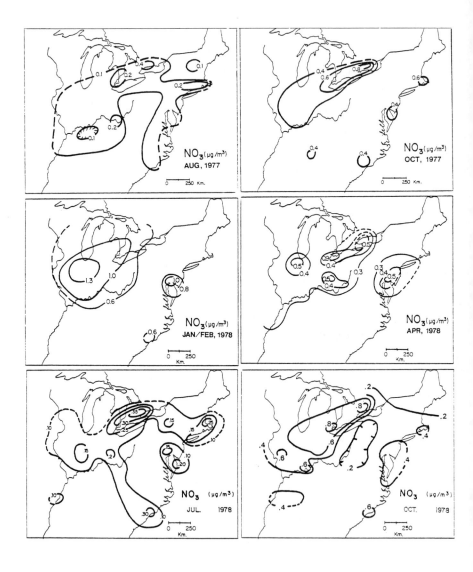

FIGURE 3.   Geographic distribution of monthly mean 24-hour nitrate values.

From this gridded data, the extent of the SURE region experiencing 24-hour average sulfate levels above 10-, 15- and 20-μg/m^3 for each of the six intensive measurements was determined and is depicted in Fig. 4. Figures 1 through 4 well illustrate the spatial and temporal variability of ground-level air quality over the rural northeast. The highest sulfate levels were observed in the two summer and winter measurement periods with much lower averages in the spring and fall (especially October 1978) months. Maximum values were observed near the Ohio River Valley, a region of especially high SO_2 source emission density (see Fig. 5). As can be seen from Fig. 4, sulfate impacts that occur during periods of length from 2 days to a week constitute the bulk of the monthly average sulfate values. In other words, during any 1 month, exposure to sulfate varied from no stations above 10 μg/m^3 to more than half the grid experiencing levels above 20 μg/m^3. On an annual basis, only 6.6% of the samples had concentrations greater than 20 μg/m^3 (3).

Unlike the sulfate distributions, maximum particulate nitrate concentrations were observed over the urbanized areas of the Great Lakes and the East Coast. Maximum levels were observed in January to February 1978 with the lowest concentrations in the two summer intensives. Maximum SO_2 levels were also observed in the January to February 1978 measurement period with lower average concentrations in the other seasons.

The seasonal variability in rural air quality can be explained by the variability in emissions and meteorological events. Table I lists seasonal average emission rates for each of five major source categories aggregated over the entire SURE region. Although there is typically less than a 15% variation in seasonal average emissions, emissions from specific source categories are highly variable from season to season. For example, NO_x emissions from commerical sources are ten times higher in winter than in summer. In general, emissions from

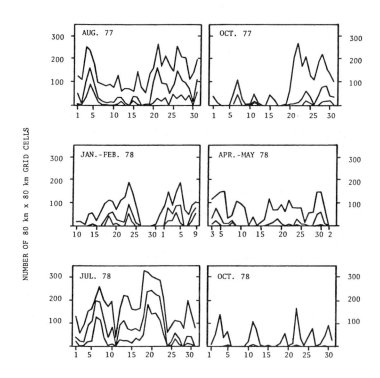

FIGURE 4. Number of 80 km x 80 km grid cells experiencing 24-hour average sulfate concentrations of 10 (top line), 15 (middle line), and 20 µg/m³ (bottom line) during intensive months.

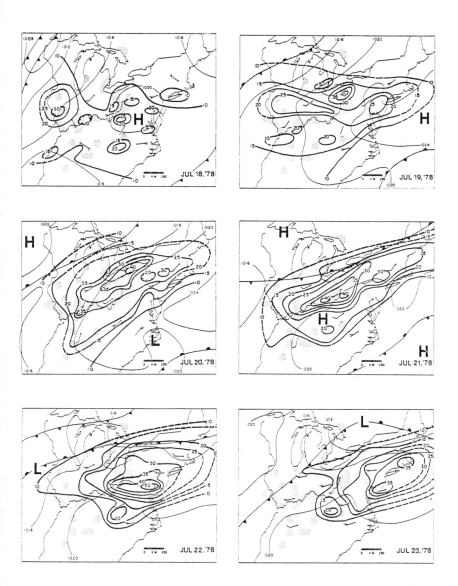

FIGURE 5. Distribution of 24-hour sulfate concentrations
for the period 18 July 1978 through 23 July 1978. Shaded areas
represent 1977-1978 summer average emission densities greater
than 1 μg/m²/sec. The 0700 EST surface pressure (mb) and 300 m
(above ground level) winds are shown.

TABLE I. *Seasonal Average SURE Grid Emissions[a]*

	Winter	Spring	Summer	Fall
SO_2				
Total	99,000[b]	86,600	86,800	88,500
Utility	52,500[b]	48,700	52,500	50,100
Industrial	36,500	33,000	31,600	33,500
Commercial	6,700	2,700	900	2,700
Residential	2,200	700	50	700
Transportation	1,300	1,500	1,600	1,500
NO_x				
Total	36,600	34,800	36,100	35,400
Utility	11,400	10,600	11,600	10,800
Industrial	7,500	6,500	6,000	6,600
Commercial	2,100	800	200	800
Residential	1,600	500	30	500
Transportation	14,000	16,500	18,200	16,700
TEP				
Total	46,300	44,500	45,000	45,200
Utility	5,800	5,400	5,700	5,100
Industrial	31,400	33,500	35,100	34,500
Commercial	4,100	1,600	400	1,500
Residential	2,000	600	25	600
Transportation	2,900	3,400	3,700	3,400

[a]*The emissions inventory was developed to represent the years 1977 and 1978; the figures represent tons per day.*

[b]*10 January 1978: Total: 101,500 tons per day; Utility: 57,500 tons per day.*

low-level sources peak in the winter, suggesting the important role these source types play in producing wintertime SO_2 and nitrate. During the winter, the combination of increased low-level emissions and low mixing heights contributes to relatively high SO_2 and nitrate. On the other hand, frequent synoptic meteorological changes during the spring and fall inhibit the regional accumulation of air pollutants. During the summer, high temperature and atmospheric moisture content and the persistence of anticyclones contribute to sulfate formation and the regional accumulation of sulfates and other air pollutants. The following information summarizes tentative conclusions regarding the seasonal variation in regional air quality:

1. Sulfate concentrations were observed to peak in summer and winter.

2. Daily SO_2 emissions vary approximately 20% on any 2 days in the year, yet sulfate levels are highly variable.

3. Sulfate was the only truly regional scale particulate pollutant.

4. Sulfur dioxide levels peaked in winter and were influenced by local sources.

5. Low-level emissions appear to be the major source of ground-level pollutants in winter.

6. Anticyclones are needed to generate regional accumulation of sulfate.

7. Long range (> 500 km) transport appears to occur only in summer.

8. The influence zone of SO_2 source-regions is limited to distances of less than about 400 km in winter.

9. Better ventilation, highly zonal flow and rapid moving weather systems inhibit regional sulfate accumulation in fall and spring.

III. METEOROLOGICAL FACTORS

The episodic nature of the regional accumulation of sulfate concentration is revealed in Fig. 4. Examination of the meteorological conditions associated with these regional events has suggested that regional sulfate accumulations can occur during two major meteorological situations, one in which regional air mass stagnation takes place under a broad high pressure area, and the other where polluted air accumulates on the west of the high between the Appalachian Mountains (or a slow moving cold front on the south), and a second quasi-stationary front across the Great Lakes northwest of the anti-cyclone. The first situation can occur in all seasons and involves light surface winds with a low mixing height; it has been discussed extensively in conjunction with urban pollution in the East (4). In such cases, no evidence has been found for transport of pollution beyond an influence zone of 200 km to 400 km from major sources. The second situation involves the "ducting of channeling" or moist, warm airflow inland over the Ohio Valley northeast across New York and New England. This case appears to occur only in summer and generally follows a stagnation event. Its behavior provides circumstantial evidence for long-range transport of sulfate over distances exceeding 1000 km from SO_2 source regions. (For additional discussion, see Refs. 5 and 6.)

An example of the ducting or channeling situation is shown in Fig. 5 for the period 18 through 23 July 1978. The six maps show isopleths of 24-hour average sulfate concentrations, the synoptic surface pressure pattern with the locations of fronts and cyclones and anticyclones, winds representative of the altitude 300 meters above ground level, and shaded areas representing SO_2 emission densities greater than 1 $\mu g/m^2/sec$.

On 18 and 19 July the areas of elevated sulfate concentrations occur within the environs of the major source region. By 20 July, the area with concentrations above 20 $\mu g/m^3$ has extended into New England, although the highest concentrations remain near the major sources. On 21 July, the sulfate pattern shifts to the east, and starts "detaching" from the zones of high emission density. This is especially well illustrated on 22 and 23 July 1978.

The significance of the combined effects of diurnal changes in mixing height and oxidation chemistry during the July event is shown in Fig. 6. The change in sulfate every 3 hours is shown at stations near Scranton, Pennsylvania, and Duncan Falls, Ohio. Throughout the period of regional accumulation, a daily sulfate maximum by mid-afternoon was observed at Scranton. At Duncan Falls, the daily sulfate maxima were more irregular, but a midday maximum generally was present. Diurnal maxima and minima in mixing height also are shown in Fig. 6, as estimated from radiosonde temperatures taken every 6 hours at Pittsburgh, Pennsylvania. The maximum in ground-level sulfate concentrations could be explained in part by downward mixing of sulfate accumulated from elevated source emissions injected above 600 meters height followed by suppression of mixing at night. At the same time, oxidation of SO_2 by photochemically related processes should be maximum in daytime. The combined effects of the two processes should enhance the daytime maximum of sulfate observed at the ground. It is interesting that the diurnal regularity in sulfate appears at Scranton far from large sources with tall stacks located in western Pennsylvania and western Ohio rather than at Duncan Falls in the center of an area of high emission density.

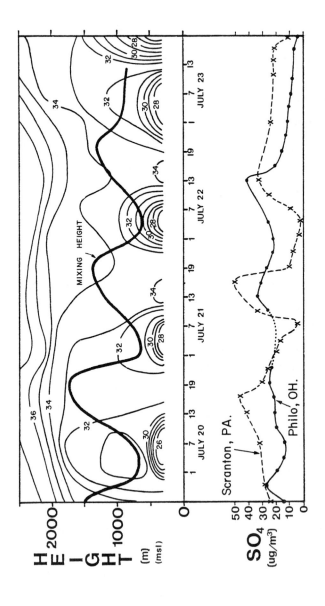

FIGURE 6. Diurnal variation in mixing height as given by radiosonde temperature from Pittsburgh, Pennsylvania, compared with ground-level 3-hour average sulfate concentrations at nearby Scranton, Pennsylvania, and Duncan Falls, Ohio. Potential temperature (C) isopleths are given as a function of altitude in the top panel.

IV.   REGIONAL GRID MODEL

To understand the relationship between SO_2 emissions and
subsequent sulfate concentrations, a regional grid model is
being developed and evaluated as part of the SURE.   In its
present version, the model is based on the numerical solution
of the mass conservation equations for SO_2 and sulfate over a
three-dimensional grid region consisting of 30 x 23 x 5 cells.
Each cell has 80 km x 80 km horizonal dimensions.   The vertical
scale is defined by five model layers: 0 to 50 km (above ground
level), 50 to 100 m, 100 to 300 m, 300 to 700 m, and 700 to
1500 m.

The model formulation follows the work of Lavery et al.,
Hidy et al., and Egan and Mahoney (7-9).   Much of the modeling
activities involve the preprocessing of various meterological
and air quality data to develop model input, including initial
conditions and boundary conditions.

The model components include algorithms to develop time and
space (three-dimensional) varying: (1) winds, (2) turbulent
diffusivities, (3) deposition velocities, (4) SO_2 to sulfate
transformation rates, and (5) SO_2 and sulfate emissions
(including the injection height of elevated buoyant plumes).

V.   CONCLUDING REMARKS

The results of the SURE have provided considerable insight
into the behavior of sulfur oxides on a regional scale over the
eastern United States.   During the study period of 1977 to 1978,
there were occurrences of widespread accumulation of particulate
sulfate concentrations centered in the Ohio Valley and western
Pennsylvania near a zone of high SO_2 emission density.   Nitrate
levels displayed a geographical distribution differing from

sulfate, with concentrations at the ground considerably lower
than sulfate. Nitrate was more prevalent along the Great Lakes
and on the East Coast compared with sulfate maximum in the rural
areas of the Midwest.

During 1977 to 1978, the maximum sulfate concentrations and
minimum SO_2 levels were observed in summer. Relatively high
sulfate and SO_2 concentrations were observed in the winter.

Regional increases in sulfates occurred periodically over
periods of 2- to 7-days. Two kinds of regional scale events
were observed; the first was associated with broad air mass
stagnation around the southern and western sides of a high
pressure over high SO_2 emission areas. The second was found in
a southwesterly directed, ducted flow of polluted air extending
over distances exceeding 1000 km. The latter appears to have a
character identifiable with long-range transport of pollution
and usually follows a stagnation event.

A regional grid model is being evaluated. Tests to date
indicate the model simulations are capable of reproducing the
observed sulfate concentrations reasonably well.

The results of a model simulation of 24-hour average sulfate
concentrations for 5 August 1977 are depicted in Fig. 7a. The
observed sulfate concentrations are given in Fig. 7b. For this
day, the model calculations accurately simulate the observations.
The linear correlation coefficient calculated from the calculations and observations is 0.94. Further model calculations will
be presented by Lavery et al.[3]

[3]Lavery, T. F., Thrasher, J. W., Lordi, N., and Lloyd, A. C.,
Development and Validation of a Regional Model to Simulate
Atmospheric Concentrations of Sulfur Dioxide and Sulfate. Paper to
be presented at the Second Joint Conference on Applications of Air
Pollution Meteorology, Boston, Massachusetts, March 24-27, 1980.

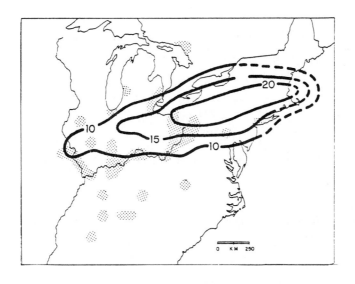

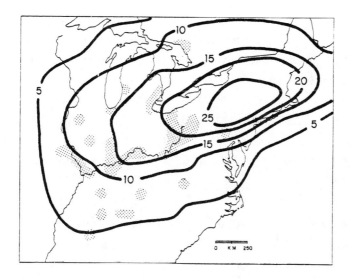

FIGURE 7.   (a) Observed 24-hour average sulfate concentra-
tions--5 August 1977.   (b) Model calculated 24-hour average
sulfate concentrations--5 August 1977.

ACKNOWLEDGMENTS

We are indebted to Dr. Ralph Perhac and Dr. Glen Hilst for their encouragement and guidance in the course of this work.

REFERENCES

1. Perhac, R., *Atmos. Environ. 12,* 641-648 (1978).

2. Hidy, G. M., Mueller, P. K., Warren, K. K., and D'Etore, K., *in* Proceedings 4th Symposium on Turbulence, Diffusion and Air Pollution, pp. 314-321. American Meteorological Society, Boston, Massachusetts (1979).

3. Mueller, P. K., Hidy, G. M., Warren, K. K., Lavery, T. F., and Baskett, R. L., The Occurrence of Atmospheric Aerosols in the Northeastern United States. Presented at the Conference on Aerosols: Anthropogenic and Natural Sources and Transport, New York Academy of Sciences, New York (1979).

4. Korshover, J. Climatology of Stagnating Anticyclones East of the Rocky Mountains, 1936-1975. ERL-ARL-55, NOAA Technical Memo, National Oceanic and Atmos. Adm., Silver Spring, Maryland (1976).

5. Lavery, T. F., Hidy, G. M., Baskett, R. L., and Thrasher, J., *in* Fourth Symposim on Turbulence, Diffusion and Air Pollution (AMS, ed.), p. 230. American Meteorological Society, Boston, Massachusetts (1979).

6. Lavery, T. F., Hidy, G. M., Baskett, R. L., Warren, K. K., Mueller, P. K., and Thrasher, J. W., *Journal of Applied Meteorology* (1980).

7. Lavery, T. F., Thrasher, J., Lloyd, A. C., Lordi, N., and Hidy, G. M. *in* Proceedings of 9th NATO/CMMS Conference on Air Pollution Modeling and Its Applications, Toronto (C. Morawa, ed.), pp. 353-362. Umweltbundefamt, Berlin (1979).

8. Hidy, G. M., Tong, E. Y., and Mueller, P. K., Design of
    the Sulfate Regional Experiment. Report EC-125, Electric
    Power Research Institute, Palo Alto, California (1976).

9. Egan, B. A., and Mahoney, J. R., *Journal of Applied
    Meterology, 11,* 1023-1039 (1972).

DISCUSSION

Kornegay: Would you say from what you showed earlier that there is very limited correlation of SO_2 versus SO_4, in SO_2 emissions and measured SO_4 concentrations?

Lavery: During at least that January and February intensive period, there was little correlation between the total emissions in the grid and total dosage estimated for the grid.

Kornegay: And it appeared even less between utilities and sulfate measurements. If that is the situation, does it appear that the tall-stack policy may have been unfounded--that indeed the assumption was that the SO_2 to sulfate transformation by alleviating the local SO_2 problem was in fact damning the rest of the country to a sulfate problem? May that have been an incorrect assumption?

Lavery: That's a possibility. That's one of the questions we are trying to answer as part of the program. And once we have the model verified, we are going to do analyses like that.

Pack: Tom, was that use of lousy coal in the winter of 1978-- January, February, March--universal? I mean the high elevated sulfates.

Lavery: I don't know the extent really.

Pack: Well, it was a loaded question because while you were showing that I was frantically looking through the rain chemistry data and it turns out that not a single station in the MAP3S network showed as high sulfate values in precipitation in the winter of 1978 as they did in the winter of 1977. They were higher in the winter of 1977. But I'm not sure that's a fair sort of thing to do because the precipitation removal is a different kind of baby than you were showing there.

Lavery: That is true, although the SURE-I data we put together for the winter of 1974-75 had much lower sulfate and much lower SO_2 values. So I don't know--there could be a meteorological explanation.

Dittenhofer: In your conclusions, you mentioned that clouds appear to enhance sulfate formation?

Lavery: I didn't say that. I said that was an area which was poorly understood. There is some evidence that clouds could enhance the oxidation of SO_2 to sulfate. I don't know. That's one of those things that has to be investigated. The July 1978 data tend to indicate that there was some enhanced oxidation, although, like I said, August 1977 showed just the opposite.

Dittenhofer: Was cloud water sampled?

Lavery: No.

Isaacs: Tom, is the SO_2 to SO_4 conversion rate a parameter or is that derived by matching the model to data?

Lavery: What we did was that in a particular simulation we looked at the measurements centered over Scranton, Pennsylvania, and then using a simple set of SO_2 to sulfate conservation equations and assuming the stuff was uniformly mixed, we actually derived a SO_2 to sulfate conversion rate based on the ground data. Then we used that in the model.

Isaacs: So in the model, that quantity is spatially homogeneous. There is one conversion rate throughout?

Lavery: In that particular simulation, yes. Although Ron Henry is looking into a statistical relationship between conversion rates and aerometric variables like ozone concentrations, temperature, and relative humidity. So we have the capability of doing it--time and space varying first order reaction rate--right now. We have not tested the model yet.

Hamill: I'm a little bit disturbed by the fact that you say there is no correlation between the SO_2 and the SO_4 concentrations that are measured and yet in your model you put in a 0.7% per hour conversion of SO_2 to SO_4, which would mean they are directly correlated and your model gives fairly good results.

Lavery: No, I did not say there was no correlation between SO_2 concentrations and sulfate concentrations. I said there was no correlation between total gridded SO_2 emissions and sulfate concentrations. In fact, between SO_2 concentrations and sulfate concentrations there is a good correlation.

Cahill: We are getting very bad correlations between oxidant and sulfate formation, for instance, in California areas on a daily basis, and very good ones on a seasonal basis. There appears to be some delay involved in terms of the chemistry occurring at higher altitude and mixing down on the ground, as was pointed out earlier. So that when you look at those correlations, sometimes they are surprisingly absent on a daily basis.

Lavery: In the Los Angeles basin, you have a much more rapid conversion of everything, in fact. We have looked at the Los Angeles data quite a bit, too. On some days the conversion rates are generally faster. In fact, in Los Angeles the correlation between sulfate and SO_2 emissions are less, even less than in the SURE region, although I think, in general, we have seen fairly good correlation between sulfate and ozone.

Cahill: We were only using the photochemically converted sulfates. We were taking out the very fine carbonous sulfate from the program.

McElroy: This question of the lifetime of SO_2 and what you can say about the lifetime of SO_2 toward conversion to sulfate, the 7% per hour, how variable is that? And you know, grossly, can you say something about what it seems to correlate with? Is the lifetime shorter with higher ozone? How does it go?

Lavery: Yes. We used 0.7% per hour over a 24-hour cycle in the simulation. In urban areas, like Los Angeles and New York City, in the summer, you can have conversion rates as high as 10% per hour pretty well correlated with ozone, relative humidity and temperature. During the winter, I would say that the SO_2 to sulfate conversion rate in rural areas is on the order of a tenth of a percent per hour.

McElroy: I misunderstood the number. So it was actually 0.7% per hour. That implies a rather long lifetime for SO_2.

Lavery: That's right. Don't forget that the whole SURE is basically designed to simulate rural regional sulfates rather than urban. If we were doing the simulation for specifically urban areas, I think that we would have to utilize much faster conversion rates.

McElroy: The number should be close to what you get with the gas phase OH reaction with the global clean OH concentration.

Lavery: Well, I think during that particular episode, namely the August 4th case, you probably could attribute most of the conversion to be due to photochemical processes.

Hsu: How do you input your wind field into your air pollution model? Do you use observed wind field or wind field generated by a dynamic model?

Lavery: We use 29 radiosonde wind observations and then do a linear interpolation and use an objective analysis procedure

based on Liu-Goodin and Eskridge. A combination of all those
things is thrown into one objective analysis procedure which
minimizes divergence. The two-dimensional wind field has no
vertical velocity at all. So we minimize the divergence
throughout the grid and we also force the model to conserve
vorticity. We feel that vorticity is important in the scale of
interest we are working on.

Hsu: I would suggest you use a dynamic model to achieve better
results.

Lavery: We have limited computer funds, here. We looked into
that and we did do some tests of a couple hours of simulation.
It just took too much computer time.

Hsu: I believe if you do that, you should have the correct
sulfur relation pattern.

Lavery: Hopefully it would do a better job if we had dynamically
predicted winds.

Kramer: What is the confidence level of your sulfate data?
What is the accuracy?

Lavery: Plus or minus 15%.

INDEX